Food Bioactive Ingredients

Series Editor

Seid Mahdi Jafari, Department of Food Materials and Process Design Engineering
Gorgan University of Agricultural Sciences and Natural Resources
Gorgan, Iran

The Food Bioactive Ingredients Series covers recent advances and research on the science, properties, functions, technology, engineering and applications of food bioactive ingredients and their relevant products. The series also covers health-related aspects of these bioactive components, which have been shown to play a critical role in preventing or delaying different diseases and to have many health-improving properties. The books in this series target professional scientists, academics, researchers, students, industry professionals, governmental organizations, producing industries and all experts performing research on functional food development, pharmaceuticals, cosmetics and agricultural crops.

Ali Imran • Ghulam Hussain
Editors

The Role of Natural Antioxidants in Brain Disorders

Springer

Editors
Ali Imran
Food Sciences
Government College University, Faisalabad
Faisalabad, Pakistan

Ghulam Hussain
Department of Physiology
Government College University Faisalabad
Faisalabad, Pakistan

ISSN 2661-8958 ISSN 2661-8966 (electronic)
Food Bioactive Ingredients
ISBN 978-3-031-41187-8 ISBN 978-3-031-41188-5 (eBook)
https://doi.org/10.1007/978-3-031-41188-5

This Springer imprint is published by the registered company Springer Nature Switzerland AG
The registered company address is: Gewerbestrasse 11, 6330 Cham, Switzerland

Preface

In *The Role of Natural Antioxidants in Brain Disorders*, the potential therapeutic use of natural antioxidants is discussed along with a review of oxidative stress in neurological illnesses and related ailments. A wide range of subjects are covered by including significant studies on naturally occurring compounds that are abundant in antioxidants. The book is a useful tool for postgraduate students and pharmaceutical professionals looking for current and crucially significant information on natural products.

This book discusses the molecular processes of neurodegenerative diseases as well as the functions of oxidative stress in brain illnesses. After that, it discusses how antioxidants, such as vitamin D, E, and C, as well as coenzyme Q10 and endogenous antioxidants like superoxide dismutase and catalase, might prevent neurodegeneration. The therapeutic potential of exogenous and endogenous antioxidants in neuroprotection is described in detail in the book.

The book contains 11 chapters written by leading researchers. For anyone who wish to learn more about the causes and available treatments for neurodegenerative illnesses, including doctors, nutritionists, and scientists who study food and nutrition, this book is a must-read.

I hope the evaluations are insightful and helpful to the readers, as this will encourage more research in the hunt for cutting-edge treatments for a variety of conditions. The editing staff's timely efforts to make this book valuable have my sincere gratitude.

Faisalabad, Pakistan

Ali Imran
Ghulam Hussain

Contents

Chapter 1
An Introduction: Overview of Nervous System and Brain Disorders

Maria Ayub and Antonello Mallamaci

1.1 Introduction to Nervous System

The nervous system is unique in the vast complexity of the processes it controls and the actions it performs. It gathers minute bits of information both externally and internally (sensory function), processes that information (integration), and sends out the required commands to the cell/tissue/organ in question to respond appropriately (motor function). The nervous system modulates every minute function of the body either directly or via sub-structures. Many of the functions that are regulated by the nervous system include higher cognitive abilities such as thought processing, and memory formation, and motor functions such as movement, posture, balance, and coordination. Sensory roles e.g., interpretation of information received from the hearing, taste, touch, homeostatic functions such as sleep, hunger, thirst, breathing, and many others. This complex array of functions regulated by the nervous system is the result of perfect coordination among its sub-structures and unique cellular organization. This system is further divided into two subsystems: the central nervous system (CNS) and the peripheral nervous system (PNS), one is responsible for voluntary control, and the other monitor involuntary actions. The brain and spinal cord make CNS and are the main relay centers of information to be processed. While cranial, spinal, and peripheral nerves together build PNS consisting of sensory and motor neurons [1].

M. Ayub (✉) · A. Mallamaci
Functional and Structural Genomics Area, Department of Neuroscience, International School for Advanced Studies (SISSA), Trieste, Italy
e-mail: mayub@sissa.it; amallama@sissa.it

A. Imran, G. Hussain (eds.), *The Role of Natural Antioxidants in Brain Disorders*, Food Bioactive Ingredients, https://doi.org/10.1007/978-3-031-41188-5_1

1.2 Nervous System Structural Organization

The nervous system is classified in general into two categories i.e., CNS and PNS, and PNS have further classification based on its autonomous and non-autonomous functions. These two divisions are somatic and autonomic, former is responsible for the voluntary actions of skeletal muscles while the latter is primarily involved in the autonomous regulation of internal organs such as cardiac myocytes, smooth muscles, and glandular epithelium. Autonomic PNS division has two subtypes, sympathetic and parasympathetic PNS, sympathetic PNS is well known for its "fight and flight response", it is activated during times of emergency and leads to increased heart rate and pupil size, the release of epinephrine/nor-epinephrine, inhibition of digestive and immune system activities and expansion of lungs. While parasympathetic PNS does the opposite of sympathetic PNS, its activation decreases heart rate and pupil size, enhances digestive and immune functions, and leads to the contraction of the lungs [2].

The structural organization of CNS builds its functional significance, it consists of the cerebrum, cerebellum, brainstem, and spinal cord. The Cerebrum, the largest part of the brain is divided into left and right hemispheres which are further classified into four lobes, each with its own set of functions: frontal, parietal, temporal, and occipital lobe. The frontal lobe is responsible for emotions, intelligence, and speech, parietal lobe regulates visual and spatial perception, constitutes language, and interprets sensory stimuli. Memory and hearing are controlled by the temporal lobe, and lastly, the occipital lobe is responsible for vision [3]. At deeper levels, these lobes are comprised of individual structures in the cerebrum such as the cortex, or deeper parts like the limbic system, hypothalamus, thalamus, pituitary gland, and pineal gland. Cerebellum is positioned at the base of the brain and monitors the balance of the body, and the brainstem acts as a relay center between the brain and spinal cord and observes sleep and breathing. The spinal cord is a cylindrical band of tissue running through the spine and connecting the brainstem and lower back of the body, carrying the information from the brain to the rest of the body via nerve bundles and cells. All this transportation of information to and from the brain via different sensory and motor routes is governed by the basic unit of CNS, called a neuron.

1.2.1 Role of CNS Individual Structures in Physiology

The cerebral cortex is the outer covering of the gray matter of the cerebrum, it has a large surface area due to its folded structure consisting of gyrus (ridges) and sulci (folds). Cortex contains 16 billion neurons arranged in such a way to give the cortex a layered structure (6 cortical layers), and the cell bodies of their neurons give the cerebrum gray color, thus called gray matter [4]. The newest part of the cortex that evolved is called the neocortex and makes up 90% of the whole cortex. While evolutionary ancient cortexes are the allocortex and paleo cortex, which compared to the neocortex contain fewer cell layers. The two hemispheres of the cortex communicate through a C-shaped structure called the corpus callosum, which consists of white matter and nerve fibers. The left hemisphere is responsible for the proper functioning of the right side of the body, while the right hemisphere controls the left side of the body.

In general, the right hemisphere is in charge of spatial ability, and creativity like artistic and musical skills, while the left hemisphere, regulates learning abilities like speech, writing, or comprehension [5]. The cerebral cortex consists of sensory and motor areas; sensory areas of the cortex receive inputs from the thalamus and process the arrived information accordingly. On the other hand, motor areas include the motor cortex and premotor cortex which are responsible for the modulation of voluntary movements.

Given the structural contribution of the cortex in the brain and its subset areas, the cortex is involved in the monitoring and manipulation of a range of physiological functions. Some of the major functions modulated by the cortex are motor functions, language processing, decision-making, intelligence, personality determination, memory storage, thinking, learning, emotions, and consciousness [6]. On a deeper level, the four lobes that comprise the cortex have individual functions as discussed in the previous section, such as the frontal lobe of the cortex is mainly considered the action cortex and is involved in the brain reward system, skeletal movements, speech control, and expression of emotions. The parietal lobe also called the sensory cortex is responsible for the integration of the sensory information received through different areas. The temporal lobe interacts with the hippocampus and modulates long-term memory formation, and lastly, the occipital lobe known as a visual processing center of the brain is responsible for the processing of visual inputs (Fig. 1.1) [3].

Brainstem also called the middle brain is the structure that connects the cerebrum with the spinal cord and has substructures midbrain, pons, and medulla oblongata. Mesencephalon commonly called the midbrain is the uppermost part of the brainstem consisting of a complex network of neurons and neural pathways which

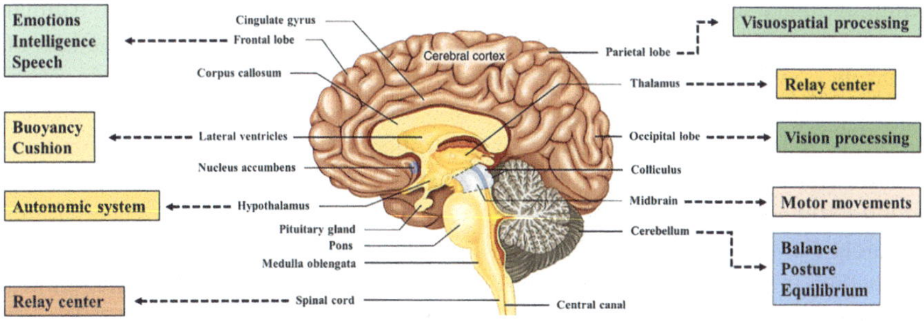

Fig. 1.1 Brain structural organization and associated physiological roles: Main structures of the brain like cerebral lobes, midbrain, limbic system, cerebellum, and spinal cord, and their associated functions in the brain

facilitates motor movements, particularly eye movements, and is involved in sensory information processing like visual and audio cues. Tectum, tegmentum, and ventral tegmental area are the subregions of the midbrain with their particular roles in the brain. The tectum forms the roof of the midbrain and processes the inputs received from the auditory cortex and peripheral brainstem nuclei. Tegmentum makes the base of the midbrain and is the house for nuclei and tracts of the reticular formation, which is a crucial brain network responsible for the survival of the being. The ventral tegmental area otherwise called VTA is involved in addiction behavior, memory, and reward system, and it does so by modulating dopaminergic neurons activity [7].

The word pons originates from the Latin word meaning "bridge", it connects the midbrain and medulla. It is the center of four of twelve cranial nerves origin, these nerves control functions such as tear production, facial expressions, blinking, chewing, and balance. Pons contain tracts carrying the sensory signals through different parts of the brain e.g., from the cerebrum to the medulla to the cerebellum, and thalamus. It coordinates the activity of cerebral hemispheres and houses the respiratory nuclei that modulates respiratory functions. At the bottom of the brainstem, a stem-like structure called medulla oblongata is positioned, which is crucial for the survival of organisms as it regulates some basic functions in the body such as breathing, rhythmic activities of the heart, blood, flow, and the body's oxygen and carbon dioxide levels. It is also responsible for sneezing, antiperistalsis movements (such as vomiting), coughing, and swallowing, altogether known as reflexes [8].

Cerebellum or little brain is a fist-size structure located at the back of the brain, above the brainstem. It consists of two hemispheres; the other outer part contains a neuronal network and the inner region communicates with the cerebral cortex. Cerebellum is mainly responsible for the posture, balance, and equilibrium of the body. It does so by monitoring the coordination of voluntary muscle movements. Cerebellum itself does not initiate any movements but contributes to coordination, precision, and timing. It receives sensory inputs from the spinal cord and other parts of the brain and integrates this information into finely tuned-motor actions. Anatomically speaking, the cerebellum is comprised of a specific set of neurons and axons i.e., Purkinje cells and granule cells, and three types of axons mossy fibers, climbing fibers, and parallel fibers that together make the cerebellar circuit. Animal and human studies have shown that damage to the cerebellum results in the loss of precisely coordinated body movements, equilibrium, posture, and motor learning [9]. Though after the damage, an individual continues to produce motor activity, precision is lost resulting in erratic, uncontrolled, and untimely movements, and actions. Before the 1990s, the sole function attributed to cerebellum was the motor activity. But advances in technology opened new avenues in the field of cerebellum research. For instance, functional imaging studies have shown the activity of the cerebellum in relation to language and attention, and correlation studies have shown the interaction between the cerebellum and non-motor parts of the cortex. Recent research is also trying to explore the role of the cerebellum in other functions such as emotions, memory, or social behaviors [10].

The limbic system is a structure buried deep within the brain underneath the cerebral cortex and consists of sub-structures, the hippocampus, hypothalamus, thalamus, and amygdala. This region is mainly involved in lower-order emotional information processing, long-term memory formation and storage, and olfaction. Hippocampus is one of the well-studied parts of the limbic system, specifically involved in cognitive functions: learning, memory formation and storage, particularly spatial learning, and memory. Morphologically speaking, the hippocampus is densely packed neurons into a seahorse-shaped structure located on the underside of the temporal lobe. Over the years, response inhibition, episodic memory, and spatial cognition have been associated with the hippocampus along with the monitoring of hypothalamic functions. In the past few decades, the hippocampus is attributed to carrying on neurogenesis in the adult stages of life, which before was considered not possible [11]. Amygdala, a small almond-like structure located next to the hippocampus is associated with the brain's reward system i.e., it regulates emotions like pleasure, fear, and anger, and is involved in fight and flight responses. Amygdala contains tracts that connect the hippocampus and entorhinal cortex [12]. Amygdala is responsible for affiliating emotional content to our memories and monitors how robustly these memories are stored [13]. Disruption of hippocampus physiology is observed in Alzheimer's disease, depression, schizophrenia, epilepsy, and many other neuropathies.

Hypothalamus is situated above the pituitary gland (master gland of the body) and is the main regulator of the autonomic system. It receives chemical stimuli from neurons of different brain regions and PNS and works to maintain an internal balance and homeostasis in the body thus called smart control of the body. It has a crucial role in monitoring some basic functions such as thirst, sleep, body temperature, blood pressure, secretion of hormones, and behaviors e.g., hunger, and sexual response [14]. And it does so by either directly influencing the autonomic system or by managing the endocrine system of the body i.e., by hormones. Hypothalamus itself synthesizes and releases some hormones to the pituitary gland such as growth hormone-releasing hormone (GHRH) which modulates the growth of long bones and muscles, gonadotrophin-releasing hormone (GnRH) that is responsible for the menstrual cycle in women, and testosterone secretion in men. It also secretes corticotropin-releasing hormone (CRH) which acts on adrenal glands during periods of stress and modulates the metabolism and immune responses, and many other hormones (Fig. 1.1) [15].

Thalamus is an egg-shaped structure present approximately in the middle of the brain, right above the mesencephalon, and its nerve fibers form multiple connections with the cerebral cortex and other parts of the brain. Thalamus consists of a bundle of nuclei mainly responsible for its function as a relay station. All the incoming information (touch, taste, hearing, and sight except smell) first pass through the thalamus and is then directed to the assigned cortical areas [16]. Thalamic nuclei are of excitatory and inhibitory nature, these neurons receive inputs from the body and present the selected information to the cortex via thalamocortical radiations. Thalamus is well connected to the hippocampus, mammillary bodies, and fornix via the mammillothalamic tract. This connection of the thalamus with the limbic structures highlights its potential involvement in learning and episodic memory and regulation of sleep and wakefulness [17].

The pituitary gland known as the master gland of the body is a pea-sized structure situated below the hypothalamus involved in the synthesis and secretion of various hormones and monitors the hormonal homeostasis of the whole body by keeping a check on other hormone-secreting glandular structures. Some of the hormones synthesized and released by the pituitary gland are growth hormone (GH), adrenocorticotropic hormone (ACTH), follicle-stimulating hormone (FSH), thyroid stimulating hormone (TSH), prolactin, and luteinizing hormone (LH) [18]. GH secretion is mediated by the hypothalamus releasing GHRH, GH acts on almost every cell of the body, but its principal targets are skeletal muscles and bones. It has a direct effect on the metabolism of proteins, fats, and carbohydrates i.e., it is anabolic and stimulates protein synthesis in the growing tissues, it decreases glucose utilization and enhances the mobilization of the fatty acids from the adipose tissues. FSH and LH are synthesized by the pituitary gland under the instruction of hypothalamic GnRH. These two hormones regulate the functions of male and female gonads i.e., ovaries and testes. FSH stimulates the growth of follicles in ovaries and LH triggers ovulation and promotes the secretion of progesterone. While in males, FSH is required for spermatogenesis, and LH fosters testosterone secretion of Leydig cells of the testes [19]. The pituitary gland also regulates the body's responses during times of stress by ACTH, which is released under the control of hypothalamic CRH. The main function of ACTH is to act on the adrenal cortex and stimulates the production of stress hormones such as glucocorticoids during stress hours [20].

1.3 Brain Barriers System and Its Physiological Significance

The concept of brain barriers refers to the evolutionary formation of physical barriers in the brain that separates the flowing blood from the neurons and regulate communication between the periphery and the brain. The brain is protected by three main brain barriers: blood-brain barrier (BBB), blood-leptomeningeal barrier (BLMB), and blood-cerebrospinal fluid barrier (BCSB). The brain barrier system is mainly responsible for the protection of the CNS from foreign toxins and chemical insults. The BBB is a monolayered structure of endothelial cells firmly held together by transmembrane proteins. BBB also called absolute or true barrier is a structure of tightly junctioned endothelial cells, pericytes, basement membrane, and astrocyte end-feet. It encompasses the blood supply of the brain thus separating the blood from the CNS surroundings. BBB is mainly responsible for the dynamic exchange of ions, nutrients, molecules, and cells between the brain and blood. Each component of BBB has its functions, such as pericytes protecting endothelial cells and are involved in vascular development, astrocytes regulating glucose transport across BBB, and endothelial cells are responsible for the structural integrity of the BBB and vaso-regulation [21]. However, during inflammatory insults or non-homeostatic conditions, BBB integrity is compromised, and it allows the unsupervised infiltration of peripheral immune components to the brain resulting in CNS dyshomeostasis [22].

BLMB is a structure made up of three layers: dura mater, arachnoid mater, and pia mater, altogether called meninges. It is also a true/absolute barrier due to its limiting nature like BBB i.e., it also restricts the entry of peripheral components into the CNS. Three layers of BLMB and subarachnoid space act as a cushion for the brain and provide support and a route for cerebrospinal fluid (CSF) circulation. The arachnoid layer of the BLMB contains CSF within the subarachnoid space and limits the invasion of activated immune cells from dural blood vessels into the CSF. Leptomeningeal cells line the stroma of the choroid plexus (ChP) and facilitate CSF drainage into the blood and lymphatics. In addition to this, leptomeningeal cells extend deep into the CNS along the outer walls of arteries and ease the CSF entry into the CNS. During homeostatic conditions, BLMB allows negligible transport of peripheral components into the brain, while in inflammatory states, BLMB's restrictive nature is disrupted leading to enhanced infiltration of leukocytes from the periphery into the CNS, resulting in neuroinflammation [23].

BCSFB on the other hand is an immunoregulatory or educational barrier i.e., it functions as a controlled junction between the brain and periphery and monitors neuroimmune communication. It is composed of ChP (structure responsible for CSF production) and CSF, together known as the ChP-CSF system (Fig. 1.2). It provides the brain a physical barrier, helps in the removal of CNS debris by CSF circulation, and acts as an immunosurveillance system owing to its diverse immune cell populations [24]. The unique anatomy of BCSFB regulates the migration of immune cells to the CNS by modulating the expression of adhesion molecules, chemokines, and cytokines at ChP and in the CSF. BCSFB physiological roles are classified as immunosurveillance and immunoprotection and the disruption of this barrier system is termed immunopathology. Immunopathology i.e., ChP-CSF system in disease, this disruption of the ChP-CSF system is highlighted in various neuropathies such as Alzheimer's disease, depression, multiple sclerosis, physiological aging, and many others.

1.3.1 Blood Supply of the Brain

Internal carotid arteries and vertebral arteries are two paired arteries that are the main supplier of blood to the brain. The internal carotid arteries give rise to two cerebral arteries (anterior and middle) which are mainly responsible for the blood flow of the cerebrum. While the cerebellum, brainstem, and base of the cerebrum get their blood intake from the vertebral arteries. The right and left vertebral arteries come together at the pons and join to form a basilar artery. At the bottom of the brain, the Circle of Willis, which is the joining area of the basilar artery and internal carotid arteries is located. This Circle of Willis is a safety feature of the brain i.e., it provides collateral blood flow to the brain in the event of blood vessel damage, thus protecting the brain from ischemia [25].

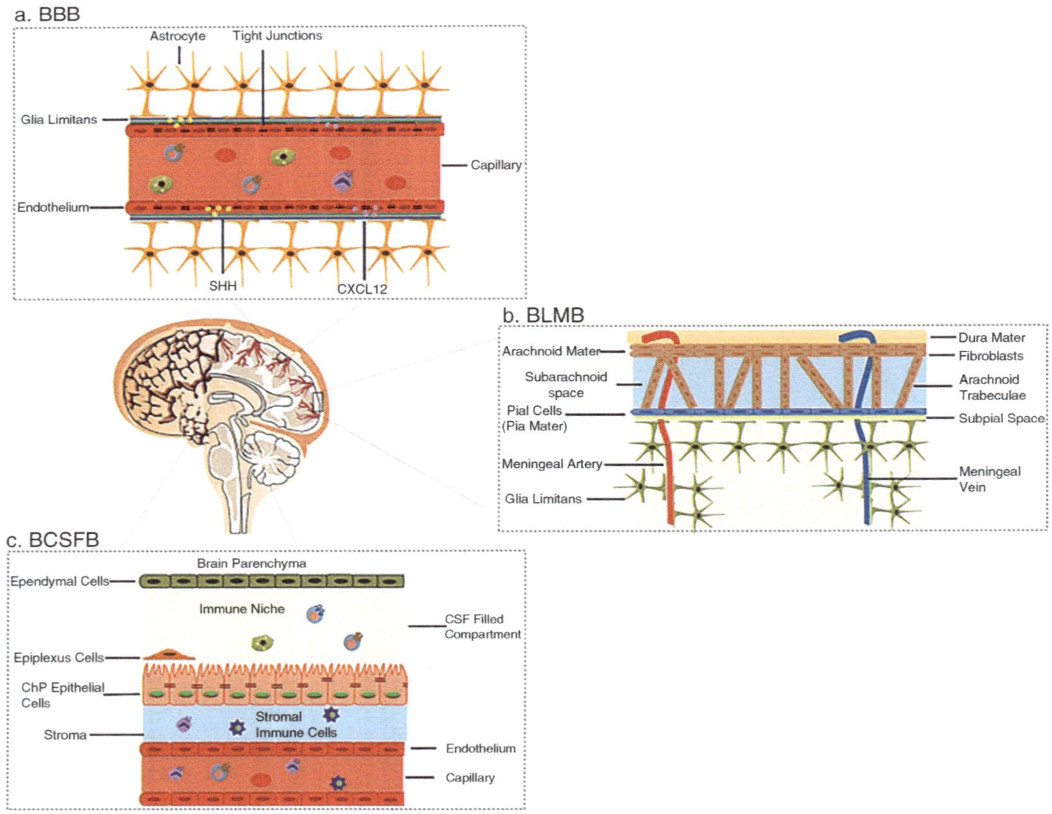

Fig. 1.2 CNS barriers system: Adapted from Ayub, Jin and Bae, 2021 from Fig. 1.1. Three main barriers of the brain i.e., Blood brain barrier, blood leptomeningeal barrier, and blood cerebrospinal fluid barrier

1.4 CNS at the Cellular Level: Structural and Functional Contributions of Various CNS Cells

1.4.1 Neuron

Neuron is the basic structural and functional unit of the nervous system. CNS is made up of more than 100 billion neurons alone excluding the supporting cells to be discussed in the coming sections. Morphologically speaking, a neuron consists of three basic substructures: soma/cell body, dendrites (receiving ends), and axons (transmitting ends). The functional classification of neurons divides them into three categories: sensory neurons, motor neurons, and interneurons [26]. Apart from this classification, neurons are also categorized based on their number of neurites i.e., unipolar, bipolar, and multipolar neurons (Fig. 1.3). A unipolar neuron has one nerve process which extends on both sides of the cell body forming one axon and one dendrite, and sensory neurons fall in this category. Bipolar neurons have two extensions i.e., one axon and one dendrite, and the majority of interneurons are bipolar. While multipolar neurons have a single axon but consist of multiple dendritic extensions and motor neurons fall into this category. There is another category of neurons not very common called anaxonic neurons which lack axons. Neurons are also subtyped based on the shape of dendrites, such as stellated neurons (star-shaped) and pyramidal neurons (of pyramid shape) [27].

Sensory neurons make the sensory part of the nervous system i.e., they collect the information from different sensory receptors across the body and carry it to the brain for processing. Such as visual receptors in the eyes to perceive visual stimuli, auditory receptors for hearing, tactile receptors for touch, and other kinds of sensory

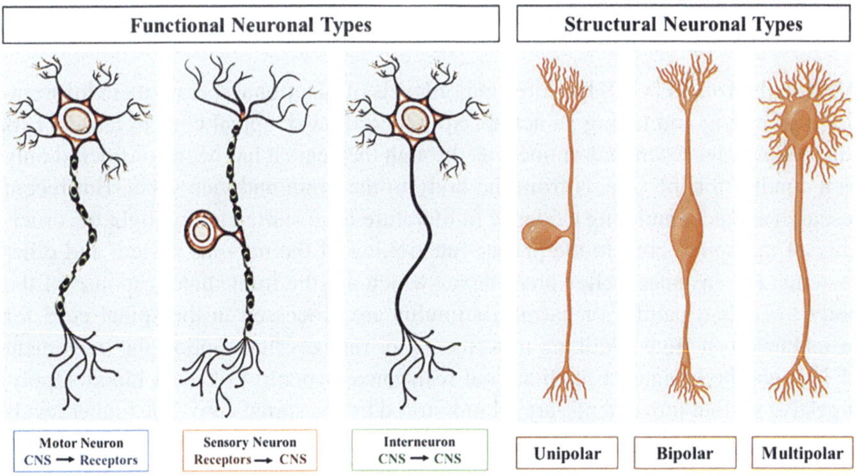

Fig. 1.3 Functional and structural neuronal types: Functional neuronal types include motor, sensory and interneurons. Structural phenotypes include unipolar, bipolar, and multipolar neurons

receptors. The information enters the brain through sensory neurons via the following pathway: the spinal cord to the brainstem, cerebellum, thalamus, and finally to the allocated regions of the cerebrum for the processing of sensory input. Integrative functions i.e., processing any kind of information entering the brain before conducting any response is the fundamental task of the nervous system. 99% of the sensory input received is discarded as unimportant and irrelevant such as continuous contact of the bodily parts with clothing or seat strain while sitting. But at the arrival of important, novel, or alarming stimuli, the information is channeled to the proper integrative areas for processing and motor regions for the desired actions. This phenomenon of channeling and processing of inputs is called the integrative function of the nervous system [28].

After the integration of the sensory inputs, the motor neurons also called effectors come into the function to elicit the desired and the best response to the received stimuli. For instance, the movement of skeletal muscles in times of physical activity or the contraction of smooth muscles in the internal organs. As sensory inputs enter the brain via a specific pathway and each structure in this pathway has its role in monitoring the input, motor output is also regulated at various levels: the spinal cord, brainstem, cerebellum, and motor cortex. Higher regions deliberate the thought process involved in controlling the movement by the lower regions involved in the motor function of the nervous system. This transition of information from sensory inputs to motor actions highlights the working of CNS at different levels for a single or multiplex of stimuli i.e., regulation at the spinal cord level, subcortical or cortical levels. These three levels of CNS have major functional significance in monitoring the input and output of information, ensuring the maximum efficacy of the system [29].

1.4.2 Functional Order in CNS

As described above, CNS has three major levels of functional specificity in information processing and taking an action. Spinal cord level: spinal cord in terms of its functionality has been undermined, as through the years it has been considered only as a conduit for the signals from the body to the brain and vice versa. But recent research and accumulating evidence in literature have started to highlight the cruciality of the spinal cord in the proper functioning of the nervous system and other systems. For instance, reflex arc/reflexes which are the immediate response of the body towards a painful or harmful stimulus are processed in the spinal cord for instantaneous actions. Walking movements, or reflexes that support the movement of legs and body against gravitational resistance or control of local blood supply, digestive system movements, are all monitored by the spinal cord. The higher levels of the nervous system often send commands to control centers in the spinal cord to take action rather than sending signals directly to the peripheral systems [30].

Subcortical or lower brain levels such as the medulla, pons, cerebellum, and basal ganglia are mainly responsible for the subconscious activities of the body. For example, body equilibrium is calibrated by the joint action of the cerebellum and reticular substance of the pons and medulla, and normal respiration and monitoring of arterial blood pressure are regulated by the pons and medulla. Feeding reflexes such as hunger are controlled by areas in the pons, medulla, amygdala, and hypothalamus. Research has also highlighted that many emotional responses such as anger, sexual responses, excitement, or reaction to pain or pleasure can still occur even after the damage to the cortex, which points out both the physiological and pathological significance of these areas of a lower order [31]. Lastly, the cortical, or higher level is responsible for thought processing and memory storage. And most importantly, lower CNS levels are precisely controlled by the actions of the cortical level, and this intercommunication between these levels is the foundation of proper and maintained functioning of the nervous system in specific and the whole body in general. All this intercommunication occurs at the neuronal level and more specifically at the synapse level, which are the junctions between two communicating neurons and responsible for intraneuronal information transfer [32].

1.4.3 Synapses of the Nervous System

The processing of information in the CNS occurs through a series of electrical and chemical reactions occurring between connected neurons. The electrical transmission of information otherwise called nerve impulse happens mainly in the form of action potentials. Each impulse reaching a neuron might be blocked, changed into repetitive impulses, or can be transformed into intricate patterns of impulses by the integration of signals from other neurons. This interneuronal communication at the chemical level is governed by the release of specific chemicals called neurotransmitters at the neuronal junction. This release of chemicals for information transmission or manipulation of electrical impulses is termed as synaptic function of neurons in the CNS.

The communicative role of synapses in the CNS is governed by their anatomical features. Morphologically speaking, a typical synapse has a presynaptic terminal, synaptic gap (space between two consecutive neurons), and postsynaptic neuronal terminal. Presynaptic terminals are also called knobs, boutons, or end feet owing to small round or oval knob-like anatomical features. This part of the synapse is responsible for the regulation of neurotransmitter release into the synaptic cleft. Presynaptic terminals are equipped with transmitter vesicles (containing required neurotransmitters) and mitochondria which provide energy in the form of adenosine triphosphate (ATP) for the release of neurotransmitters from the vesicles into the synaptic cleft (Fig. 1.4). Depending on the type of nerve impulse received at the presynaptic neuron, either excitatory or inhibitory neurotransmitters are synthesized in the transmitter vesicles. So far, more than 40 neurotransmitters have been reported and some of the best knows are glutamate, epinephrine, nor-epinephrine (excitatory transmitters), gamma-aminobutyric acid, serotonin, and glycine (inhibitory neurotransmitters). These synthesized transmitters are then emptied into the synaptic

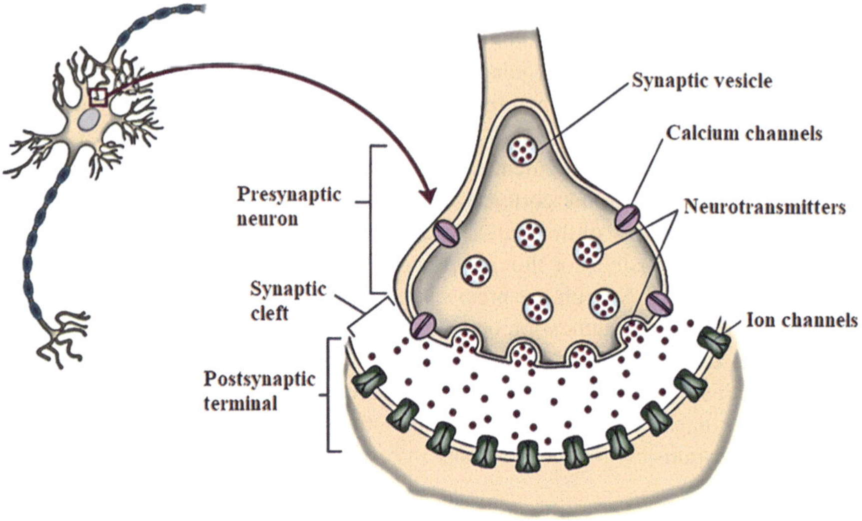

Fig. 1.4 Structural organization of a synapse: Illustration of a typical chemical synapse, consisting of presynaptic terminal containing neurotransmitters in synaptic vesicles, voltage gated calcium channels, synaptic cleft where neurotransmitters are released, and postsynaptic terminal

cleft, which is then recognized by the receptors expressed at the postsynaptic neuronal terminal and leads to either excitatory or inhibitory responses [33].

Based on their functionality, synapses are classified into two categories named electrical and chemical synapses. The majority of synapses in the human CNS are of chemical origin. As described in the paragraph above, these synapses pass signals from one neuron to the other via neurotransmitter release at the presynaptic terminals which in turn will either excite, inhibit, or modify the sensitivity of the signal at the postsynaptic terminal. While electrical synaptic communication occurs via open fluid channels which act as electrical conductors. Most of these channels are the tubular structure of proteins called gap junctions, which make the free movement of ions from one neuron to the other possible. The best examples of electric synapses are the transmission of action potentials between smooth and cardiac muscles [34].

1.4.4 Supporting Cells of CNS

Though neurons are largely responsible for the accurate functioning of the nervous system, the performance of neurons is highly dependent on supporting cells called neuroglia or glial cells. This term was coined in 1907 by Emilio Lugaro, who suggested that neuroglia control the neuronal environment by monitoring the exchange of substances with the extracellular fluid. Advancements over time in glial biology have found evidence that neurons and glia exist in a 1:1 ratio, and are easily distinguishable as they don't possess axons, these cells do not form synapses and can divide throughout the lifespan of an individual. Six types of neuroglia i.e., microglia, astrocytes, oligodendrocytes, and ependymal cells exist in CNS, satellite glial cells, and Schwann cells are present in the PNS (Fig. 1.5).

Astrocytes is a name derived from the Greek word Astron meaning star. These star-shaped cells are the most abundant type of neuroglia and have various physiological roles ranging from making BBB to migration of cells, from regulating metabolic activities in the CNS to facilitating the formation of synaptic connections. The most studied role of astrocytes is their contribution to the formation and maintenance of BBB. The astrocytic-end feet or foot processes, wrap around the blood capillaries, forming the physical barrier which separates the circulating blood and neurons. They regulate the exchange of substances at the BBB and are effective in filtering out inflammatory or harmful substances, preventing the onset of pathology [35]. Metabolic activities governed by astrocytes ensure the steady supply of energy to neurons for proper functioning, for instance, they serve as reservoirs of glycogen, which can be utilized by the brain in the hours of need. Apart from this, these star-shaped cells also help in the removal of byproducts generated as the result of cellular metabolism. Piling evidence has reported that defective astrocytic metabolism can lead to many neurological abnormalities such as depression, anxiety, bipolar disorder, or neurodegenerative pathologies like Alzheimer's disease [36]. Astrocytes guide the migration of progenitor cells in the CNS, which then differentiate into different cell types. Literature also suggests astrocytes ensheath the neuronal synapses to form tripartite synapses, where they tend to influence synaptic communication and neuronal plasticity. Some reports also point out their role in the release of specific neurotransmitters also called gliotransmitters such as glutamate, ATP, D-serine, and many others. These released transmitters then act on both pre and post-synaptic neuronal terminals to regulate synaptic transmission [37].

Microglia also known as brain macrophages or immune cells of the CNS are the second most abundant neuroglia. They have small oval-shaped nuclei and

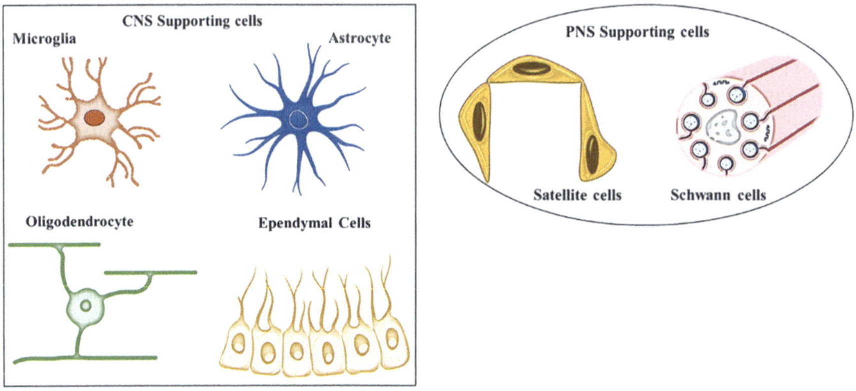

Fig. 1.5 CNS and PNS supporting cells: CNS supporting cells also known as neuroglia include microglia, astrocytes, oligodendrocytes, and ependymal cells. PNS supporting cells are Schwann cells and satellite cells

projections from their cell soma, which help their chemotactic movement in the CNS. These cells enter the CNS during embryonic development and attain the role of brain immune cells. Their origin differs from other neuroglia, while other glial cells originate from neuroectoderm (embryonic layer which gives rise to nervous tissue), microglia are derived from mesoderm, which forms blood and immune cells. During homeostatic conditions, microglia patrol the brain, and clear out any damaged or toxic substances. Microglia have different physiological states, and each type possesses a particular function in a particular condition. For instance, ameboid microglia also called surveilling microglia are prevalent during brain development and have scavenger properties. Ramified microglia have long branches, and maintain an immunologically stable environment in the adult brain. While activated microglia are the type of microglia present in neuroinflammatory conditions, these microglia are in a fully active phagocytic form, have short thick branches, are neurotoxic, and act as antigen-presenting cells. Apart from this classification, microglia also have M1 and M2 forms, former proinflammatory and later anti-inflammatory. In different pathological environments, such as inflammatory or degenerative, the polarization of microglia from M2 to M1 form is commonly observed, which exacerbates the ongoing pathology. Microglia also regulate the infiltration of peripheral immune cells such as monocytes into the brain via the release of specific chemokines, influencing BBB permeability [38]. Recent literature has pointed out the critical role of microglia specifically in depression and Alzheimer's disease, in the former it regulates the inflammatory response while in the latter it is involved in the phagocytosis of amyloid-beta (Aβ) plaques. Microglia also monitor synaptic pruning, which is the elimination process of extra synapses during early childhood and adulthood. Studies in mice lacking microglia have shown delayed synaptic pruning and synaptic abnormalities resulting in developmental disorders [39].

Oligodendrocytes are myelin sheath-synthesizing glial cells, which are derived from oligodendrocyte precursor cells. The term oligodendrocyte is derived from a Greek word meaning a cell with few branches, and these few branches of oligodendrocytes are wrapped around the adjacent neuronal processes to provide myelin sheath protection, while soma and nucleus remain separated. The myelin-containing cells in the PNS are called Schwann cells, which serve the same function of providing myelin protection to the nerves in the PNS. Unlike Schwann cells, oligodendrocytes can wrap around multiple axons at the same time, for example, one oligodendrocyte can cover almost 50 axons at a time while Schwann cells can wrap around only one neuronal process. The main purpose of this cellular ensheathing is to ensure the effective and fast transfer of nerve impulses without the loss of signal intensity by preventing the leakage of ions and decreasing the cell membrane capacitance.

The ideal example of this type of fast travel of nerve impulse is observed at the nodes of Ranvier (myelin sheath gaps) called saltatory conduction (which is the propagation of action potential along the myelinated axons) at the speed of 150 m/s [40]. Apart from increasing the signal velocity, myelin sheath also ensures an energy-efficient process, by reducing the energy expenditure over the axonal membrane. With advances in the studies in this field, oligodendrocytes have been reported to also secrete some neuronal trophic factors such as glial cell line-derived neurotrophic factor (GDNF), brain-derived neurotrophic factor (BDNF), or insulin-like growth factor-1 (IGF-1). Any injury, infection, or autoimmune action which results in demyelination can result in serval pathological states as the most common disorder associated with demyelination of neurons in multiple sclerosis. Demyelination results in the loss of effective signal transmission over the neurons, which can lead to various signs and symptoms such as muscle weakness, blindness, double vision, and disrupted coordination [41].

Ependymal cells are ciliated cuboidal to columnar epithelial cells that develop from radial glia and line the brain ventricles and central canal in the spinal cord. Given their location in the brain, these cells are primarily responsible for monitoring CSF homeostasis. Their apical surface is provided with cilia which facilitate the CSF circulation in the CNS, and microvilli which serve the function of CSF absorption. In addition to the lining of the ventricles, within brain ventricles, a modified form of ependymal cells exists in conjunction with capillaries called tela choroidea, which form ChP, the structure responsible to produce CSF. The tight junctions between these ependymal cells regulate the exchange of nutrients, ions, and other factors between ventricles and peripheral blood. These epithelial cells also secrete some chemokines to attract immune cells such as monocytes, memory $CD4^+$ T cells, or peripheral macrophages into the CNS, hence explaining the complex immune profile of the CSF. The clinical significance of these cells can be highlighted by the fact that any abnormality or dysregulation of their function will ultimately affect the functioning of ChP and CSF. Ependymoma, the tumor of ependymal cells is the most common pathology associated with these cells, dysregulation of their functionality can affect the physiology of ChP which can lead to several neurodevelopmental disorders such as autism, schizophrenia, intellectual disability, or in some cases neurodegenerative pathologies [42].

Satellite glial cells (SGCs) are exclusively found in PNS ganglia, particularly in sensory, sympathetic, and parasympathetic ganglia, where they form neuronal envelopes by covering the cell bodies of neurons. SGCs are equipped with potassium channels and glutamate transporters via which they monitor the microenvironment around neurons. SGCs share many molecular markers with other neuroglia such as they express glutamine synthetase, potassium channels, and glutamate aspartate transporter like astrocytes. SGCs express transcription factor Sox10 like oligodendrocytes, they contain cadherin19, a Schwann cell marker, and share many other morphological and molecular similarities. Closely located SCGs tend to form strong interactions, otherwise called SGCs-SGCs coupling, or with neurons, neuron-SGCs coupling. This neuronal-SGCs and SGCs-SGCs coupling plays a critical role during injury to the PNS and pain sensation. Recent research has reported the contribution of SGCs in sensory glia during pain pathology. Research has observed that inflammation or peripheral nerve damage can activate SGCs in sensory ganglions. This results in the upregulation of astrocyte marker glial fibrillary acidic protein (GFAP), an increase in SGCs-SGCs coupling, and enhanced reactivity of SGCs to the ATP (pain mediator). In addition to that, SGCs upon activation also release proinflammatory cytokines such as interleukin-1β (IL-1β), interleukin-6 (IL-6), tumor necrosis factor, and fractalkine, which in turn can act on the neurons and result in hyperexcitability. The working mechanism of SGCs has been explored in many clinical pathologies related to pain such as systemic inflammation, chronic post-surgical pain, diabetic neuropathic pain, and post-herpetic neuralgia. Though the potential contribution of SGCs in pain has been explored in rodent models, a lot is still needed to be explored about the physiological significance of SGCs in homeostatic states [43].

Schwann cells are the oligodendrocytes of PNS except they only wrap myelin sheath around a single neuron, unlike oligodendrocytes which can wrap multiple axons at a time. These cells are derived from the neural crest, neural crest cells differentiate into Schwann cell precursors, which migrate to the periphery along the axonal tracts and proliferate. These cells play a significant role in PNS nerve development, maintenance, functioning, and regeneration. Schwann cells are categorized into two types based on their myelinating properties: myelinating and non-myelinating cells. Myelinating Schwann cells associate and ensheath a single axon, and like oligodendrocytes, this process is not continuous. Myelinating Schwann cells help the fast and energy-efficient transmission of nerve impulses over the axons. While non-myelinating Schwann cells associate with several smaller axons such as those involved in transmitting pain and temperature stimuli and make Remak bundle (group C nerve fiber). And this non-myelinating gap exists between the myelinated axons and make-up nodes of Ranvier as described in the oligodendrocyte section. Schwann cells can lose their myelinating properties in demyelinating neuropathies, autoimmune reactions, or external toxin attacks. These pathological states can displace the Schwann cells away from the nerve fibers which can ultimately lead to the loss of proper nerve conduction or in some cases complete blockade of nerve impulse transmission down the axons as in Guillain-Barré syndrome and diphtheria [44].

The comprehensive description of supporting cells of CNS and PNS highlights their critical physiological contribution and how dysfunction of any of these can lead to neuropathies, and not only this but how this can affect the proper functioning of neurons. The following section underlines the several pathologies ranging from neurodevelopmental disorders to autoimmune disorders to neurodegeneration and how all cellular populations play a role in either contributing to or alleviating these pathologies.

1.5 Brain in Disease: Overview of Various Brain Pathologies

A wide range of pathologies can affect the brain, and their incidence is increasing with the increase in the aging population and genetic and environmental alterations. Not every brain disorder can be categorized into a particular classification but there are some established main classes of brain diseases that encompass a variety of pathologies: neurodevelopmental disorders, autoimmune brain diseases, infections, epilepsy, stroke, traumatic brain injuries, mental illness, and neurodegenerative pathologies. The subsequent section shed light on some of these disorders which will then be discussed in detail in the coming chapter.

1.5.1 Neurodevelopmental Disorders

Neurodevelopmental disorders are a group of disorders that affect the normal development of the CNS, resulting in abnormal brain functions such as emotional, social, and memory deficits, difficulty in learning and processing, and self-control and persist through the lifespan of an individual. Autism spectrum disorder (ASD), Attention-Deficit/Hyperactivity disorder (ADHD), and intellectual disability (ID) are the main pathologies included in this category. The pathophysiology of these disorders is very complex and ranges from genetic mutations to environmental factors or in most cases the combined effect of these two and other biological stimulants such as psychosocial risks. For instance, intellectual disability is specifically associated with mutations in specific genes, and some variants of ASD and ADHD also result from genetic alteration. While the majority of ASD and ADHD cases result from the mutual effect of genetic and environmental factors such as during pregnancy abuse of alcohol, tobacco, or drugs by the mother, poor socioeconomic state, childhood traumas, or environmental contaminants (lead, methylmercury), low birth weight, or preterm birth.

ASD, previously known as autism or pervasive disorder is a lifelong condition characterized by deficits in social interactions and communication, repetitive behaviors like flapping arms, poor motor coordination, learning problems, and enhanced sensitivity to sensory stimuli like loud noises. According to the world health organization (WHO), about 1 in 100 children is affected by autism. Literature reports that specific genetic causes account for up to 10–20% of ASD cases, while the rest of them are the result of the combined effect of genetic and environmental risk factors. There are no specific diagnostic techniques to determine ASD, but healthcare workers receive specific training in behavioral screenings and evaluations to assess the disorder. No drugs have been produced so far that can cure the pathology completely, and the medicine given to the affected individuals is used only to alleviate some of the signs and symptoms. Autistic children and adults mainly go through behavioral therapies aimed to address the core behavioral abnormalities and to reduce their severity. Apart from the affected individuals, the entire family is also advised to undergo specific training that helps in educating the parents and other members of the family about autism, which can help in improving living conditions for both family and the autistic person [45].

ADHD also a neurodevelopmental disorder is characterized by inattentiveness, impulsivity, and hyperactivity more pronounced in affected persons than in individuals of the same age. ADHD symptoms usually appear around the age of 7 years, and children suffering from ADHD exhibit specific behavioral activities like distractedness, diminish academic performance, social problems, and difficulty in following instructions or completing a task. Systematic reviews report that about 2–7% of people are suffering from ADHD globally. ADHD share somewhat the same pathophysiology as ASD, but with different disease outcome and different brain regions involved. ADHD in children, teens, and adults is diagnosed via an established protocol i.e., identification of ADHD symptoms, ruling out of alternative causes of symptoms, and identification of comorbid conditions such as depression or anxiety. Unlike ASD, ADHD is treatable and some drugs have proved effective in the treatment of ADHD. Psychostimulants (drugs used to stimulate CNS), methylphenidates (often known as Ritalin), and dextroamphetamine are the most used psychostimulants for ADHD. These drugs help individuals with ADHD focus their attention and ignore unnecessary thoughts and stimulants. Data reports that stimulant medicines are 70–90% effective in patients with ADHD [46].

ID also called mental retardation is characterized by diminished intellectual functioning in a person before the age of 18, or put in other words IQ of less than 70. These individuals also display impaired communicative, interpersonal, and social skills, difficulty with self-care, and in some cases severe retardation. Genetic mutations, traumatic brain injuries, maternal use of alcohol or drugs, or infection during pregnancy are observed to be the major causes of ID according to the reports. However, the main causes of intellectual disability are still unknown in 30–50% of the cases, but the identification of triggers in severe retardation (IQ less than 50) is more frequent than in individuals with mild retardation (IQ between 50 and 70) which remains unknown in more than 75% of cases. To diagnose an intellectual disability, a series of tests are performed which include IQ tests, psychological testing, special education tests, hearing, speech, and vision tests, and many others to formulate a conclusive diagnosis. Currently, there is no treatment for this disorder, but there is occupational therapy (self-care training, employment activities, skills, domestic activities), speech therapy (to improve communication skills, vocabulary, speech articulations, and language skills), and physical therapy (improves sensory integration, quality of life, and educating the individuals to adaptive solutions to mobility problems) [47].

1.5.2 Neurodegenerative Diseases

Neurodegeneration refers to the progressive loss of either structure or function of a neuron, ultimately resulting in neuronal death, and pathologies occurring because of neurodegeneration are called neurodegenerative diseases. These diseases occur at several brain levels i.e., from molecular to systemic, and affect memory, motor movements, coordination, speech, and breathing and so far, these are incurable. Neurodegenerative diseases include Alzheimer's disease (AD), Parkinson's disease (PD), Amyotrophic lateral sclerosis (ALS), Huntington's disease including many others. Many of these disorders are of genetic origin and result from alterations at the genetic level, but in some reported cases infections, alcoholism, toxins, tumor, or stroke can result in the diseased outcome.

AD is the most common neurodegenerative disease mainly occurring in people over the age of 60 and is the prevalent cause of dementia (memory loss). It is characterized by loss of memory, affected learning, difficulty in managing daily routine tasks and coping with unfamiliar situations, and disorientation. Abnormal protein aggregation especially of Aβ plaques and tau neurofibrillary tangles lead to neuronal damage and loss in the cerebral cortex and some subcortical areas resulting in the temporal and parietal lobe and some parts of the frontal cortex atrophied. Research reports that the first symptoms of AD can appear after 10 years of the disease onset, thus making an early and timely diagnosis difficult. The current diagnostic methods used are brain imaging, mental status testing, neuropsychological, and blood testing. As neurodegeneration is an irreversible process, no cure exists at the moment for the treatment of AD or other neurodegenerative diseases, but some drugs are assigned to the patients to alleviate symptoms [48].

PD is the second most common chronic neurodegenerative disease affecting 1% of the population over the age of 60. The signs and symptoms of PD are categorized into motor and non-motor, former include bradykinesia (slow movements due to muscle weakness), tremors, loss of body coordination, rigidity, and stiffness, and later includes dysphagia (trouble with swallowing), drooling or monotonous facial expressions. PD shares some similar pathological features with AD but the affected area in question is different. In PD mainly dopaminergic neurons are lost in substantia nigra (a structure in basal ganglia that regulates body movements) presumably due to the accumulation of aggregated alpha-synuclein complexes in Lewy bodies within the affected neurons. PD like AD is both familial and sporadic i.e., it can be hereditary or occur at later stages in life due to some non-genetic factors. PD is diagnosed based on blood tests, computerized tomography (CT) scans, magnetic resonance imaging (MRI), and genetic testing. There are some drugs available to treat the symptoms of PD but no medication for a cure exists for now [49].

ALS or Lou Gehrig's disease is a neurodegenerative disease that particularly affects the neuromuscular system causing muscle weakness and leading to difficulty in talking, eating, or moving. It is a progressive disorder and throughout the disease, the symptoms get worse, and with time muscle atrophy and wasting occur. As the symptoms get worse, the individual starts to experience breathing problems and most ALS patients die of respiratory failure because of weakened muscle contraction. ALS affects both upper-level neurons (in the brain) and lower levels (spinal cord and brainstem) and both genetic and environmental factors can lead to the disease. Sporadic cases of ALS have reported the aggregation of TAR DNA-binding protein 43 (TDP-43) and RNA-binding FUS proteins in the affected motor neurons. ALS is diagnosed by a couple of tests including an electromyogram (EMG), MRI, nerve conduction study, spinal fluid test, and muscle and/or nerve biopsy [50].

1.5.3 Autoimmune Brain Diseases

Autoimmune brain diseases occur when the body's immune system mistakenly attacks the healthy cells and tissues of the nervous system resulting in the inflammation of the affected area. This inflammation of the affected tissue or cells can lead to neurological and psychological signs and symptoms. Affected individuals can develop symptoms like seizures, loss of vision, muscle weakness, sleep problems, depression, hallucinations, and an overall decline in the quality of life. Autoimmune brain diseases include Multiple sclerosis (MS), autoimmune encephalitis, autoimmune-related epilepsy, and many others.

In MS, the immune system attacks the healthy myelin-producing cells in the brain and spinal cord, resulting in the disruption of proper nerve signal transmission. This impaired signal transduction can lead to optic neuritis (blurriness or pain in the eye), muscle weakness or spasms particularly in arms and legs, fatigue, loss of coordination, changes in the gait, and memory problems. Epidemiological studies suggest that MS is more prevalent in women than men and affected individuals fall in the age group of 20–40 years. Many factors can trigger the onset of MS like viral or bacterial infections, living in an environment with less sun exposure can lead to vitamin D deficiency (a risk factor for MS), how your immune system works, and genetic mutations. How genetics play a role in the onset of MS is still unclear but having a family member with MS increases the risk of MS development. MS is diagnosed by multiple tests which particularly include MRI to look for the lesions in the brain or spinal cord resulting due to damage to the myelin sheath, physical examination, and some blood and urine tests. Currently, there is no cure for MS, the treatment focuses on managing symptoms and slowing the disease progression [51].

1.5.4 Neuropsychiatric Disorders

Neuropsychiatric disorders are a class of mental, behavioral, and emotional diseases that occur both at cellular and systemic levels and affect a person's ability to carry out daily life tasks and diminish the quality of life. The pathologies in this category mostly result from environmental factors such as psychological traumas, poor socioeconomic status, infections, comorbidity with other diseases, and in some cases genetic predisposition. At the cellular level, microglia and neuronal communication are highlighted in the pathology of these orders, and their contributions are now explored in the research to establish some mechanisms and pathways. Neuropsychiatric disorders include anxiety, depression, bipolar disorder, schizophrenia, and many others.

Anxiety, depression, and bipolar disorder: These three disorders share some similar signs and symptoms, and pathophysiology. The shared symptoms include nervousness, panic, anhedonia (sadness), mood swings, mania, and changes in thinking patterns. These disorders also have some common causative initiators like traumas, chemical imbalances in the brain, medical conditions, long-lasting stress, or sometimes genetics or hereditary. The diagnosis of any of these disorders requires an extensive observation of affected individuals' behavior, for instance, in case of depression, symptoms of sadness or hopelessness lasting for more than 2 weeks, physical examination, blood tests, and thorough examination of family medical history are required. The common treatment for these disorders is a combination of medication (anxiolytics, anti-depressants), psychotherapy, and counseling. Establishing a healthy lifestyle such as exercise, meditation or yoga has also been observed to improve the quality of life for individuals [52].

Schizophrenia refers to a spectrum of conditions that fall in the range of psychosis and involve a disconnection from reality like suffering from hallucinations or delusions. The common symptoms of schizophrenia include slow/unusual movements, incoherent/disorganized speech, delusions, hallucinations, depression/anxiety, paranoia, and alcohol or drug abuse. According to statistics, about 2.27 million new cases are reported each year globally, and about 22.1 million people are suffering from this disease at the moment. For schizophrenia diagnosis, a person must have at least two of five main symptoms lasting for no less than a month. Other medical tests are performed to rule out other problems like stroke, injuries, tumors, or other diseases. Schizophrenia is treatable but for now, no full cure exists for this disease. This treatment includes medication (anti-psychotic drugs), psychotherapy, and in some cases electroconvulsive therapy [53].

1.6 Conclusion

This chapter discussed nervous system structural organization and shed light on the cellular composition of the nervous system and the various roles performed by specific parts and cells. And lastly, an overview of some of the common brain disorders is provided, detail of which will be discussed in the coming chapter. This discussion laid the foundation for the coming section as the basic knowledge of brain anatomy and physiology is essential to understanding and comprehending neuropathies in the light of structural, molecular, and behavioral abnormalities.

References

1. Thau L, Reddy V, Singh P (2023) Anatomy, central nervous system. StatPearls Publishing, Treasure Island (FL)
2. Catala M, Kubis N (2013) Gross anatomy and development of the peripheral nervous system. Handb Clin Neurol 115:29–41. https://doi.org/10.1016/B978-0-444-52902-2.00003-5
3. Firat RB (2019) Opening the "Black Box": functions of the frontal lobes and their implications for sociology. Front Sociol 4:3. https://doi.org/10.3389/fsoc.2019.00003
4. Briggs F (2010) Organizing principles of cortical layer 6. Front Neural Circuits 4:3. https://doi.org/10.3389/neuro.04.003.2010

5. Güntürkün O, Ströckens F, Ocklenburg S (2020) Brain Lateralization: A Comparative Perspective. Physiol Rev 100:1019–1063. https://doi.org/10.1152/physrev.00006.2019
6. Jawabri KH, Sharma S (2022) Physiology, cerebral cortex functions. Treasure Island (FL)
7. Puelles E, Martínez-de-la-Torre M, Watson C, Puelles L (2012) Chapter 10 - Midbrain. In: Paxinos G, Puelles LBT-TMNS (eds) Watson C. Academic, San Diego, pp 337–359
8. Caminero F, Cascella M (2022) Neuroanatomy, mesencephalon midbrain. Treasure Island (FL)
9. Lara-Aparicio SY, Laureani-Fierro AJ, Morgado-Valle C et al (2022) Latest research on the anatomy and physiology of the cerebellum. Neurol Perspect 2:34–46. https://doi.org/10.1016/j.neurop.2021.12.002
10. Reeber SL, Otis TS, Sillitoe RV (2013) New roles for the cerebellum in health and disease. Front Syst Neurosci 7:83. https://doi.org/10.3389/fnsys.2013.00083
11. Semënov MV (2019) Adult hippocampal neurogenesis is a developmental process involved in cognitive development. Front Neurosci 13:159. https://doi.org/10.3389/fnins.2019.00159
12. Anand KS, Dhikav V (2012) Hippocampus in health and disease: an overview. Ann Indian Acad Neurol 15:239–246. https://doi.org/10.4103/0972-2327.104323
13. Baxter MG, Croxson PL (2012) Facing the role of the amygdala in emotional information processing. Proc Natl Acad Sci U S A 109:21180–21181. https://doi.org/10.1073/pnas.1219167110
14. Shahid Z, Asuka E, Singh G (2022) Physiology, hypothalamus. StatPearls Publishing, Treasure Island (FL)
15. Saper CB, Lowell BB (2014) The hypothalamus. Curr Biol 24:R1111–R1116. https://doi.org/10.1016/j.cub.2014.10.023
16. Kosif R (2016) The thalamus: a review of its functional anatomy. Arch Med Res 4:1–9. https://doi.org/10.18103/mra.v4i8.740
17. Fama R, Sullivan EV (2015) Thalamic structures and associated cognitive functions: relations with age and aging. Neurosci Biobehav Rev 54:29–37. https://doi.org/10.1016/j.neubiorev.2015.03.008
18. Barkhoudarian G, Kelly DF (2017) Chapter 1 – the pituitary gland: anatomy, physiology, and its function as the master gland. In: Laws ERBT-CD (ed). Academic, pp 1–41
19. Oduwole OO, Huhtaniemi IT, Misrahi M (2021) The roles of luteinizing hormone, follicle-stimulating hormone and testosterone in spermatogenesis and folliculogenesis revisited. Int J Mol Sci 22. https://doi.org/10.3390/ijms222312735
20. Aguilera G (1994) Regulation of Pituitary ACTH Secretion during Chronic Stress. Front Neuroendocrinol 15:321–350. https://doi.org/10.1006/frne.1994.1013
21. Kadry H, Noorani B, Cucullo L (2020) A blood–brain barrier overview on structure, function, impairment, and biomarkers of integrity. Fluids Barriers CNS 17:69. https://doi.org/10.1186/s12987-020-00230-3
22. Rosenberg GA (2012) Neurological diseases in relation to the blood-brain barrier. J Cereb Blood Flow Metab Off J Int Soc Cereb Blood Flow Metab 32:1139–1151. https://doi.org/10.1038/jcbfm.2011.197
23. Derk J, Jones HE, Como C et al (2021) Living on the edge of the CNS: meninges cell diversity in health and disease. Front Cell Neurosci 15:703944. https://doi.org/10.3389/fncel.2021.703944
24. Ayub M, Jin HK, Bae J-S (2021) The blood cerebrospinal fluid barrier orchestrates immunosurveillance, immunoprotection, and immunopathology in the central nervous system. BMB Rep 54:196–202. https://doi.org/10.5483/BMBRep.2021.54.4.205
25. Brodal P (2016) 120The Blood Supply of the CNS. Cent. Nerv. Syst. 0
26. Sharpee TO (2014) Toward functional classification of neuronal types. Neuron 83:1329–1334. https://doi.org/10.1016/j.neuron.2014.08.040
27. Hernáth F, Schlett K, Szücs A (2019) Alternative classifications of neurons based on physiological properties and synaptic responses, a computational study. Sci Rep 9:13096. https://doi.org/10.1038/s41598-019-49197-8
28. Sah P, Stuart GJ, Egan GF (2020) Editorial: integrative brain function down under. Front Neural Circuits 14:48

29. Tozzi A (2015) Information processing in the CNS: a supramolecular chemistry? Cogn Neurodyn 9:463–477. https://doi.org/10.1007/s11571-015-9337-1
30. Khatibi A, Vahdat S, Lungu O et al (2022) Brain-spinal cord interaction in long-term motor sequence learning in human: an fMRI study. NeuroImage 253:119111. https://doi.org/10.1016/j.neuroimage.2022.119111
31. Min B-K, Hämäläinen MS, Pantazis D (2020) New cognitive neurotechnology facilitates studies of cortical-subcortical interactions. Trends Biotechnol 38:952–962. https://doi.org/10.1016/j.tibtech.2020.03.003
32. Zagha E (2020) Shaping the cortical landscape: functions and mechanisms of top-down cortical feedback pathways. Front Syst Neurosci 14:33. https://doi.org/10.3389/fnsys.2020.00033
33. Shao Z, Yang Y, Hu Z (2022) Editorial: Regulation of synaptic structure and function. Front Mol Neurosci 15:1060367
34. Fröhlich F (2016) Chapter 3 – synaptic transmission. In: Fröhlich FBT-NN (ed). Academic, San Diego, pp 29–45
35. Abbott NJ, Rönnbäck L, Hansson E (2006) Astrocyte–endothelial interactions at the blood–brain barrier. Nat Rev Neurosci 7:41–53. https://doi.org/10.1038/nrn1824
36. Ricci G, Volpi L, Pasquali L et al (2009) Astrocyte-neuron interactions in neurological disorders. J Biol Phys 35:317–336. https://doi.org/10.1007/s10867-009-9157-9
37. Verkhratsky A, Parpura V, Vardjan N, Zorec R (2019) Physiology of Astroglia. Adv Exp Med Biol 1175:45–91. https://doi.org/10.1007/978-981-13-9913-8_3
38. Saijo K, Glass CK (2011) Microglial cell origin and phenotypes in health and disease. Nat Rev Immunol 11:775–787. https://doi.org/10.1038/nri3086
39. Wolf SA, Boddeke HWGM, Kettenmann H (2017) Microglia in Physiology and Disease. Annu Rev Physiol 79:619–643. https://doi.org/10.1146/annurev-physiol-022516-034406
40. Nie D-Y, Ma Q-H, Law JWS et al (2006) Oligodendrocytes regulate formation of nodes of Ranvier via the recognition molecule OMgp. Neuron Glia Biol 2:151–164. https://doi.org/10.1017/S1740925X06000251
41. Kipp M (2020) Oligodendrocyte physiology and pathology function. Cell 9
42. Jiménez AJ, Domínguez-Pinos M-D, Guerra MM et al (2014) Structure and function of the ependymal barrier and diseases associated with ependyma disruption. Tissue Barriers 2:e28426. https://doi.org/10.4161/tisb.28426
43. Hanani M, Spray DC (2020) Emerging importance of satellite glia in nervous system function and dysfunction. Nat Rev Neurosci 21:485–498. https://doi.org/10.1038/s41583-020-0333-z
44. Griffin JW, Thompson WJ (2008) Biology and pathology of nonmyelinating Schwann cells. Glia 56:1518–1531. https://doi.org/10.1002/glia.20778
45. Yenkoyan K, Grigoryan A, Fereshetyan K, Yepremyan D (2017) Advances in understanding the pathophysiology of autism spectrum disorders. Behav Brain Res 331:92–101. https://doi.org/10.1016/j.bbr.2017.04.038
46. Albrecht B, Uebel-von Sandersleben H, Gevensleben H, Rothenberger A (2015) Pathophysiology of ADHD and associated problems-starting points for NF interventions? Front Hum Neurosci 9:359. https://doi.org/10.3389/fnhum.2015.00359
47. Lee K, Cascella M, Marwaha R (2022) Intellectual disability. StatPearls Publishing, Treasure Island (FL)
48. Tiwari S, Atluri V, Kaushik A et al (2019) Alzheimer's disease: pathogenesis, diagnostics, and therapeutics. Int J Nanomedicine 14:5541–5554. https://doi.org/10.2147/IJN.S200490
49. Kouli A, Torsney KM, Kuan W-L (2018) Parkinson's disease: etiology, neuropathology, and pathogenesis. In: Stoker TB, Greenland JC (eds). Codon Publications, Brisbane (AU)
50. van den Bos MAJ, Geevasinga N, Higashihara M et al (2019) Pathophysiology and diagnosis of ALS: insights from advances in neurophysiological techniques. Int J Mol Sci 20. https://doi.org/10.3390/ijms20112818
51. Huang W-J, Chen W-W, Zhang X (2017) Multiple sclerosis: pathology, diagnosis and treatments. Exp Ther Med 13:3163–3166. https://doi.org/10.3892/etm.2017.4410

52. Fisar Z (2013) Pathophysiology of Mood disorders and mechanisms of action of antidepressants and mood stabilizers. Endocannabinoid Regul Monoamines Psychiatr Neurol Disord 103–134. https://doi.org/10.1007/978-1-4614-7940-6_6
53. Deng C, Dean B (2013) Mapping the pathophysiology of schizophrenia: interactions between multiple cellular pathways. Front Cell Neurosci 7:238. https://doi.org/10.3389/fncel.2013.00238

Chapter 2
Pathophysiological Mechanisms of Brain Disorders

Maria Ayub and Antonello Mallamaci

2.1 Neurodevelopmental Disorders: Etiologies and Management

Autism spectrum disorder (ASD), Attention-Deficit/Hyperactivity disorder (ADHD), and intellectual disability are the most reported neurodevelopmental disorders. These pathologies disrupt normal brain development and affect an individual's emotional, learning, memory, and self-control abilities and diminish the quality of life. The cause of any of these neuropathies is complex as most of them result from a synergy of genetic alterations, and environmental and immunological factors (Fig. 2.1).

2.1.1 Autism Spectrum Disorder (ASD)

2.1.1.1 Synaptic Dysfunction in ASD

One of the most observed disrupted molecular functions in ASD pathology is synaptic dysfunction. Individuals with ASD tend to have more synapses due to disrupted synaptic pruning, abnormal dendritic spine morphology, and dysregulated synaptic transmission during development. This surplus of synapses in the brain results in hyperexcitation of the neurons. The published genomic reports of autistic patients and animal models have reported several genes that are particularly involved

M. Ayub (✉) · A. Mallamaci
Functional and Structural Genomics Area, Department of Neuroscience, International School for Advanced Studies (SISSA), Trieste, Italy
e-mail: mayub@sissa.it; amallama@sissa.it

A. Imran, G. Hussain (eds.), *The Role of Natural Antioxidants in Brain Disorders*, Food Bioactive Ingredients, https://doi.org/10.1007/978-3-031-41188-5_2

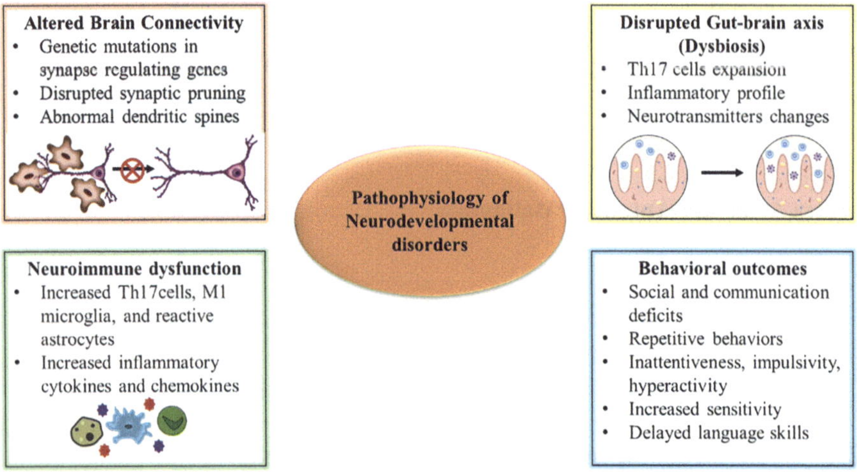

Fig. 2.1 Etiology of neurodevelopmental disorders and phenotypic outcomes: Neurodevelopmental disorders like Autism spectrum disorder or attention deficit hyperactivity disorder share some similar mechanistic changes like altered brain activity, neuroinflammatory profile both in CNS and periphery, and dysbiosis which can lead to disease specific behavioral outcomes like social and communication deficits and repetitive behaviors in autism and inattentiveness, impulsivity, and hyperactivity in case of ADHD

in homeostatic synaptic transmission and diseased conditions, disrupting normal synaptic communication. Normal synaptic function is governed by the synergy between different factors like adhesion molecules, scaffold proteins, ion channels, and neurotransmitter receptors. And some of the genes reported via genomic studies are responsible for keeping in check the expression level of the above-mentioned genes to ensure proper synaptic functioning.

Synaptic transmission is a transcription and translation-dependent neuronal activity, and mutations in many of the transcription factors associated with this translation and transcription processes are reported in ASD. For example, postsynaptic scaffolding proteins (SH3 and multiple ankyrin repeat domains protein (SHANK)), glutamate receptors, chromodomain-helicase-DNA-binding protein 8 (CHD8), and many others. Disorders in any of these synaptic adhesion molecules, neuroligins (NLGNs), scaffolding proteins, or receptors translation or transcription can result in the altered synthesis of synaptic proteins and thus disrupted neuronal plasticity. Changes in the expression of any of these molecular factors can eventually affect the number and strength of synapses leading to alterations in neuronal connectivity [1]. Neuropathological studies in past years have reported that in ASD, abnormal dendritic spine morphology like increased spine density and aberrant structures are observed [2]. Post-synaptic scaffolding proteins Shank1,2 and 3 are encoded by SHANK genes directly in the post-synaptic densities of excitatory synapses. Shank mutations especially Shank3 have been well studied in both humans and its counterpart mutant mice models. Individuals with Shank3 mutations exhibit faulty dendrite development and morphological features along with defective axonal growth cone mobility. Similar alterations are observed in Shank3 knockout mice together with the reduction in the number of cortico-striatal connections [3]. Shank3 mutant mice display disrupted long-term potentiation due to aberrant synaptic signaling in the hippocampus. Shank2 knockout mice studies have reported defects in excitatory neurotransmission and synaptic plasticity. Shank proteins regulate dendritic spine size, Shank1 overexpression in hippocampal neurons leads to oversized dendritic spines as well as their early maturation. On the other hand, Shank1 deletion results in reduced spine size, thinner PSDs, and weakened synaptic communication [4].

2.1.1.2 Immune Dysfunction and Neuroinflammation in ASD

Immune dysfunction and neuroinflammation have gained the spotlight in the past few years in the field of ASD research. Both patients and animal models have been identified with persistent immune dysregulations. For instance, one of the earliest studies in this field reported 150 differentially expressed genes in ASD individuals compared to controls, 85% of these upregulated genes were involved in the immune response pathways [5]. These changes in the immune landscape occur both in the CNS and periphery, such as ASD patients have more circulating inflammatory Th17 cells, increased M1 microglia (inflammatory) and reactive astrocytes, and their respective signaling pathways. This enhancement in inflammatory molecular signaling pathways in CNS and periphery can affect normal brain functions like synaptic transmission, social behaviors, and other physiological actions [6, 7]. These reactive microglia and astrocytes can lead to increased release of inflammatory cytokines and chemokines such as IL-6 (interleukin-6), TNFα (tumor necrosis factor α), IFNγ (interferon gamma), IL-17 (interleukin-17), and many others. On the other hand, there is a reduction of anti-inflammatory factors like IL-10 (interleukin-10) and TGFβ (transforming growth factor β) [8, 9]. These studies have helped in establishing the diagnosis of immune dysfunction and neuroinflammation via monitoring the levels of inflammatory factors mentioned above as a biomarker tool.

Similar findings have been reported in the maternal immune activation (MIA) and valproic acid (VPA) induced autism mice models. Upregulation of IL-6, IL-17, and TNFα is observed in MIA and VPA mice models both in CNS and periphery [10]. Some reports also provide evidence of infiltration of Th-17 cells from the periphery into the brain along with the migration of IL-1α (interleukin-1 alpha), IL-1β (interleukin-1 beta), TNFα, and IL-6 [11]. This increase in the inflammatory profile of the brain enhances the microglia and astrocyte crosstalk, which can affect blood-brain barrier (BBB) permeability, thus leading to unsupervised communication between CNS and periphery. Maternal autoimmune disorders such as fever or infection or exposure of the mother to external toxins during pregnancy can result in exaggerated immune responses, increasing the risk of ASD in progeny.

One of the well-established mice models to explore ASD pathology from this perspective is MIA, which is created by viral infection molecules (poly(I: C)), bacterial mimics (lipopolysaccharide (LPS)), and some inflammatory cytokines injection at mid-gestation stage. Treatment of mice with either poly(I: C), LPS, or cytokines disturbs the maternal immune and cytokine profile, such as an increase in IL-6, IL-17, and TNFα levels. These cytokines and inflammatory Th-17 cells can cross the blood-placenta barrier and enter fetal blood, resulting in the alteration of the immune profile of the fetus. These alterations result in the development of ASD phenotypes such as anxiety-like behaviors, deficits in social interactions, and repetitive behaviors [12].

2.1.2 Attention-Deficit/Hyperactivity Disorder (ADHD)

ADHD called hyperkinetic disorder is a persistent neurodevelopmental syndrome that affects 5% of school-age children and 2.5% of adults worldwide [13]. ADHD is characterized by distractibility, inattentiveness, impulsivity, and locomotor hyperactivity. This disorder can enhance the developmental risk for other neuropsychiatric disorders, accidents, and social deficits, which can lead to addictions or educational or professional failures throughout the lifespan of an individual. No single risk factor is associated with the development of ADHD, both genetic and environmental factors together lead to the initiation of this disorder. External factors such as childhood traumas, exposure of the mother to toxins or drugs during pregnancy, and exposure of children to environmental toxins such as lead can trigger ADHD lasting for an individual's life.

2.1.2.1 Genetics of ADHD

Genetics have a strong role in the development of ADHD, about 74% of ADHD cases have mutations at the gene level. Genes impact the onset of ADHD, its persistence, and remission thought the lifespan of an individual. Having said that, reports have also suggested that, living with ADHD individuals can increase their risk of developing ADHD two-ten folds compared to the general population [14]. ADHD occurrence is affected by both stable genetic factors like ADHD-associated genes or the alterations that arise at later time points of an individual's life. Studies have also shown that ADHD shares genetic influences with other neurodevelopmental and psychological disorders including ASD, cognitive impairments, and mood disorders.

Copy number variants (CNVs) such as deletion, insertions, or mutations in the genes have been reported to have a potential connection with ADHD pathology. Some studies have found that a significant number of ADHD cases carry large CNVs of >500,000 base pairs in length compared to non-ADHD individuals [15]. Some of these CNVs occur in genes that encode for neuronal nicotine receptors, key receptors of synaptic transmission in the nervous system. These CNVs also impact the activity of glutamate receptor encoding genes, which are essential for excitatory neuronal activity, and neuropeptide Y encoding gene, this peptide regulates food intake by monitoring signaling in the brain and autonomic nervous system [16, 17].

ADHD cases are presented with lower dopaminergic neuronal functioning (essential for voluntary body movements and a wide array of behavioral processes). ADHD individuals carry mutations for the genes (D4 and D1B) involved in the expression of dopaminergic neurons and the SlC6A3 dopamine transporter. Other genes in ADHD pathology are associated with the dysregulation of the monoamine system include SLC64 (serotonin transporter), HTR1B (gene encoding for serotonin receptor 1B), and SNAP25 (which encodes synaptosomal protein 25) [18, 19]. Some studies also report neuronal structural dysfunction such as neurite outgrowth in ADHD pathology.

2.1.2.2 Environmental Factors in ADHD Pathogenesis

Environmental risk factors are strong contributors to neurodevelopmental disorders, and many of these developmental disorders share the same external causative variables such as alcohol intake during pregnancy, exospore of mother to the toxins or inflammatory substances, difficult childhood, premature birth, or lower birth weight. Traumatic brain injury can increase ADHD risk by 30% [20], and certain infections such as measles or enteroviral infection can lead to an enhanced incidence of ADHD [21].

Animal studies have contributed a lot to our understanding of the association between environmental factors and ADHD. Some of these studies have found a strong correlation between exposure to tobacco or alcohol during pregnancy and ADHD, these external toxins can affect normal CNS development. Prenatal exposure to nicotine is one of the main causative agents of ADHD which can affect early brain development. Nicotinic acetylcholine receptor proteins are expressed early in the brain, indicating their importance in modulating dendrite outgrowth during the developmental phase. Prenatal or perinatal exposure to nicotine can affect neurite growth, influence the glutamate release and uptake by neurons, and may produce changes in the catecholaminergic system (it regulates diverse cognitive, motor, and endocrine functions). Early exposure to nicotine can affect locomotor activity, and impair cognitive functions, principally working memory, reported in human and animal studies [22].

2.1.3 Behavioral Outcomes of ASD and ADHD and Current Therapies

Clinical diagnosis of ASD is based on three main symptoms observed in the individuals i.e., social and communication deficits, repetitive behaviors, and general lack of interest or resistance to change. Along with these main symptoms, there are other behavioral alterations observed in the affected individuals that help in the diagnosis of ASD. For example, avoiding eye contact, reduced interest in children or people, showing more attention towards toys or objects, increased sensitivity to various stimuli like loud noises, smell, or touch, repetitive movements such as the flapping of hands or spinning in circles, delayed language skills, or repetition of words.

ADHD individuals display three types of core behavioral phenotypes i.e., difficulty in focusing on one thing for a longer time (inattentiveness), exhibiting strange behaviors on sudden urges such as throwing things, talking out in a class, or spending too much money (impulsivity), and restlessness, inability to sit in a place, or climbing when it's not appropriate (hyperactivity). Based on the data collected from the behavioral assessment of the ADHD individuals, the clinician can diagnose the person with one of these ADHD subtypes: predominately hyperactive/impulsive type, in this type a person exhibits the hyperactive or impulsive behaviors for at least six months, but do not meet the criteria for inattention. Predominately inattentive type, an individual must exhibit constant inattentive behavior for six months but do not show hyperactivity for a longer duration, and lastly, combined type, an individual diagnosed with this ADHD type displays both hyperactive and inattentiveness for at least six months, and this is the most common form of ADHD in children [23].

ASD and ADHD management includes psychological counseling, psychoeducational programs, and special schools designed to improve the social and communicative skills, attentive span, and cognitive abilities of individuals. Special education programs and early behavioral therapies can help individuals to acquire communication, self-care, or specific employment skills. Psychoeducational programs are designed to improve cognitive abilities, and social and communication skills. While pharmacological interventions such as psychoactive drugs, anticonvulsants, or antidepressants focus on alleviating one or more symptoms of ASD and ADHD to improve the overall quality of life for the individual.

2.2 Neurodegenerative Diseases: Pathophysiology of Neurodegeneration and Resulting Pathologies

Alzheimer's disease (AD), Parkinson's disease (PD), Amyotrophic lateral sclerosis (ALS), Huntington's disease, and many others fall into the neurodegenerative disorders category. These disorders in general affect the neuronal structure and function, leading to neuronal death, resulting in memory deficits, dysregulated motor coordination, and affecting speech and breathing. These pathologies share a somewhat similar working hypothesis i.e., these diseases arise from the progressive degeneration of neurons, share strong genetic and aging factors, but affect different brain regions (Fig. 2.2).

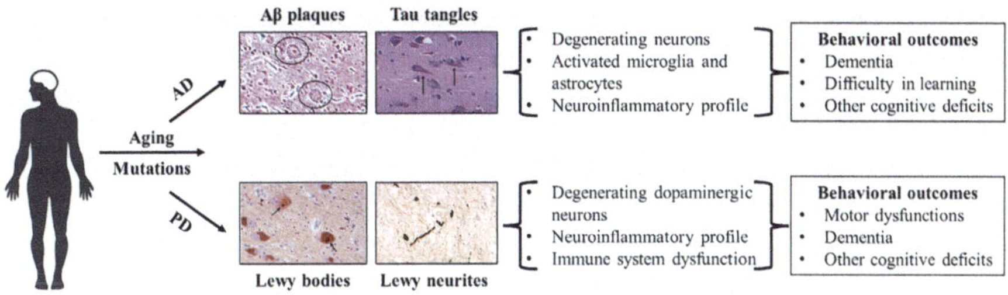

Fig. 2.2 Pathophysiological hallmarks of Alzheimer's disease and Parkinson's disease and behavioral outcomes: Pathological hallmarks of Alzheimer's disease (AD) include deposition of amyloid beta (Aβ) plaques and tau neurofibrillary tangles particularly in the cortical areas which lead to the degeneration of neurons and initiation of inflammatory cascades leading to the associated behavioral outcomes like dementia. Though Parkinson's disease (PD) shares the same concept of degenerating neurons with AD, the affected neurons in PD are mainly dopaminergic neurons in substantia nigra region of the brain. The atrophy of these neurons leads to poor neurotransmission and motor dysfunctions

2.2.1 AD Pathology

AD is an age-associated neurodegenerative disorder, resulting in dementia, affecting approximately 45 million individuals globally, and the fifth leading cause of death worldwide [24]. AD is characterized by two main histopathological hallmarks: amyloid beta (Aβ) aggregation into the extracellular senile plaques, and the formation of intracellular neurofibrillary tangles (NFTs) by hyperphosphorylated tau protein (pTau). One of the well-established mechanisms dysregulated in AD pathology is Aβ accumulation in the brain and how it leads to neuronal degeneration and behavioral outcomes. Recent studies have also identified immune system dysregulation and mitochondrial dysfunction as strong contributors to AD pathology.

2.2.1.1 Genetics and Aβ Pathway

Aβ is a 4kDa downstream molecule of amyloid precursor protein (APP), widely produced in the brain by both neuronal and non-neuronal cells. APP in homeostatic conditions is catalyzed by two proteolytic enzymes, β-secretase (β-APP- cleaving enzyme-1 (BACE1)) and γ-secretase subsequently, generating Aβ fragments. Pathomechanistic studies indicate that dysregulation in the production of Aβ and its clearance from neurons leads to Aβ dyshomeostasis, resulting in amyloid protein misfolding and aggregation into the neuronal senile plaques. Neuroimaging studies have identified a gradual spreading of Aβ plaques in the brain, senile plaque accumulation initially starts in cerebral regions and spreads from there to the brainstem and lower brain regions. This spreading phenomenon can explain the different phases observed in AD pathologies such as the pre-clinical stage in which Aβ continues to accumulate in the same brain areas and the clinical stage in which protein plaques start spreading to the other brain regions, leading to behavioral outcomes [25].

AD pathology has a strong correlation with genetic mutations in the genes responsible for Aβ homeostasis in the brain, particularly in early-onset AD (EOAD). Genome-wide association studies (GWAS) of EOAD have identified mutations in the APP gene and presenilin 1 and 2 (PSEN1 and PSEN2), these genes govern Aβ synthesis and removal in the brain. In mouse models of EOAD, a mutation in any of these genes result in dysregulation of Aβ homeostasis, leading to misfolded proteins, aggregated plaques, and their buildup in the brain parenchyma. EOAD accounts for 1% of all AD cases, and most of these incidences have reported mutations in the above-mentioned genes. These studies of one-gene-one mutation-one misfolded protein further strengthen the Aβ hypothesis of AD pathology [26].

In LOAD, which is late-onset AD, no causal relation is observed so far between genetic mutations and AD pathology, but several genetic risk factors have been suggested to increase the susceptibility of LOAD development. And many of these genetic factors are linked to the homeostasis of Aβ in the brain such as APP, PSEN1, and PSEN2 responsible for Aβ expression, apolipoprotein E family (especially APOE4) involved in Aβ trafficking, and genes responsible for Aβ degradation. Several genes associated with LOAD pathogenesis are involved in other processes as well which can contribute to AD pathogenesis such as inflammatory and immune responses, cellular trafficking, lipid metabolism, cholesterol transport, endocytosis, and ubiquitination (mechanisms crucial for protein clearance) [27].

Current studies have found a synergy between Aβ aggregation and tau tangles formation, it is observed that Aβ might functions as a facilitator of tau dyshomeostasis, including tau protein misfolding, accumulation in tangles, spreading to the different areas, and associated neuronal degeneration. Most of the studies have identified tau markers as drivers of neurodegeneration and cognitive impairments in AD, suggesting that Aβ pathophysiology might trigger downstream pathways such as tau-spreading and tau-induced toxicity [28]. AD mouse models also show that modulation of tau accumulation in the brain irrespective of Aβ plaques levels can result in reduced neurodegeneration and cognitive deficits [29]. Similar findings have been observed in *in-vitro* studies, such as treating healthy neuronal cultures with AD cortex-derived Aβ oligomer results in neuronal dystrophy and tau hyperphosphorylation. However, no pathology is observed if tau is knockdown before the treatment with Aβ oligomers [30].

2.2.1.2 Neuro-immune Crosstalk in AD Pathology

Recent studies in identifying the critical significance of glial cells in particular and the immune system in general coupled with human genetic studies in AD led to rediscovery and re-evaluate the importance of the immune system in AD pathology. AD *in-vitro* and mice models studies also provide evidence that neuroinflammation is one of the key pathogenic events occurring in AD etiology [31]. Microglia, also known as immune cells of the brain are chief cells involved in the clearance of cellular waste, metabolic by-products, and protein debris. Many experimental AD mouse model studies have reported that microglia tend to surround Aβ plaques and tau fibrils, to prevent further spread and phagocytosed them. Microglia may aid in reducing the overall Aβ brain burden by facilitating the removal of Aβ and tau.

Furthermore, Aβ can lead to the priming of the microglia to the inflammatory type, these primed microglia also known as activated microglia, secret pro-inflammatory factors, increasing the inflammatory load of the brain. Moreover, dysregulation of microglia phagocytic activity can increase the Aβ accumulation, as primed microglia fail to clear the accumulated protein plaques and neurofibrils. Both human and animal studies have reported an abnormal increase in the levels of TNFα, by activated microglia, neurons, and astrocytes, and Aβ triggers its release. TNFα is reported to stimulate γ-secretase activity, increasing the Aβ production, which further increases TNFα release, thus working in a vicious loop. In mice models of AD, blocking the TNFα pathway resulted in the reduction of activated microglia and Aβ accumulation [32].

Parallel to microglial activation, the human postmortem brain, and animal model studies have reported that astrocytes tend to surround Aβ, and become reactive like microglia [33]. These reactive astrocytes release a plethora of pro-inflammatory molecules such as interleukins, nitric oxide, complement system components, and many other cytotoxic elements [34]. Human and rodent studies have also suggested the presence of Aβ plaques in the astrocytes, reflecting the phagocytic ability of astrocytes to engulf and phagocytize Aβ [35].

2.2.2 *Parkinson's Disease (PD)*

PD is the second most common progressive neurodegenerative disease primarily distinguished by motor system dysfunction i.e., it affects muscle movements, control, and balance. PD progresses for 10-15 years before the manifestation of clinical signs and symptoms, making the early diagnosis challenging. As age progresses, non-motor or autonomic symptoms of PD start to appear such as anxiety, depression, dementia, and sleep issues. Aging is the greatest risk factor for the development of PD, and apart from aging, genetic and environmental factors are also major initiators of PD pathogenesis. PD shares several aspects of pathology with other neurodegenerative disorders linked to alpha-synuclein (αSyn) aggregation such as synucleinopathies. The main histological features of PD is Lewy bodies (intracellular inclusions (LBs)) that are aggregated αSyn in the neuronal cell bodies and Lew neuritis (LNs) and dopaminergic neuronal loss in substantia nigra [36].

2.2.2.1 Pathophysiology of PD

Genetics of PD: About 5-10% of PD cases are due to mutations in the specific genes associated with PD, these mutations may not lead to the PD onset but can increase the risk of the disease onset and progression combined with environmental and age factors. Both autosomal dominant and autosomal recessive mutations have been reported in PD patients, at least 11 genes have been identified to be associated with autosomal dominant phenotype and 9 genes with autosomal recessive traits. The risk of developing PD increases by 20–30 folds if the mutation is present in any of the associated genes. Autosomal dominant genes include SCNA, PARK3, UCHL1, EIF4G1, RICE3, and many others, among which the most common mutations are observed in SCNA. Genes associated with autosomal recessive phenotype are PRKN, PINK1, ATP13A2, and PARK7, including several others [37, 38].

Studies of familial forms of PD have identified mutations in SCNA, that are responsible for the αSyn expression, mutations result in the overexpression of αSyn, a pathological hallmark of PD. Human postmortem brain studies have suggested that mutations or duplications in the SCNA gene are sufficient to cause PD or Lewy body dementia. Macroscopically speaking, mild frontal cortex atrophy and distinctive histological changes are observed in substantia nigra pars compacta (SNpc) and locus querulous of the PD brain. Almost all the reported cases present with dopaminergic neuronal degeneration in SNpc and loss of noradrenergic neurons in locus calculus. This neuronal damage can lead to denervation of the nigrostriatal pathway (dopaminergic signaling pathway that connects SNpc with striatum), which is critical for the movement. Denervation of this pathway leads to diminished dopamine levels in the striatum, which is cardinally responsible for the motor deficits observed in PD [39].

Lewy body pathology: LBs are the inclusion bodies of abnormally aggregated proteins inside the nerve cells formed during the PD pathology. The pathological characteristic of PD is the development of unchecked accumulation of αSyn in neuronal cell bodies, that aggregate to form LBs, resulting in the dystrophy of neuronal projections both in axons and dendrites accompanied by cellular loss. In PD and other synucleinopathies, αSyn is abnormally phosphorylated, acquires an amyloid-like filamentous structure, and forms clumps of LBs. Several mechanisms have been proposed for this structural transformation of αSyn into an amyloid phenotype, among which the phosphorylation at serine 129 is mostly observed in reported cases and animal studies [40]. Apart from αSyn serine subunit phosphorylation, dysfunctional post-translational modifications of αSyn have also been seen such as ubiquitination or C-terminal truncation [41].

LB is mainly comprised of αSyn along with other proteins such as heat shock proteins, tau, ubiquitin, proteasomal and lysosomal elements, and many others [42]. αSyn apart from individually contributing to PD pathology has been documented to interact with other proteins in the brain and affect several molecular pathways. One such protein is tau; exaggerated levels of hyperphosphorylated tau in the striatum of both PD and PD-induced dementia patients are observed in postmortem studies [43]. This has been backed by animal model studies as well, which also report that increasing the αSyn levels can trigger the hyperphosphorylation of tau both *in-vivo* and *in-vitro* [44, 45]. Moreover, GWAS found a strong link between MAPT (gene encoding tau protein) and PD onset and progression risk. αSyn is reported to interact with Aβ; in a subgroup of PD patients, αSyn associated Aβ aggregates are deposited in the cortical regions. Moreover, PD patients with cognitive symptoms seem to have a widespread accumulation of tau tangles and Aβ plaques [46].

Neuroinflammation in PD: Whether neuroinflammation independently triggers PD or it is the consequence of PD pathology is still debatable. But postmortem brain studies have identified dysregularities in the immune system of the brain in particular and of the whole body in general. These studies have reported an increased number of activated microglia, complement system, peripheral T lymphocyte infiltration, and subsequent release of proinflammatory cytokines by these cells, especially in the SNpc and striatum of patients with PD [47, 48]. PD rodent model studies have reported that diminishing the microglial activity pre- and post-neurotoxic insult with minocycline significantly reduces the dopaminergic neuronal death in the SNpc, suggesting that microglia-triggered neuroinflammation might be responsible for the neuronal degeneration [49, 50]. On the other hand, evidence also shows that αSyn can trigger microglia activation into inflammatory phenotype, initiating the inflammatory processes. *In-vitro* studies have shown that αSyn treated primary cortical cultures mediate microglial activation in a dose-dependent manner [51].

Genetic studies suggest a strong association between immune system components and PD; studies have reported human leukocyte antigen (HLA) class II region (a key molecule of the immune system) dysregulation and PD onset risk [52]. Moreover, based on recent research, PD patients are also screened for a pro-inflammatory immune profile i.e., enhanced levels of inflammatory components of the immune system are considered to be associated with accelerated motor system dysfunction and severe cognitive impairment [53]. Epidemiological data suggests a reduced PD risk in individuals taking the non-steroidal anti-inflammatory drug ibuprofen on regular basis [54].

2.2.3 Common Behavioral Alterations in AD and PD and Current Medications

The main symptoms of AD include memory loss or dementia which is usually one of the first symptoms to appear. PD phenotype is characterized by different stages depending on the signs and symptoms that are observed in the affected individuals i.e., motor, and non-motor abnormalities. Hand and muscle weakness leading to slower movements also called bradykinesia, tremors, rigidity or stiffness, unstable gait, drooling, and difficulty in swallowing (dysphagia) are generalized motor dysfunctions observed in the motor stage of PD. While the non-motor stage of PD shares similar phenotypic outcomes as observed in AD such as cognitive deficits e.g., difficulty in learning or remembering new information, trouble in carrying out daily life activities like self-care, sleep disturbances, difficulty reading, speaking, walking, anxiety, depression, trouble recognizing familiar faces like friends and family, and many others.

Currently, no medication exists for the complete cure of AD, but available medications slow down the progression of pathology and help with behavioral problems observed throughout the disease. There are four medicines from two classes of drugs currently prescribed to AD patients: cholinesterase inhibitors which include donepezil, rivastigmine, and galantamine used to treat mild to moderate AD symptoms, and NMDA antagonist memantine for the treatment of AD moderate to severe symptoms. Medications currently prescribed to PD individuals are majorly targeting to improve the dopamine levels in the brain, and the most commonly used drug is levodopa which increases the available dopamine levels in the brain and can ease some of the motor symptoms. Apart from medication, rehabilitation such as physical therapy and exercise can help improve the motor symptoms in the individuals, e.g., improved mobility, gait, flexibility, strength, and quality of life. Occupational therapies are also recommended for patients, the idea of which is to engage people in activities related to daily life and enhance their health and quality of life.

2.3 Neuropsychiatric Pathologies: Recent Mechanistic Findings and Treatments

Neuropsychological diseases/disorders fall in the wide spectrum of mental illnesses that mainly target our emotional states and result in anxious behaviors, depression-like symptoms, mood issues, social deficits, and poor quality of life. The most common diseases in this category are anxiety, depression, bipolar disorder, and schizophrenia. Anxiety, depression, and bipolar disorder share several similar phenotypes and to some extent similar pathophysiology and will be discussed together in the coming section (Fig. 2.4).

2.3.1 Pathophysiology of Neuropsychiatric Disorders

2.3.1.1 Disruption of Neuronal Circuitry in Anxiety, Depression, and Bipolar Disorder

The generalized hypothesis addressing the changes in the brain circuitry in depression and anxiety hypothesizes that there is a dysfunction in the brain areas responsible for governing emotional and cognitive functions. Disruption of neural circuitry involving the amygdala and hippocampus is reported to have a critical role in the onset and progression of anxiety i.e., people who suffer from anxiety tend to show higher activity in the amygdala in response to emotional stimulus. Following the research on individuals through infancy to adolescence has provided the evidence that nucleus accumbens of these individuals are more sensitive when it comes to making decisions compared to other people [55]. Structural MRI studies of depressive patients indicate a reduction in the thalamus, basal ganglia, hippocampus, and prefrontal cortex (PFC), and some other studies have also reported this reduction in the amygdala and anterior cingulate cortex [56]. Meta-analysis across a variety of experimental paradigms found a generalized hypoactivity of dorsal PFC, temporal cortex, insular cortex, cerebellum, and hyperactivity in the thalamus, visual cortex, and anterior PFC associated with depression [57].

The neurogenic model of the bipolar disorder suggests that emotional circuitry in bipolar disorder consists of two substructures i.e., the ventral system (main regulator of emotional behaviors) comprised of the amygdala, insular cortex, ventral striatum, and ventral anterior cingulate cortex, and dorsal system consisting of the hippocampus, dorsal anterior cingulate cortex, and some other parts of the prefrontal cortex. This neurogenic model suggests that in the case of bipolar disorder, generalized hyperactivity is observed in the ventral system while dorsal system activity is diminished. Bipolar disorder is distinguished by the episodes of mania and hypomania; the former encompasses elevated mood levels often called psychosis, abnormally energetic, happy, and irritable, and later covers the episodes of severe depression, suicidal thoughts, and self-harm. Reports have suggested that manic episodes in bipolar disorder are associated with the overactivity of right vPFC and depressive episodes are linked to the under activity of left vPFC [58].

2.3.1.2 Neuroinflammation as a Driving Force in Neuropsychiatric Pathologies

A well-established triggering mechanism in the pathology of anxiety, depression, and bipolar disorder is the development of neuroinflammation. Studies have suggested that during these diseases, microglial phenotypic remodeling occurs in the brain, and the proliferation of proinflammatory phenotype i.e., M1 is increased. These M1 microglia increase the synthesis and release of inflammatory molecules such as IL-6, IL-1β, TNFα, IFNβ, and many others [59]. Microglia also secrete chemokines to facilitate the peripheral immune cells, particularly IL-1β producing monocyte infiltration into the brain, thus exacerbating the ongoing inflammation [60]. There is evidence reporting that sterile inflammation (inflammation occurring in the absence of microorganisms) can affect neuronal functioning, resulting in the release of damage-associated molecules (DAMPs) which are then recognized by microglia and can result in the activation of inflammatory cascades [61]. Anti-inflammatory factors such as TGFβ, IL-2, IL-10, and neurotrophic factors are severely diminished in both humans and animal model studies [62].

There is a generalized shift in the immunological responses in the individuals affected by anxiety, depression, or bipolar disorder, these patients tend to have higher levels of circulating inflammatory Th17 and Th1 cells, along with their immune profiles demonstrating increased expression of proinflammatory cytokines. These cells and factors can cross the compromised BBB and enter CNS to further contribute to an ongoing inflammatory cascade. Neuroinflammation affects the proper functioning of the hypothalamic-pituitary axis also called the HPA axis, which is the main axis that monitors the stress and energy levels in the body and regulates the activities of the hypothalamus, pituitary, and adrenal glands. Studies have provided evidence that neuroinflammation interferes with the serotonin levels in the brain, as affected individuals have lower levels of serotonin both in the brain and circulation.

2.3.1.3 Gut-Brain Axis in Mood Disorders

Gut-brain axis and microbiota have garnered much attention in the last few years, as how the changes in the intestinal microbiota during stress conditions modulate behavioral responses and vice versa. Studies have found that microbiota-free rats and specific pathogens-free rats exhibit increased anxiety and depression-like behavior, particularly through the overactivation of the HPA axis. On contrary, studies have shown that *Lactobacillus and Bifidobacterium* can reduce cortisone levels and reverse the HPA regulation [63]. Intestinal microbiotas can synthesize and secrete several neurotransmitters as *Lactobacillus* subspecies produce acetylcholine, *Candida, Streptococcus, Escherichia coli,* and *Enterococcus* can synthesize and release serotonin, and *Bacilli and Serratia* secrete dopamine. Literature reports that lower levels of dopamine, and increased kynurenine/tryptophan plasma ratio are associated with depression resulting due to disrupted tryptophan metabolism. Germ-free rats administered with microbiota from depressed patients showed a similar trend of increased kynurenine/tryptophan ratio and reduced dopamine levels. Studies are reporting a reduction in the levels of serotonin in germ-free mice and rats particularly in males, pointing to the potential role of microbiota in keeping the balance of several neurotransmitters in the body [64]. Apart from neurotransmitters, the microbiota is also reported to influence the expression of neurotrophic factors such as brain-derived neurotrophic factor (BDNF), which is essential for neuronal cell growth, differentiation, and maturation. Intestinal microbiota, especially *Bifidobacterium* and *Lactobacillus* can increase the BDNF levels in the hippocampus. Similarly, levels of BDNF are reported to be reduced in germ-free mice [65] (Fig. 2.3).

2.3.2 Disease Management and Therapeutic Interventions

Anxiety is characterized by panic, extreme nervousness, uncontrollable thoughts, repeated thoughts of traumatic experiences, inability to stay calm, trouble with sleep, cold or sweaty hands, nausea, muscle tension, and heart palpitations. Major symptoms of depression include anhedonia (feelings of sadness), abnormal eating habits such as eating too much or too little, difficulty in concentration, not enjoying the things that used to give joy, easy irritation, frustration, headaches, and stomachache. Bipolar disorder symptoms include changes in energy levels such as low energy and general tiredness, irritability, lack of motivation, suicidal thoughts, problems with concentration, increased sleeping, and changes in appetite.

Though neuropsychiatric diseases are curable, relapse of the disease can happen at any stage. The management of these diseases involves medication,

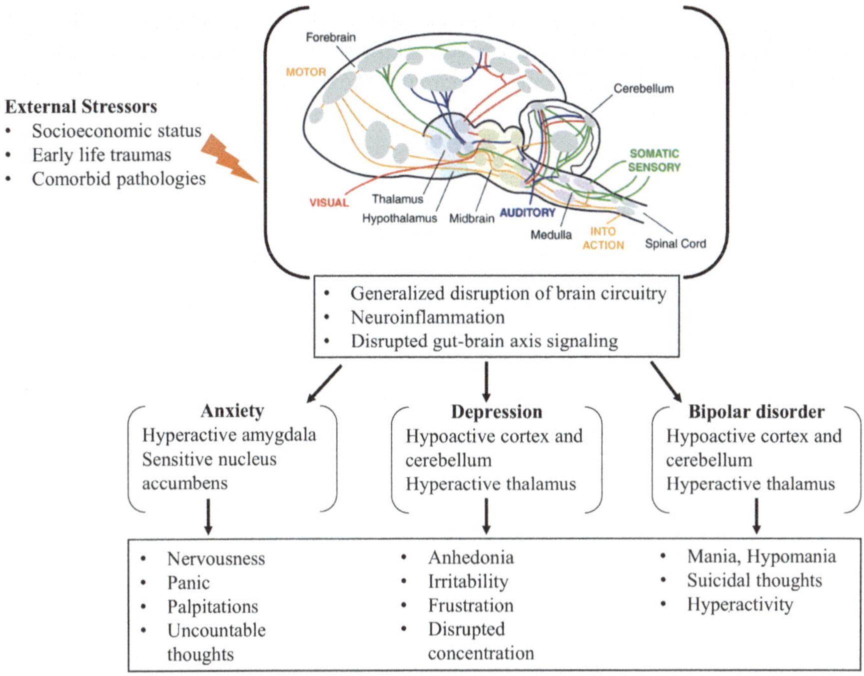

Fig. 2.3 Pathogenesis of neuropsychiatric disorders and their respective phenotypic outcomes: Generalized disruption of brain circuitry is observed in neuropsychiatric disorders along with changes in the immune profile of brain and periphery and alterations in gut-brain axis communication. These changes lead to disease specific behavioral outcomes

psychotherapy, and counseling. Commonly prescribed medicines are anxiolytics such as benzodiazepines, which can alleviate anxiety and panic, and antidepressants, and beta-blockers which can help with physical symptoms like palpitation, shaking, or trembling. Psychotherapy is recommended and essential for these individuals, it helps affected individuals in thought processing, and emotional management, and aid in developing coping skills. Regular exercise and getting enough sleep have proven to alleviate some of the symptoms and overall improve the life and health quality of individuals.

2.4 Traumatic Brain Injuries and Their Management

Traumatic brain injury (TBI) or intracranial injury, is an injury to the brain caused by external blunt traumas or physical force. TBI can be categorized based on the severity of the injury i.e., mild TBI also called concussion to severe TBI, also the injury can be penetrating or closed or affect a specific part of the brain. TBI can result in several signs and symptoms e.g., limitation of physical activity, cognitive and emotional abnormalities, and social and other behavioral changes, and the TBI outcome can range from complete recovery of the individual to permanent disability or death. Causes of TBI are all environmental and external such as gunshot wounds, falls, assaults, vehicle collisions, or domestic violence.

2.4.1 Mechanisms of TBI Progression

TBI pathology is generally categorized into primary and secondary brain injury. The damage that occurs at the time of trauma falls in the category of primary injury and includes compression, stretching, or tattering of brain tissue or blood vessels. After days or weeks of the initial trauma, a complex cascade of intracellular processes and biochemical alterations are observed, called secondary injury. Secondary brain injury complications are the main cause of death in a large population of people killed by brain trauma. The secondary brain injury complications include disruption of BBB integrity which results in the infiltration of peripheral system components into the brain both regulatory and inflammatory resulting in alterations, and in most cases causing inflammation in the brain. Reactive oxygen species (ROS) and free radical levels are elevated in the brain after TBI, which further contribute to an ongoing inflammatory insult. Elevated serum levels of several molecular factors such as IL-1β, IL-6, and CCL2 are reported in patients after TBI. These cytokines and chemokines can cross the damaged BBB and enter the brain, where they can lead to the priming of microglia and astrocytes to their activated state, resulting in further synthesis and release of inflammatory substances by these cells [8].

The dysregulation in the neurotransmitter system particularly glutamate levels and transmission are reported as well, for instance, studies have found excessive synthesis and release of glutamate resulting in the excitotoxicity of the neurons [66]. Changes in the neuronal membrane potential leading to hyperexcitation of cells due to increased sodium and calcium ions influx are reported. Mitochondrial bioenergetics are also disrupted, resulting in the excessive release of free radicals and ROS. Physical injury to the neurons can also take place i.e., neuronal death due to the disconnection of axons from their cell bodies [67]. Ischemia, hypoxia, edema, or intracranial pressure are also observed in TBI patients. Intracranial pressure may arise due to hemorrhage, due to which blood supply to the brain is diminished, resulting in ischemia. On the other hand, too much pressure in the skull can lead to the crushing of the brain by the skull, resulting in brain herniation or death [68] (Fig. 2.4).

2.4.2 TBI Management and Therapeutic Interventions

The common signs of TBI are behavior or mood changes, memory problems, seizures, headaches, nausea, dizziness and fatigue, sleep problems, and slurred speech. All the cognitive, social, and other behavioral alterations along with changes in the

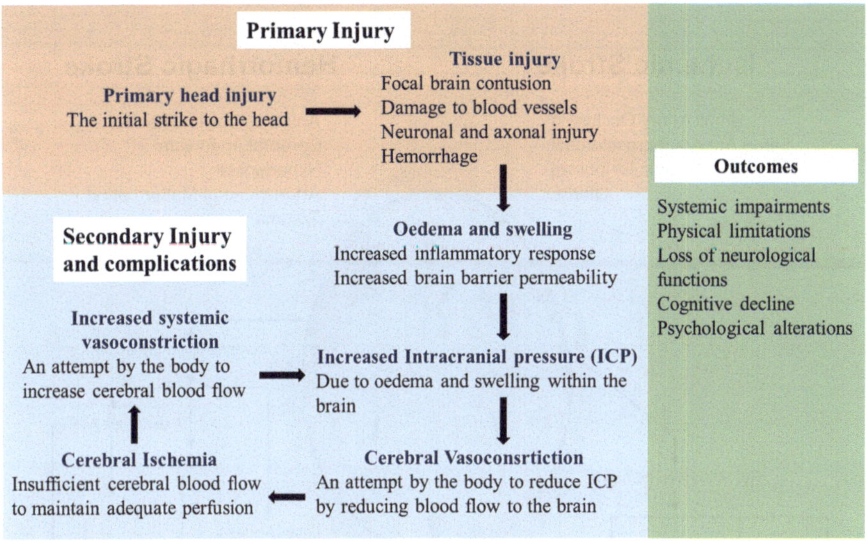

Fig. 2.4 Traumatic brain injury pathogenesis: Following blunt trauma to head also called primary head injury or primary injury can result in blood vessels or neuronal which is then followed by secondary injury. Secondary injury is the set of physiological alterations happening in the brain and systemically, which can lead to various phenotypic outcomes

proper functioning of other systems are the result of secondary brain injury complications. TBI is manageable, and individuals with TBI are usually recommended for counseling to address the cognitive and emotional problems, and rehabilitation including physical, occupational, and speech therapy to address the physical limitations resulted due to TBI. Surgery is often done, if there is internal brain bleeding to prevent the further pooling of the blood and to reduce brain swelling. Complete rest is advised for days to weeks till the individual is capable of self-care and management of professional and occupational activities. In case of physical disability, continuous physical therapy is recommended, and follow-ups are done.

2.5 Stroke, Classifications, and Its Management

A stroke or brain attack occurs when the blood supply to the brain is restricted or diminished due to blood vessel blockage or bursting. As the brain is entirely dependent on the blood supply for its nutrition and oxygenation, stroke can cut off both nutrition and oxygen supply which can result in the death of the surrounding nerve cells. Depending on the site of the stroke, both pooling and lack of blood supply can affect the specific functions, for example, one of the most common symptoms of stroke is contralateral paralysis, meaning if the brain is affected on the right side, it will result in the paralysis of the left side of the body and vice versa. High blood pressure and cholesterol levels, excessive tobacco intake, diabetes, end-stage renal

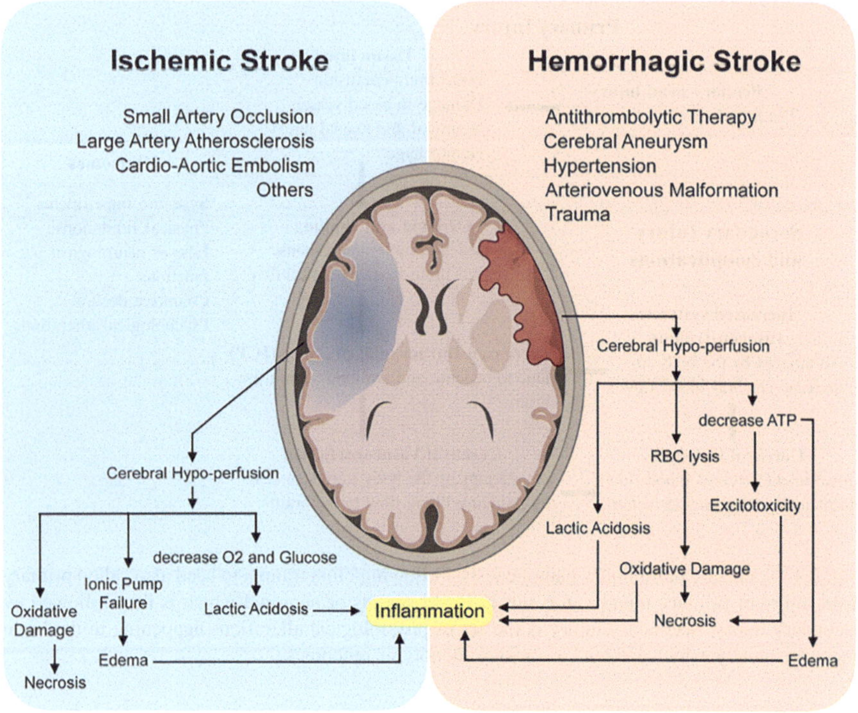

Fig. 2.5 Pathophysiology and Mechanism involved in Ischemic and Hemorrhagic stroke: Adapted from Peng and Jiang, 2022 from Figure 1. Mechanistic alterations taking place in ischemic and hemorrhagic strokes involve cerebral hypo-perfusion leading to lower oxygen supply to the brain. This reduction in the oxygen supply can switch the cellular metabolism from aerobic to anaerobic, resulting in the lactic acid accumulation which can directly damage the neurons and glial cells. Changes in ionic pumps, ecotoxicity, oxidative damage, and edema all lead to inflammatory responses further exacerbating the pathology

disease, or previous transient ischemic attack are the main contributing initiators of stroke (Fig. 2.5).

2.5.1 Classification of Stroke and Their Pathophysiology

2.5.1.1 Ischemic Stroke

In ischemic stroke, a diminished blood supply to some parts or a particular part of the brain is observed, which results in the damage and death of that tissue and dysfunction of that area. Primarily, four causes can lead to ischemic stroke: blockage of a vessel by a locally formed clot (thrombosis), impairment of a blood vessel by a clot formed somewhere else in the body (embolism), a generalized reduction in the blood supply to the brain (hypoperfusion), and the presence of a blood clot in cerebral veins resulting in cerebral venous sinus thrombosis. One of the main initiating factors in stroke is atherosclerosis i.e., the buildup of fats, cholesterol, and other substances inside and outside of the arterial wall. This build-up of substances in and around the arterial wall can lead to the tightening of the blood vessels, thus disrupting the normal blood flow to the brain. In addition to this, atherosclerosis can lead to the formation of multiple blood clots (emboli) by fragmenting the larger atherosclerotic plaques (emboli infarction) [69].

As the brain is fully dependent for its nutritional supply on blood flow, this disruption due to blocked vessels leads to diminished energy levels in the brain, shifting the neuronal metabolism from aerobic to anaerobic. Anaerobic metabolism produces less energy and releases lactic acid as a by-product, which is a neuronal irritant and can potentially destroy the cells and disrupts the homeostatic acid balance of the brain. And the area affected as a result of this ischemic cascade is known as the ischemic penumbra. This shift of cellular metabolism towards the anaerobic can result in the initiation of interrelated processes and pathways that can result in cell injury and death. For example, neuronal injury can result due to abnormally elevated levels of glutamate (excitatory neurotransmitter). Glutamate concentration outside the cells is kept normally low by uptake carriers, their activity is driven by the ion gradient concentration established across the membrane (particularly of sodium). Oxygen and glucose are the main mediators powering ion pumps and keeping the gradients across the neuronal membrane, but during stroke pathology, poor oxygen, and glucose supply lead to the loss of this ion gradient. This loss of transmembrane ion gradients reverses the activity of glutamate transporters, resulting in higher extracellular glutamate levels, thus excitotoxicity and eventually neuronal death.

2.5.1.2 Hemorrhagic Stroke

Hemorrhagic strokes make up 13% of all the reported cases of stroke. Hemorrhagic strokes have two subclasses based on the site of bleeding i.e., rupturing of an artery leading to blood pooling and flooding in the brain (intracerebral hemorrhage) and subarachnoid hemorrhage. In subarachnoid hemorrhage, bleeding happens outside the brain but still within the skull, precisely in the meninges. Both Intraparenchymal hemorrhages which is the accumulation of blood within the brain tissue and intraventricular hemorrhage which is the pooling of blood in the brain ventricles can result in intracerebral hemorrhage.

Hemorrhagic strokes may occur due to alterations in the architecture of the brain vessels, such as cerebral amyloid angiopathy in which amyloid beta plaques deposit inside the blood vessels and meninges of the brain, abnormal connection between the arteries and veins of the brain (cerebral arteriovenous malformation), and intracranial aneurysm that is dilatation or ballooning of the blood vessels due to weakness in the walls of a cerebral artery. Moreover, hemorrhagic strokes often cause specific symptoms, for instance, a severe form of headache known as thunderclap headache results due to subarachnoid hemorrhage. Hemorrhagic stroke can result in both primary and secondary brain injury. For instance, tissue injury can occur due to compression of the surrounding areas from expanding hematoma, this developing pressure can affect the blood supply to the affected area resulting in the loss of blood supply and leading to infarction. In addition, blood released by hemorrhage is reported to have direct toxic effects on the brain tissue and vasculature, and not only this, pooling of blood and cutting off of local blood supply result in secondary brain injury by initiating inflammatory cascades. This inflammatory response is pronounced in several hemorrhagic types but particularly in intracerebral hemorrhage. The inflammatory response in intracerebral hemorrhage is characterized by the activation and accumulation of immune system components, both in the brain and periphery. Intracerebral hemorrhage allows the infiltration of peripheral immune cells such as leukocytes and macrophages and their mediators into the brain, which lead to the activation of local immune cells, microglia as well as astrocytes. The infiltrated cells and activated microglia and astrocytes further contribute to ongoing inflammatory insult by releasing proinflammatory mediators such as IL-1β, TNFα, IL-6, and IFNγ [70].

2.5.2 *Stroke Management and Current Therapeutics*

The first few hours of stroke are critical in managing the following outcomes. In ischemic strokes, definitive treatment is targeted to remove the blood vessel blockage by breaking the blood clot by thrombolysis (by medication) or removing it mechanically by thrombectomy (surgical removal of clots from cerebral arteries). Aspirin also known as a blood thinner and recombinant tissue plasminogen activator (to break down the clots) is reported to have beneficial effects in patients. In cases when a larger portion of the brain is affected, hemicraniectomy i.e., the temporary surgical removal of the skull from one side of the brain to reduce the intracranial pressure is done. Patients with hemorrhagic strokes are monitored for blood pressure, and their oxygenation and blood sugar levels are kept at optimum. In some patients, accumulated blood is removed by surgical interventions to reduce the compression of the surrounding tissue and increased pressure. Stroke rehabilitation is aimed to improve the quality of life for the patients affected by stroke and includes medications, routine follow-ups with clinicians, psychotherapies, physical therapies to aid individuals to return to their normal physical activities, occupational therapies, speech-language therapies, and orthotics (which is the designing of devices to support neuromuscular and skeletal system of the affected individuals).

2.6 Conclusion

This subunit highlighted already established structural changes, molecular alterations, and cellular contributions in various neurological disorders. The recently published reports and literature discussed in this chapter shed light on how diversified diseases can be even if they fall into the same category such as neurodegenerative disorders. The structural and molecular findings discussed have helped a lot in designing various diagnostic tools and therapeutics, though for many disorders a complete cure does not exist yet. But these studies highlight the importance of a pathophysiological understanding of the disease, to design better tools for timely diagnosis and medication for timely treatment.

References

1. Rylaarsdam L, Guemez-Gamboa A (2019) Genetic causes and modifiers of autism spectrum disorder. Front Cell Neurosci 13:385. https://doi.org/10.3389/fncel.2019.00385
2. Pagani M, Barsotti N, Bertero A et al (2021) mTOR-related synaptic pathology causes autism spectrum disorder-associated functional hyperconnectivity. Nat Commun 12:6084. https://doi.org/10.1038/s41467-021-26131-z
3. Pagani M, Bertero A, Liska A et al (2018) Deletion of autism risk gene Shank3 disrupts prefrontal connectivity. bioRxiv:409284. https://doi.org/10.1101/409284
4. Monteiro P, Feng G (2017) SHANK proteins: roles at the synapse and in autism spectrum disorder. Nat Rev Neurosci 18:147–157. https://doi.org/10.1038/nrn.2016.183
5. Garbett K, Ebert PJ, Mitchell A et al (2008) Immune transcriptome alterations in the temporal cortex of subjects with autism. Neurobiol Dis 30:303–311. https://doi.org/10.1016/j.nbd.2008.01.012
6. Vargas DL, Nascimbene C, Krishnan C et al (2005) Neuroglial activation and neuroinflammation in the brain of patients with autism. Ann Neurol 57:67–81. https://doi.org/10.1002/ana.20315
7. Li X, Chauhan A, Sheikh AM et al (2009) Elevated immune response in the brain of autistic patients. J Neuroimmunol 207:111–116. https://doi.org/10.1016/j.jneuroim.2008.12.002
8. Ashwood P, Krakowiak P, Hertz-Picciotto I et al (2011) Elevated plasma cytokines in autism spectrum disorders provide evidence of immune dysfunction and are associated with impaired behavioral outcome. Brain Behav Immun 25:40–45. https://doi.org/10.1016/j.bbi.2010.08.003
9. Masi A, Quintana DS, Glozier N et al (2015) Cytokine aberrations in autism spectrum disorder: a systematic review and meta-analysis. Mol Psychiatry 20:440–446. https://doi.org/10.1038/mp.2014.59
10. Choi GB, Yim YS, Wong H et al (2016) The maternal interleukin-17a pathway in mice promotes autism-like phenotypes in offspring. Science 351:933–939. https://doi.org/10.1126/science.aad0314
11. Banks WA, Kastin AJ, Broadwell RD (1995) Passage of cytokines across the blood-brain barrier. Neuroimmunomodulation 2:241–248. https://doi.org/10.1159/000097202
12. Ponzio NM, Servatius R, Beck K et al (2007) Cytokine levels during pregnancy influence immunological profiles and neurobehavioral patterns of the offspring. Ann N Y Acad Sci 1107:118–128. https://doi.org/10.1196/annals.1381.013
13. Polanczyk G, Rohde LA (2007) Epidemiology of attention-deficit/hyperactivity disorder across the lifespan. Curr Opin Psychiat 20:386–392. https://doi.org/10.1097/YCO.0b013e3281568d7a
14. Faraone SV, Larsson H (2019) Genetics of attention deficit hyperactivity disorder. Mol Psychiatry 24:562–575. https://doi.org/10.1038/s41380-018-0070-0
15. Williams NM, Zaharieva I, Martin A, et al (2010) Rare chromosomal deletions and duplications in attention-deficit hyperactivity disorder: a genome-wide analysis. Lancet (London, England) 376:1401–1408. https://doi.org/10.1016/S0140-6736(10)61109-9
16. Elia J, Glessner JT, Wang K et al (2011) Genome-wide copy number variation study associates metabotropic glutamate receptor gene networks with attention deficit hyperactivity disorder. Nat Genet 44:78–84. https://doi.org/10.1038/ng.1013

17. Lesch K-P, Selch S, Renner TJ et al (2011) Genome-wide copy number variation analysis in attention-deficit/hyperactivity disorder: association with neuropeptide Y gene dosage in an extended pedigree. Mol Psychiatry 16:491–503. https://doi.org/10.1038/mp.2010.29
18. Gizer IR, Ficks C, Waldman ID (2009) Candidate gene studies of ADHD: a meta-analytic review. Hum Genet 126:51–90. https://doi.org/10.1007/s00439-009-0694-x
19. Franke B, Vasquez AA, Johansson S et al (2010) Multicenter analysis of the SLC6A3/DAT1 VNTR haplotype in persistent ADHD suggests differential involvement of the gene in childhood and persistent ADHD. Neuropsychopharmacol Off Publ Am Coll Neuropsychopharmacol 35:656–664. https://doi.org/10.1038/npp.2009.170
20. Eme R (2012) ADHD: an integration with pediatric traumatic brain injury. Expert Rev Neurother 12:475–483. https://doi.org/10.1586/ern.12.15
21. Millichap JG (2008) Etiologic classification of attention-deficit/hyperactivity disorder. Pediatrics 121:e358–e365. https://doi.org/10.1542/peds.2007-1332
22. Tiesler CMT, Heinrich J (2014) Prenatal nicotine exposure and child behavioural problems. Eur Child Adolesc Psychiatry 23:913–929. https://doi.org/10.1007/s00787-014-0615-y
23. Wilens TE, Spencer TJ (2010) Understanding attention-deficit/hyperactivity disorder from childhood to adulthood. Postgrad Med 122:97–109. https://doi.org/10.3810/pgm.2010.09.2206
24. Prince M, Wimo A, Guerchet M, et al (2015) World Alzheimer report 2015. The global impact of dementia. An analysis of prevalence, incidence, cost and trends
25. Thal DR, Rüb U, Orantes M, Braak H (2002) Phases of A beta-deposition in the human brain and its relevance for the development of AD. Neurology 58:1791–1800. https://doi.org/10.1212/wnl.58.12.1791
26. Jansen IE, Savage JE, Watanabe K et al (2019) Genome-wide meta-analysis identifies new loci and functional pathways influencing Alzheimer's disease risk. Nat Genet 51:404–413. https://doi.org/10.1038/s41588-018-0311-9
27. Sims R, Hill M, Williams J (2020) The multiplex model of the genetics of Alzheimer's disease. Nat Neurosci 23:311–322. https://doi.org/10.1038/s41593-020-0599-5
28. He Z, Guo JL, McBride JD et al (2018) Amyloid-β plaques enhance Alzheimer's brain tau-seeded pathologies by facilitating neuritic plaque tau aggregation. Nat Med 24:29–38. https://doi.org/10.1038/nm.4443
29. Roberson ED, Scearce-Levie K, Palop JJ et al (2007) Reducing endogenous tau ameliorates amyloid beta-induced deficits in an Alzheimer's disease mouse model. Science 316:750–754. https://doi.org/10.1126/science.1141736
30. Jin M, Shepardson N, Yang T et al (2011) Soluble amyloid beta-protein dimers isolated from Alzheimer cortex directly induce Tau hyperphosphorylation and neuritic degeneration. Proc Natl Acad Sci U S A 108:5819–5824. https://doi.org/10.1073/pnas.1017033108
31. Heneka MT, Carson MJ, El Khoury J et al (2015) Neuroinflammation in Alzheimer's disease. Lancet Neurol 14:388–405. https://doi.org/10.1016/S1474-4422(15)70016-5
32. Decourt B, Lahiri DK, Sabbagh MN (2017) Targeting tumor necrosis factor alpha for Alzheimer's disease. Curr Alzheimer Res 14:412–425. https://doi.org/10.2174/1567205013666160930110551
33. Wyss-Coray T, Loike JD, Brionne TC et al (2003) Adult mouse astrocytes degrade amyloid-beta in vitro and in situ. Nat Med 9:453–457. https://doi.org/10.1038/nm838
34. Brosseron F, Krauthausen M, Kummer M, Heneka MT (2014) Body fluid cytokine levels in mild cognitive impairment and Alzheimer's disease: a comparative overview. Mol Neurobiol 50:534–544. https://doi.org/10.1007/s12035-014-8657-1
35. Söllvander S, Nikitidou E, Brolin R et al (2016) Accumulation of amyloid-β by astrocytes result in enlarged endosomes and microvesicle-induced apoptosis of neurons. Mol Neurodegener 11:38. https://doi.org/10.1186/s13024-016-0098-z
36. Spillantini MG, Schmidt ML, Lee VM et al (1997) Alpha-synuclein in Lewy bodies. Nature 388:839–840
37. Deng H, Wang P, Jankovic J (2018) The genetics of Parkinson disease. Ageing Res Rev 42:72–85. https://doi.org/10.1016/j.arr.2017.12.007

38. Puschmann A (2017) New genes causing hereditary parkinson's disease or parkinsonism. Curr Neurol Neurosci Rep 17:66. https://doi.org/10.1007/s11910-017-0780-8
39. Ledonne A, Mercuri NB (2017) Current concepts on the physiopathological relevance of dopaminergic receptors. Front Cell Neurosci 11:27. https://doi.org/10.3389/fncel.2017.00027
40. Fujiwara H, Hasegawa M, Dohmae N et al (2002) α-Synuclein is phosphorylated in synucleinopathy lesions. Nat Cell Biol 4:160–164. https://doi.org/10.1038/ncb748
41. Barrett PJ, Timothy Greenamyre J (2015) Post-translational modification of α-synuclein in Parkinson's disease. Brain Res 1628:247–253. https://doi.org/10.1016/j.brainres.2015.06.002
42. Shahmoradian SH, Lewis AJ, Genoud C et al (2019) Lewy pathology in Parkinson's disease consists of crowded organelles and lipid membranes. Nat Neurosci 22:1099–1109. https://doi.org/10.1038/s41593-019-0423-2
43. Wills J, Jones J, Haggerty T et al (2010) Elevated tauopathy and alpha-synuclein pathology in postmortem Parkinson's disease brains with and without dementia. Exp Neurol 225:210–218. https://doi.org/10.1016/j.expneurol.2010.06.017
44. Duka T, Rusnak M, Drolet RE et al (2006) Alpha-synuclein induces hyperphosphorylation of Tau in the MPTP model of parkinsonism. FASEB J Off Publ Fed Am Soc Exp Biol 20:2302–2312. https://doi.org/10.1096/fj.06-6092com
45. Duka T, Sidhu A (2006) The neurotoxin, MPP+, induces hyperphosphorylation of Tau, in the presence of alpha-Synuclein, in SH-SY5Y neuroblastoma cells. Neurotox Res 10:1–10. https://doi.org/10.1007/BF03033329
46. Goris A, Williams-Gray CH, Clark GR et al (2007) Tau and alpha-synuclein in susceptibility to, and dementia in, Parkinson's disease. Ann Neurol 62:145–153. https://doi.org/10.1002/ana.21192
47. McGeer PL, Itagaki S, Boyes BE, McGeer EG (1988) Reactive microglia are positive for HLA-DR in the substantia nigra of Parkinson's and Alzheimer's disease brains. Neurology 38:1285 LP – 1285. https://doi.org/10.1212/WNL.38.8.1285
48. Loeffler DA, Camp DM, Conant SB (2006) Complement activation in the Parkinson's disease substantia nigra: an immunocytochemical study. J Neuroinflammation 3:29. https://doi.org/10.1186/1742-2094-3-29
49. Wu DC, Jackson-Lewis V, Vila M et al (2002) Blockade of microglial activation is neuroprotective in the 1-methyl-4-phenyl-1,2,3,6-tetrahydropyridine mouse model of Parkinson disease. J Neurosci Off J Soc Neurosci 22:1763–1771. https://doi.org/10.1523/JNEUROSCI.22-05-01763.2002
50. He Y, Appel S, Le W (2001) Minocycline inhibits microglial activation and protects nigral cells after 6-hydroxydopamine injection into mouse striatum. Brain Res 909:187–193. https://doi.org/10.1016/S0006-8993(01)02681-6
51. Su X, Maguire-Zeiss KA, Giuliano R et al (2008) Synuclein activates microglia in a model of Parkinson's disease. Neurobiol Aging 29:1690–1701. https://doi.org/10.1016/j.neurobiolaging.2007.04.006
52. Saiki M, Baker A, Williams-Gray CH et al (2010) Association of the human leucocyte antigen region with susceptibility to Parkinson's disease. J Neurol Neurosurg Psychiat 81:890–891. https://doi.org/10.1136/jnnp.2008.162883
53. Williams-Gray CH, Wijeyekoon R, Yarnall AJ et al (2016) Serum immune markers and disease progression in an incident Parkinson's disease cohort (ICICLE-PD). Mov Disord 31:995–1003. https://doi.org/10.1002/mds.26563
54. Gao X, Chen H, Schwarzschild MA, Ascherio A (2011) Use of ibuprofen and risk of Parkinson disease. Neurology 76:863–869. https://doi.org/10.1212/WNL.0b013e31820f2d79
55. Bar-Haim Y, Fox NA, Benson B et al (2009) Neural correlates of reward processing in adolescents with a history of inhibited temperament. Psychol Sci 20:1009–1018. https://doi.org/10.1111/j.1467-9280.2009.02401.x
56. Nugent AC, Davis RM, Zarate CAJ, Drevets WC (2013) Reduced thalamic volumes in major depressive disorder. Psychiatry Res 213:179–185. https://doi.org/10.1016/j.pscychresns.2013.05.004

57. Zhang F-F, Peng W, Sweeney JA et al (2018) Brain structure alterations in depression: psychoradiological evidence. CNS Neurosci Ther 24:994–1003. https://doi.org/10.1111/cns.12835
58. Strakowski SM, Adler CM, Almeida J et al (2012) The functional neuroanatomy of bipolar disorder: a consensus model. Bipolar Disord 14:313–325. https://doi.org/10.1111/j.1399-5618.2012.01022.x
59. Thibaut F (2017) Neuroinflammation: new vistas for neuropsychiatric research. Dialogues Clin Neurosci 19:3–4
60. McKim DB, Weber MD, Niraula A et al (2018) Microglial recruitment of IL-1β-producing monocytes to brain endothelium causes stress-induced anxiety. Mol Psychiatr 23:1421–1431. https://doi.org/10.1038/mp.2017.64
61. Nie X, Kitaoka S, Tanaka K et al (2018) The innate immune receptors TLR2/4 mediate repeated Social defeat stress-induced social avoidance through prefrontal microglial activation. Neuron 99:464–479.e7. https://doi.org/10.1016/j.neuron.2018.06.035
62. Dhabhar FS, Burke HM, Epel ES et al (2009) Low serum IL-10 concentrations and loss of regulatory association between IL-6 and IL-10 in adults with major depression. J Psychiatr Res 43:962–969. https://doi.org/10.1016/j.jpsychires.2009.05.010
63. Zhang C, Franklin CL, Ericsson AC (2021) Consideration of gut microbiome in murine models of diseases. Microorganisms 9
64. Legan TB, Lavoie B, Mawe GM (2022) Direct and indirect mechanisms by which the gut microbiota influence host serotonin systems. Neurogastroenterol Motil Off J Eur Gastrointest Motil Soc 34:e14346. https://doi.org/10.1111/nmo.14346
65. Bercik P, Denou E, Collins J et al (2011) The intestinal microbiota affect central levels of brain-derived neurotropic factor and behavior in mice. Gastroenterology 141:599–609, 609.e1–3. https://doi.org/10.1053/j.gastro.2011.04.052
66. Chen Y-H, Huang EY-K, Kuo T-T et al (2017) Impact of traumatic brain injury on dopaminergic transmission. Cell Transplant 26:1156–1168. https://doi.org/10.1177/0963689717714105
67. Hernandez ML, Chatlos T, Gorse KM, Lafrenaye AD (2019) Neuronal membrane disruption occurs late following diffuse brain trauma in rats and involves a subpopulation of NeuN negative cortical neurons. Front Neurol 10:1238. https://doi.org/10.3389/fneur.2019.01238
68. Rostami E, Engquist H, Enblad P (2014) Imaging of cerebral blood flow in patients with severe traumatic brain injury in the neurointensive care. Front Neurol 5:114. https://doi.org/10.3389/fneur.2014.00114
69. Deb P, Sharma S, Hassan KM (2010) Pathophysiologic mechanisms of acute ischemic stroke: an overview with emphasis on therapeutic significance beyond thrombolysis. Pathophysiology 17:197–218. https://doi.org/10.1016/j.pathophys.2009.12.001
70. Wang J (2010) Preclinical and clinical research on inflammation after intracerebral hemorrhage. Prog Neurobiol 92:463–477. https://doi.org/10.1016/j.pneurobio.2010.08.001

Chapter 3
Physiological Significance of Oxidative Stress and Anti-oxidative System

Saddam Hussain, Azhar Rasul, Ghulam Hussain, Majeeda Rasheed, Maria Manan, Komal Riaz, Saba Riaz, Muhammad Asif Khalil, Ayesha Sadiqa, and Sevki Adem

3.1 Introduction

Two valence electrons have parallel spins in each of their two anti-bonding orbitals in molecular oxygen. This spin restriction allows it to accept a pair of electrons from a donor. A redox reaction is a fundamental metabolic activity in living organisms [1]. The movement of a single electron may result in the formation of free radicals and other issues [2]. Free radicals generally show a high level of reactivity. These radicals are extremely unstable and reactive with other chemicals. Guyton de Morveau coined the term "radical" in 1786, and later, Gay-Lussac, Berzelius, and Liebig used it to refer to unaltered atomic groups in numerous substances [3].

S. Hussain · A. Rasul (✉) · K. Riaz · S. Riaz · M. A. Khalil · A. Sadiqa
Department of Zoology, Faculty of Life Sciences, Government College University, Faisalabad, Pakistan
e-mail: azharrasul@gcuf.edu.pk

G. Hussain
Department of Physiology, Government College University Faisalabad, Faisalabad, Pakistan

M. Rasheed
Department of Life Sciences, Khwaja Fareed University of Engineering and Information Technology, Rahim Yar Khan, Pakistan

M. Manan
Department of Pharmacology, Faculty of Pharmaceutical Sciences, Government College University, Faisalabad, Pakistan

S. Adem
Department of Biochemistry, Çankırı Karatekin Üniversitesi, Çankırı, Turkey

A. Imran, G. Hussain (eds.), *The Role of Natural Antioxidants in Brain Disorders*, Food Bioactive Ingredients, https://doi.org/10.1007/978-3-031-41188-5_3

Free radicals not only take part in pathogenic processes but are also essential for many physiological activities of living organisms, such as healthy aging [4]. Lipid peroxidation was reported to have both negative and positive consequences [5]. Free radicals can cause numerous diseases in humans by damaging lipids, proteins, and DNA. ROS and RNS are responsible for cellular damage by substituting macromolecules [6]. There are numerous antioxidants, both natural and artificial. Endogenous antioxidants are characterized as enzymatic or non-enzymatic [7].

3.2 Roots of Oxidative Stress

An imbalance in the production of reactive oxygen species results in the oxidative stress and capacity of an organism's antioxidative defense mechanisms to lessen the harm due to oxidants. As a byproduct of normal aerobic metabolism, ROS may provide a fundamental health concern when the amount increases in response to stress [8]. The mitochondrion is a primary organelle which is taking part in the production of ROS. ATP is produced by numerous processes including the electron transport chain. Only one or two electron of oxygen are reduced instead of four electrons during this process, which is responsible for the formation of O_2 or H_2O_2, which then changes into other ROS [9]. Free radicals may be created by both internal and external processes. Infection, inflammation, ischemia, immune cell activation, cancer, mental stress, and aging contribute to endogenous free radical formation [10].

Numerous studies show that excessive macronutrient intakes might increase oxidative stress. An excessive amount of high caloric intake will increase the number of substrates entering mitochondrial respiration. As a consequence, the number of contributed electrons to the electron transport chain will be surged [11]. When superoxide concentrations cross a certain point, extra electrons may gather at complex III and donate more electrons to molecular oxygen [12].

ROS generation is fundamentally dependent on enzymatic and non-enzymatic processes. Superoxide radical is produced by xanthine oxidase, peroxidases, and NADPH oxidase [10]. The sole class of enzymes with the specific purpose of producing ROS is the NADPH oxidases, which differentiates it from other enzymes producing ROS as the byproduct of their activity [13]. Free radicals may also be created by non-enzymatic processes like oxygen's interactions with organic materials or the radiation that is exposed to cells. Non-enzymatic free radicals production may also take place during mitochondrial respiration [14].

3.2.1 *Endogenous Sources of ROS Production*

Different cellular organelles with high oxygen consumption rates, including the endoplasmic reticulum, mitochondria, and peroxisomes are examples of endogenous generators of ROS.

3.2.1.1 Production of Oxidative Stress in Mitochondria

Mitochondria generate the majority of the intracellular ROS. Oxidizing radicals are generated at complex-I and complex-III in oxidative phosphorylation [15]. Alongside cytochrome c oxidase, monoamine oxidase, glycerol phosphate dehydrogenase, a-ketoglutarate dehydrogenase, and p66shc also take part in ROS generation within mitochondria [16].

3.2.1.2 Generation of Oxidative Stress in Peroxisomes

The respiratory pathway in peroxisomes involves the transport of electrons from different metabolites to O_2, which ultimately causes the generation of hydrogen peroxides [17]. β-oxidation of fatty acids produces hydrogen peroxide in peroxisomes. $OH^{\bullet}$, H_2O_2, and $O^{2\bullet-}$ are also produced in peroxisome [18].

3.2.1.3 Generation of Oxidative Stress in the Endoplasmic Reticulum

Diamine oxidase, cytochrome b5, and cytochrome P-450 play role in ROS production [19]. Erop1p is a thiol oxidase that leads to the production of H_2O_2 [20]. Auto oxidation of the prostaglandin synthesis, immune cell activation, adrenaline, cytochrome P-450, phagocytic cells, flavin mononucleotide (FMNH2), flavin adenine dinucleotide (FADH2), inflammation, anxiety, mental stress [21], infection, excessive exercise, aging, ischemia, and cancer are other endogenous sources of ROS [19].

3.2.2 *Production of Oxidative Stress by Exogenous Sources*

Various synthetic products are causing oxidative stress directly or via producing by-products. Some of the major exogenous sources for the generation of oxidative stress are given below (Fig. 3.1).

I. **Smoke-generated oxidative stress**

Smoke from cigarettes comprises a variety of extremely unstable free radicals that increase the generation of ROS and RNS and cause oxidative stress [22]. Lung

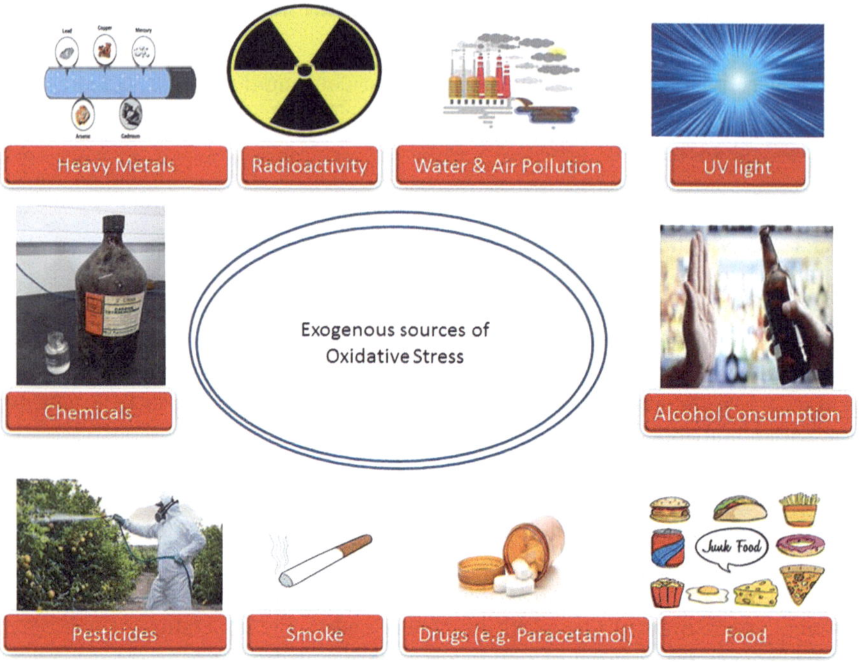

Fig. 3.1 Exogenous sources of oxidative stress

inflammatory cells (macrophages, epithelium, and neutrophils) are affected by cigarette smoke, due to the activation of NADPH oxidase 2, which produces superoxide radicals [23].

II. **Ultra-violet generated oxidative stress**

There are two ways that UV light might harm cellular components. The first method involves the cell and its constituent parts directly absorbing incoming light. This results in the production of an excited state of the molecules following chemical reactions. The second mechanism is photosensitization. Incoming radiation is absorbed by photosensitizers such as bilirubin. As a result, the sensitizers are excited to triple states [24].

III. **Other exogenous sources of oxidative stress**

Other factors like air and water pollution are involved in the production of oxidative stress in the body. Radiations and radioactivity also take part in the production of oxidative stress. Drugs like halothane, bleomycin, paracetamol, doxorubicin, and metronidazole have a record of generation of oxidants. Industrial solvents, pesticides, chemicals like carbon tetrachloride, transition metals, heavy metals, alcohol consumption, and cooking (smoked meat, fat, and junk foods) are also recorded as the sources of oxidative stress [25].

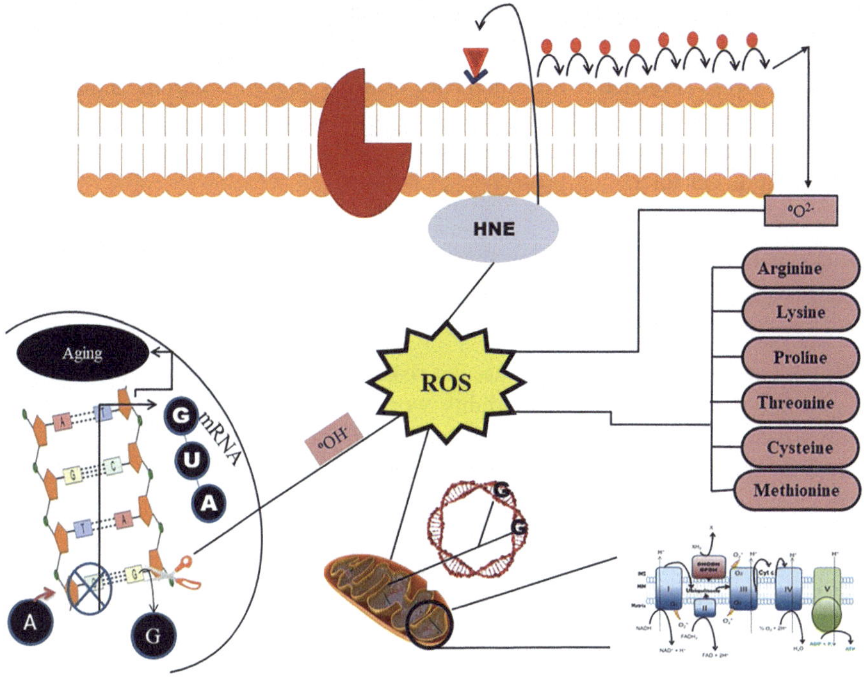

Fig. 3.2 Involvement of ROS in the pathophysiology of cell (**a**) ROS produces lipid peroxides in the cell membrane, inducing lipid peroxidation chain reaction or the generation of aldehydes such as 4-Hydroxy-2-nominal (HNE), that are detrimental to cellular activities via Ca^{2+} signaling and thus cause diseases such as inflammation [26], (**b**) Proteins are primary targets of ROS with reversible or irreversible modifications to the amino acid residues like Cys, Met, Arg and Tyr [26], (**c**) ROS are produced by electron transport chain in mitochondria to a large extent. ROS are generated via single electron leakage in the following situations: (i) during normal ETC function, at complex-I and complex-III; (ii) during conditions of high NADH/ NAD^+ ratio and low electron transport chain activity, (iii) during conditions of a high pool of reduced ubiquinone and transmembrane H^+ gradient, at complex I and (iv) during hypoxic conditions, at complex III [27], (**d**) The oxidative modifications of guanine base is one of the most common forms of DNA damage. Nuclear DNA is far less susceptible to ROS than mitochondrial DNA, which contributes to age-related mitochondrial malfunction. The fact that guanine is quickly oxidized could have important physiological consequences [26].

3.3 Molecular Targets of Free Radicals

Increased generation of RNS and ROS and decreased antioxidant defense result in nitrosative and oxidative stress. Major components of cells (mitochondria, plasma membrane, and DNA molecule) are damaged as shown in Fig. 3.2, leading to multiple disorders [14].

3.4 Role of Oxidative Stress in Health Illness

Oxidative stress is related to the emergence of many acute and chronic ailments in addition to speeding up aging and generating acute illnesses. The impact of oxidative stress on hypertension, Alzheimer's disease, and some malignancies will be covered in this chapter.

3.4.1 Oxidative Stress and Hypertension

The intricate and prevalent cardiovascular risk factor is hypertension [28], which is responsible for morbidity and mortality worldwide [29]. Hypertension is linked with inflammatory processes but is not confirmed whether inflammation is the consequence or cause of hypertension [30]. Tissue damage and remodeling in hypertension ensure its central role in hypertension and its side effects [31].

3.4.1.1 Sources of ROS

Research on ROS sources in hypertension is extensive. These are NADPH oxidase, uncoupled eNOS, xanthine oxidases, and mitochondria [32].

3.4.1.2 Oxidative Stress as a Mediator of Hypertension

In the year of 1991, Nakazono and his colleagues described that the blood pressure of spontaneously hypertensive rats (SHR) was reduced by intravenous injection of a fusion protein composed of human Cu/ Zn SOD and COOH terminal basic peptides with enhanced attraction for heparan sulfate. This result indicated that oxidative stress could be a mediator of hypertension in SHR. Additionally, they discovered that the xanthine oxidase inhibitor oxypurinol decreased the blood pressure in male SHR, correlating hypertension in male SHR to oxidative stress [33].

After 5 years of Nakazono's findings, Rajagopalan and his colleagues reported that administering large doses of angiotensin to rats raised their blood pressure and increased vascular superoxide, which was mediated by NADPH oxidase [34]. Superoxide levels were unaffected by norepinephrine, which raised blood pressure to comparable levels. Vascular dysfunction and constriction were eliminated when researchers administered a liposome-encapsulated superoxide dismutase [35]. These researchers later demonstrated that superoxide probably degraded vascular NO to raise blood pressure [36].

Superoxide binds to the NO generated by endothelial NO synthase (eNOS), forming peroxynitrite. This decreases NO bioavailability which results in vasoconstriction. Furthermore, in the presence of ROS, the eNOS cofactor, tetrahydrobiopterin (BH4) is converted to dihydrobiopterin leading eNOS to synthesize superoxide [37]. The instability was only partially reversed by the addition of BH4. These researchers hypothesized that peroxynitrite can inactivate eNOS by oxidizing BH4, as well as by damaging the enzyme's heme/heme core [38]. Antioxidants such as vitamins E and C; tempol, apocynin, allopurinol, N-acetylcysteine, and BH4 reduced depression according to a study performed in male animals [39].

3.4.2 *Oxidative Stress and Alzheimer's Disease*

Clinical symptoms of Alzheimer's disease include a gradual decline in memory and cognitive abilities and severe dementia. Over the next few decades, people with Alzheimer's disease are expected to rise upto 15 million from the present number of over 4 million [40, 41]. When hyperphosphorylated tau protein aggregates bind to Fe3+, neurofibrillary tangles are produced [42]. The amyloid-peptide may form a chelation complex with transition metal ions, which then catalyzes the production of H_2O_2 and the poisonous OH radical [43]. In AD patients, there is significant lipid peroxidation, which could lead to neuronal loss by a variety of pathways, gathered with impaired activity of glucose transporters, ion pumps, and glutamate transporters. Patients with AD have been found to have additional oxidative protein damage indicators like 3-nitrotyrosine and protein carbonyls [44].

3.4.3 *Oxidative Stress and Cancer*

Cancer ranks among the main causes of mortality in people. Free radicals alter DNA chemically in many ways, make them potentially mutagenic, and contribute to the development of cancer [45, 46]. Cancer cells exhibit increased levels of oxidative stress due to the activation of the oncogenes and loss of tumor suppressors [47]. ROS changes the gene expression and growth signals, which leads to cancer cell proliferation [48].

3.4.3.1 Colorectal Cancer (CRC)

CRC is one of the important types of cancer with 608,000 fatalities per year [49]. ROS from internal and external sources are continually exposed to the gastrointestinal system, especially the colon, and rectum [50]. Epithelial cells are sites where colon cancer begins to develop. These cells have high metabolic rate and divide quickly [51]. This exposure eventually leads to a disrupted intestinal metabolic equilibrium that results in cancer [52].

3.4.3.2 Breast Cancer

ROS damages the breast epithelium which results in hyperplasia of epithelium, breast cancer, and fibroblast proliferation [53]. Thymidine phosphorylase produces oxygen radicals in the carcinoma cell when proteins are quickly glycated. It can be overexpressed in a majority of breast cancer which might cause oxidative stress [54].

3.4.3.3 Prostate Cancer

Cellular growth of prostate cancer is caused by ROS production [55]. Prostate cancer first appears when the protein NADPH oxidase 1 (Nox1) is overexpressed. ROS and Nox1 levels are noticeably greater in prostate cancer [56].

3.4.3.4 Lung Cancer

Among the main global causes of cancer mortality in males, lung cancer has been increasing at a steady rate in recent decades. Approximately 30% of all cancer deaths are caused by lung cancer. Lung inflammation and cancer are two conditions that oxidative stress contributes to significantly [57]. The significant environmental risk factor for lung cancer is cigarette smoking. The particulate matter from cigarette smoke is a complicated combination of several stable ROS and carcinogens with very long half-lives [49].

3.5 Antioxidants and Classification of Antioxidants

Antioxidants may be synthetic or natural. The natural antioxidant system has two categories, enzymatic antioxidants, and non-enzymatic antioxidants as shown in Fig. 3.3 [58]. Free radicals may be stabilized or inactivated by antioxidant enzymes before they damage cellular components. Synthetic antioxidants are chemically prepared substances [58]. Natural antioxidants are further divided into two categories. They may be endogenous and exogenous antioxidants [59]. Exogenous are those antioxidants that we take through food and supplements that are high in antioxidants [60].

Examples of exogenous antioxidants include vitamins, minerals, carotenoids, beta carotene, lycopene, lutein, zeaxanthin, organic sulfur compounds, allium, allyl sulfide, indoles, uric acid, glutathione and polyphenols which are phenolic acids and flavonoids. Flavonoids may be anthocyanidins cyanidin, pelargonidin, isoflavonoids, genistein, flavonols, catechin, EGCG, flavonols quercetin kaempferol, and flavanones. Endogenous antioxidants are the primary defense system including glutathione peroxidase, superoxide dismutase, catalase, and the secondary defense system which includes glucose-6 phosphate dehydrogenase and glutathione reductase.

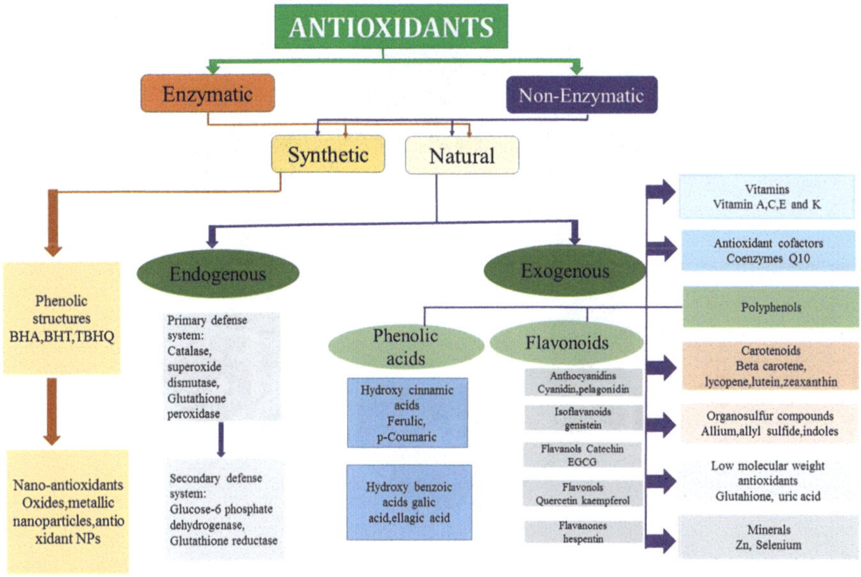

Fig. 3.3 Classification of antioxidants based on enzymatic and non-enzymatic categories

Synthetic antioxidants are also categorized as enzymatic and non-enzymatic antioxidants. They are phenolic structures, nano-antioxidants, oxides, and metallic nanoparticles [61].

3.6 Sources of Antioxidants

Antioxidants are found in natural foods and can also be synthesized. Antioxidants are mostly found in plants [62]. Phenolic structures are endogenous. A brief description of sources of antioxidants is elaborated in Fig. 3.4. We get phenolic structures from apples, grapes, pomace, pomegranate, berries, oranges, tomatoes, olive oil, coffee, and tea. Exogenous may be polyphenols, minerals, carotenoids, vitamins, and organosulfur compounds [63].

Polyphenols are found in spices, berries, nuts, herbs, cocoa powder, flaxseeds, olives, vegetables, coffee, and tea. Polyphenols may trigger apoptosis, inhibit tumor development and increase cell survival since they are prooxidants and antioxidants. However, polyphenols' biological impacts could go well beyond just reducing oxidative stress [64]. Minerals are found in meat, dairy foods, cereals, fish, nuts milk, fruits, and vegetables [65]. Other sources of antioxidants are vitamins which are found in potatoes, citrus fruits, red and green peppers, strawberries, green leafy vegetables, blueberries, blackberries, carrots, and kale [66]. Carotenoids are also the type of antioxidants that are found in spinach, yams, cantaloupe, kale, watermelon, tomatoes, bell peppers, and carrots [67]. Organosulfur compounds are found in

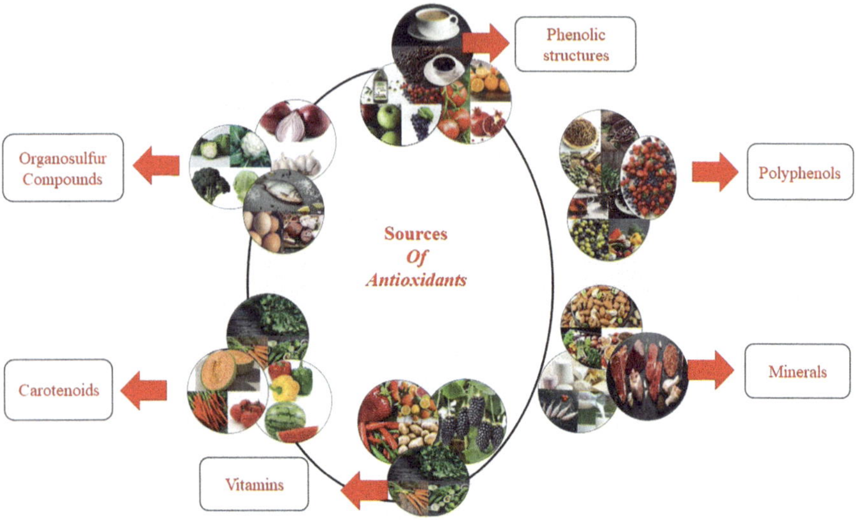

Fig. 3.4 Natural sources of antioxidants enriched with phenolic compounds, polyphenols, minerals, vitamins, carotenoids, and organosulphur compounds

cabbage, broccoli, cauliflower, brussels sprouts, garlic, onion, meat, eggs, and fish [68]. Additionally, there are excellent sources of certain particular antioxidants, such as the allium sulfur compounds found in garlic, onions, and leeks [69]. Anthocyanins are found in berries, grapes, and eggplant [70]. Beta carotene is found in apricots, pumpkins, carrots, mangoes, parsley, and spinach [71]. Flavonoids are found in different fruits, onions, tea, green tea, and apples [72].

3.7 Mechanism of Action of Antioxidants

Reactive intermediates are produced both endogenously and exogenously. Concerning the mechanism of antioxidants, there are five basic ways by which antioxidants work namely (1) radical-scavenging mechanisms (2) H• species donation, (3) oxidant enzyme inhibition, (4) metal chelation, and lastly (5) repair of damaged cell components [73]. Several physical, chemical, and enzymatic factors promote oxidative reactions that result in the loss of an electron from the outermost shell of a given substance [74, 75]. This series of damage is prevented when there are enough antioxidants present in the body through the five mechanisms which are illustrated in Fig. 3.5. The first one employs the free radical scavenging mechanism thus interrupting the chain reactions by inhibiting further oxidation Fig. 3.5 Part 1. The second way of the antioxidant system involves the donation of H^0 species to unstable molecules thus producing a more stable radical which does not contribute to further propagation and is stable comparatively to Fig. 3.5 part 2 [76–87]. The

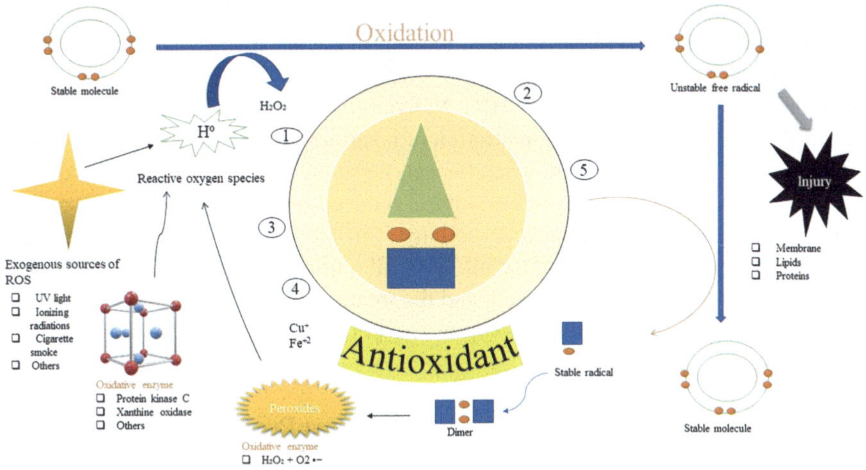

Fig. 3.5 Mechanism of action of antioxidants

third path of the antioxidant system involves the inhibition or deactivation of oxidative enzymes Fig. 3.5 part 03 [88–90]. The fourth mechanism refers to the chelation of different metals such as Fe^2 which results in the production of highly aggressive $HO^•$ radical which, in turn, prevents metal-induced free radical formation Fig. 3.5 part 4 [91–95]. The last mechanism, the fifth one, employs the repairing of damaged components of the cell such as proteins, membrane, lipids, and deoxyribonucleic acid (DNA) [83, 96–99]. Depending upon the structure and nature of the antioxidant agents, the said mechanisms may act alone or in association with one another [74, 100, 101].

3.8 Role of Antioxidants in the Treatment of Different Diseases

Antioxidants play a major role in the treatment of different ailments by scavenging free radicals and eliminating them from the body through different processes. Some of them are enlisted below;

3.8.1 Antioxidants and Hypertension

Hypertension is an important cardiovascular issue that contributes to almost half of the prevalent coronary heart diseases and associated disorders like chronic kidney diseases (CKD) [102–107]. In addition, hypertension ranks third among the list of six major factors which cause global diseases [108].

Antioxidant treatment appears to be an effective method for reestablishing a healthy equilibrium between oxidants and antioxidants in hypertensive patients. Antioxidants have promising potential to relieve hypertension in animal models. In spontaneously hypertensive rats (SHR), NO viability was enhanced and blood pressure was lowered after oral administration of lazaroid, the ROS scavenger medicine [109]. Similar results were seen when N-acetylcysteine (NAC), another antioxidant, was used to treat high blood pressure. NAC reduced blood pressure by preventing ROS production and increasing NOS activity [110]. A similar pattern was observed in SHR given the xanthine oxidase inhibitor allopurinol [111]. The bioavailability of nitric oxide was greatly increased, and the treatment blunted the progressive and time-dependent rise in systolic blood pressure [112].

3.8.1.1 Anti-hypertensive Drugs with Antioxidant Properties

Several molecules with anti-hypertensive and antioxidant properties have been discovered so far. Among these celiprolol, nebivolol, propranolol, and carvedilol got major focus [113]. Tissue lipid peroxidation and oxidative stress are both decreased by propranolol [113, 114]. Patients with heart failure can benefit from carvedilol's free radical scavenging properties which decrease lipid peroxidation [113, 115]. However, not all beta-blockers have these antioxidant properties; for example, atenolol has been demonstrated to possess no affect on ROS generation in lining cells [116].

3.8.2 Antioxidants and Aging

Aging is a universal, inevitable, biological phenomenon affecting almost all living organisms from multicellular to unicellular life [117–119]. When we talk about the process behind the oxidative stress associated with aging, we can't find clear data despite the presence of many different hypotheses, most probably elevated levels of RONS, a process that inhibits the proliferation that results due to damage during replication [120].

Several antioxidants are available which have anti-aging properties such as retinoids [121–123], vitamin C [124–129], tea extracts [130–132], grapes seed extracts [133], peptides, and hydroxy acids have anti-aging character. The interesting thing is that almost all of these are antioxidants [134–144].

3.8.3 Antioxidants and Cancer

Antioxidants have the ability to avoid harmful and sometimes carcinogenic effects. Mice that have been exposed to carcinogens or have lost tumor suppressor genes got benefit from many isoforms of glutathione S-transferases (GSTs) which work together to keep the liver, skin, and colon cancer-free [145–147]. Glutathione Peroxidases (GPXs) can also protect against carcinogen and ROS-induced malignancies initiation in a variety of animals. In colon cancer mouse models, GPX3 inhibits tumor initiation [32]. Similarly, animals with reduced SOD2 expression, either alone or in combination with GPX1 loss, exhibited higher DNA damage and tumor incidence [148, 149].

Catechins, especially epigallocatechin-3-gallate (EGCG), are abundant in green tea (*Camellia sinensis*). Animal studies on carcinogenesis have revealed that EGCG and green tea can reduce tumor growth. Polyphenols found in tea are potent radical scavengers due to the presence of dihydroxy and trihydroxy groups. NRF2-antioxidant response element-dependent upregulation of glutamate cysteine ligase, glutamyl transferase, and heme oxygenase-1 gene expression in EGCG-treated mice [150]. Berberine has been shown to suppress the growth of a wide variety of cancers by binding to oligonucleotides, stabilizing DNA triplexes or G-quadruplexes, and blocking the enzymes telomerase and topoisomerase. Berberine can scavenge reactive oxygen species (ROS), inhibit lipid peroxidation, and decrease metal ion concentrations associated with lipid peroxidation [151].

3.9 Conclusion and Future Perspectives

Oxidative stress arises when the balance between the rate at which oxygen-reactive species are produced and accumulated in cells and tissues and the rate at which the body can eliminate them is disturbed. Mainly ROS is generated as a byproduct of normal cellular reactions. ROS production that is necessarily produced at a limited level is easily diminished, but certain chemicals, drugs, and other sources become responsible for high ROS production. Oxidative stress has a vital role in different diseases including cancer. Antioxidants are substances that counteract oxidative stress. Although a lot of research work regarding the mechanism of product and action of ROS was discovered, more investigations should be done to find out a link between disease and ROS level, food and antioxidant production, and the role of ROS in normal cellular activities. There should be educational seminars and public awareness campaigns that emphasize the importance of antioxidants and encourage antioxidants-enriched diets.

References

1. Gulcin İ (2020) Antioxidants and antioxidant methods: an updated overview. Arch Toxicol 94(3):651–715. https://doi.org/10.1007/s00204-020-02689-3
2. Singh A, Kukreti R, Saso L, Kukreti S (2019) Oxidative stress: a key modulator in neurodegenerative diseases. Molecules 24(8). https://doi.org/10.3390/molecules24081583
3. Di Meo S, Venditti P (2020) Evolution of the knowledge of free radicals and other oxidants. Oxidative Med Cell Longev 2020:9829176. https://doi.org/10.1155/2020/9829176

4. Linnane AW, Kios M, Vitetta L (2007) The essential requirement for superoxide radical and nitric oxide formation for normal physiological function and healthy aging. Mitochondrion 7(1-2):1–5. https://doi.org/10.1016/j.mito.2006.11.009
5. Niki E (2009) Lipid peroxidation: physiological levels and dual biological effects. Free Radic Biol Med 47(5):469–484. https://doi.org/10.1016/j.freeradbiomed.2009.05.032
6. Circu ML, Aw TY (2010) Reactive oxygen species, cellular redox systems, and apoptosis. Free Radic Biol Med 48(6):749–762. https://doi.org/10.1016/j.freeradbiomed.2009.12.022
7. Neha K, Haider MR, Pathak A, Yar MS (2019) Medicinal prospects of antioxidants: a review. Eur J Med Chem 178:687–704. https://doi.org/10.1016/j.ejmech.2019.06.010
8. Rahal A, Kumar A, Singh V, Yadav B, Tiwari R, Chakraborty S, Dhama K (2014) Oxidative stress, prooxidants, and antioxidants: the interplay. Biomed Res Int 2014:761264. https://doi.org/10.1155/2014/761264
9. Inoue M, Sato EF, Nishikawa M, Park AM, Kira Y, Imada I, Utsumi K (2003) Mitochondrial generation of reactive oxygen species and its role in aerobic life. Curr Med Chem 10(23):2495–2505. https://doi.org/10.2174/0929867033456477
10. Pizzino G, Irrera N, Cucinotta M, Pallio G, Mannino F, Arcoraci V, Squadrito F, Altavilla D, Bitto A (2017) Oxidative stress: harms and benefits for human health. Oxidative Med Cell Longev 2017:8416763. https://doi.org/10.1155/2017/8416763
11. Teodoro JS, Duarte FV, Gomes AP, Varela AT, Peixoto FM, Rolo AP, Palmeira CM (2013) Berberine reverts hepatic mitochondrial dysfunction in high-fat fed rats: a possible role for SirT3 activation. Mitochondrion 13(6):637–646. https://doi.org/10.1016/j.mito.2013.09.002
12. Brownlee M (2001) Biochemistry and molecular cell biology of diabetic complications. Nature 414(6865):813–820. https://doi.org/10.1038/414813a
13. Selemidis S, Sobey CG, Wingler K, Schmidt HH, Drummond GR (2008) NADPH oxidases in the vasculature: molecular features, roles in disease and pharmacological inhibition. Pharmacol Ther 120(3):254–291. https://doi.org/10.1016/j.pharmthera.2008.08.005
14. Dröge W (2002) Free radicals in the physiological control of cell function. Physiol Rev 82(1):47–95. https://doi.org/10.1152/physrev.00018.2001
15. Finkel T, Holbrook NJ (2000) Oxidants, oxidative stress and the biology of ageing. Nature 408(6809):239–247. https://doi.org/10.1038/35041687
16. Starkov AA (2008) The role of mitochondria in reactive oxygen species metabolism and signaling. Ann N Y Acad Sci 1147:37–52. https://doi.org/10.1196/annals.1427.015
17. De Duve C, Baudhuin P (1966) Peroxisomes (microbodies and related particles). Physiol Rev 46(2):323–357. https://doi.org/10.1152/physrev.1966.46.2.323
18. Schrader M, Fahimi HD (2006) Peroxisomes and oxidative stress. Biochim Biophys Acta 1763(12):1755–1766. https://doi.org/10.1016/j.bbamcr.2006.09.006
19. Cheeseman KH, Slater TF (1993) An introduction to free radical biochemistry. Br Med Bull 49(3):481–493. https://doi.org/10.1093/oxfordjournals.bmb.a072625
20. Gross E, Sevier CS, Heldman N, Vitu E, Bentzur M, Kaiser CA, Thorpe C, Fass D (2006) Generating disulfides enzymatically: reaction products and electron acceptors of the endoplasmic reticulum thiol oxidase Ero1p. Proc Natl Acad Sci U S A 103(2):299–304. https://doi.org/10.1073/pnas.0506448103
21. Wang L, Muxin G, Nishida H, Shirakawa C, Sato S, Konishi T (2007) Psychological stress-induced oxidative stress as a model of sub-healthy condition and the effect of TCM. Evid Based Complement Alternat Med 4(2):195–202. https://doi.org/10.1093/ecam/nel080
22. Padmavathi P, Raghu PS, Reddy VD, Bulle S, Marthadu SB, Maturu P, Varadacharyulu N (2018) Chronic cigarette smoking-induced oxidative/nitrosative stress in human erythrocytes and platelets. Mol Cell Toxicol 14(1):27–34
23. Gloire G, Legrand-Poels S, Piette J (2006) NF-kappaB activation by reactive oxygen species: fifteen years later. Biochem Pharmacol 72(11):1493–1505. https://doi.org/10.1016/j.bcp.2006.04.011
24. de Jager TL, Cockrell AE, Du Plessis SS (2017) Ultraviolet light induced generation of reactive oxygen species. Adv Exp Med Biol 996:15–23. https://doi.org/10.1007/978-3-319-56017-5_2

25. Pham-Huy LA, He H, Pham-Huy C (2008) Free radicals, antioxidants in disease and health. Int J Biomed Sci 4(2):89–96
26. Radak Z, Zhao Z, Goto S, Koltai E (2011) Age-associated neurodegeneration and oxidative damage to lipids, proteins and DNA. Mol Asp Med 32(4-6):305–315. https://doi.org/10.1016/j.mam.2011.10.010
27. Payen VL, Zampieri LX, Porporato PE, Sonveaux P (2019) Pro- and antitumor effects of mitochondrial reactive oxygen species. Cancer Metastasis Rev 38(1-2):189–203. https://doi.org/10.1007/s10555-019-09789-2
28. Ventura HO, Taler SJ, Strobeck JE (2005) Hypertension as a hemodynamic disease: the role of impedance cardiography in diagnostic, prognostic, and therapeutic decision making. Am J Hypertens 18(S2):26S–43S
29. Kearney PM, Whelton M, Reynolds K, Muntner P, Whelton PK, He J (2005) Global burden of hypertension: analysis of worldwide data. Lancet 365(9455):217–223
30. Agita A, Alsagaff MT (2017) Inflammation, immunity, and hypertension. Acta Medica Indonesiana 49(2):158
31. Castro MM, Tanus-Santos JE (2013) Inhibition of matrix metalloproteinases (MMPs) as a potential strategy to ameliorate hypertension-induced cardiovascular alterations. Curr Drug Targets 14(3):335–343
32. Pinheiro LC, Oliveira-Paula GH (2020) Sources and effects of oxidative stress in hypertension. Curr Hypertens Rev 16(3):166–180
33. Nakazono K, Watanabe N, Matsuno K, Sasaki J, Sato T, Inoue M (1991) Does superoxide underlie the pathogenesis of hypertension? Proc Natl Acad Sci 88(22):10045–10048
34. Rajagopalan S, Kurz S, Münzel T, Tarpey M, Freeman BA, Griendling KK, Harrison DG (1996) Angiotensin II-mediated hypertension in the rat increases vascular superoxide production via membrane NADH/NADPH oxidase activation. Contribution to alterations of vasomotor tone. J Clin Investig 97(8):1916–1923
35. Ojeda NB, Hennington BS, Williamson DT, Hill ML, Betson NE, Sartori-Valinotti JC, Reckelhoff JF, Royals TP, Alexander BT (2012) Oxidative stress contributes to sex differences in blood pressure in adult growth-restricted offspring. Hypertension 60(1):114–122
36. Laursen JB, Rajagopalan S, Galis Z, Tarpey M, Freeman BA, Harrison DG (1997) Role of superoxide in angiotensin II–induced but not catecholamine-induced hypertension. Circulation 95(3):588–593
37. Chen D-D, Chen L-Y, Xie J-B, Shu C, Yang T, Zhou S, Yuan HF, Chen A (2014) Tetrahydrobiopterin regulation of eNOS redox function. Curr Pharm Des 20(22):3554–3562
38. Chen W, Druhan LJ, Chen C-A, Hemann C, Chen Y-R, Berka V, Tsai A-L, Zweier JL (2010) Peroxynitrite induces destruction of the tetrahydrobiopterin and heme in endothelial nitric oxide synthase: transition from reversible to irreversible enzyme inhibition. Biochemistry 49(14):3129–3137
39. Montezano AC, Dulak-Lis M, Tsiropoulou S, Harvey A, Briones AM, Touyz RM (2015) Oxidative stress and human hypertension: vascular mechanisms, biomarkers, and novel therapies. Can J Cardiol 31(5):631–641
40. Katzman R, Saitoh T (1991) Advances in Alzheimer's disease. FASEB J 5(3):278–286
41. Salmon DP, Thomas R, Pay M, Booth A, Hofstetter C, Thal L, Katzman R (2002) Alzheimer's disease can be accurately diagnosed in very mildly impaired individuals. Neurology 59(7):1022–1028
42. Chang Y, Kong Q, Shan X, Tian G, Ilieva H, Cleveland DW, Rothstein JD, Borchelt DR, Wong PC, Lin C-lG (2008) Messenger RNA oxidation occurs early in disease pathogenesis and promotes motor neuron degeneration in ALS. PLoS One 3(8):e2849
43. Uttara B, Singh AV, Zamboni P, Mahajan R (2009) Oxidative stress and neurodegenerative diseases: a review of upstream and downstream antioxidant therapeutic options. Curr Neuropharmacol 7(1):65–74
44. Butterfield DA, Perluigi M, Sultana R (2006) Oxidative stress in Alzheimer's disease brain: new insights from redox proteomics. Eur J Pharmacol 545(1):39–50

45. Goldstein BD, Witz G (1990) Free radicals and carcinogenesis. Free Radic Res Commun 11(1-3):3–10
46. Dreher D, Junod AF (1996) Role of oxygen free radicals in cancer development. Eur J Cancer 32(1):30–38
47. Cairns R, Harris I, McCracken S, Mak T Cancer cell metabolism. In: Cold Spring Harbor symposia on quantitative biology, 2011. Cold Spring Harbor Laboratory Press, pp 299–311
48. Valko M, Leibfritz D, Moncol J, Cronin MT, Mazur M, Telser J (2007) Free radicals and antioxidants in normal physiological functions and human disease. Int J Biochem Cell Biol 39(1):44–84
49. Jemal A, Bray F, Center MM, Ferlay J, Ward E, Forman D (2011) Global cancer statistics. CA Cancer J Clin 61(2):69–90
50. Blau S, Rubinstein A, Bass P, Singaram C, Kohen R (1999) Differences in the reducing power along the rat GI tract: lower antioxidant capacity of the colon. Mol Cell Biochem 194(1):185–191
51. Foksinski M, Rozalski R, Guz J, Ruszkowska B, Sztukowska P, Piwowarski M, Klungland A, Olinski R (2004) Urinary excretion of DNA repair products correlates with metabolic rates as well as with maximum life spans of different mammalian species. Free Radic Biol Med 37(9):1449–1454
52. Guz J, Foksinski M, Siomek A, Gackowski D, Rozalski R, Dziaman T, Szpila A, Olinski R (2008) The relationship between 8-oxo-7, 8-dihydro-2′-deoxyguanosine level and extent of cytosine methylation in leukocytes DNA of healthy subjects and in patients with colon adenomas and carcinomas. Mutat Res/Fundament Mol Mech Mutagen 640(1-2):170–173
53. Murrell T (1991) Epidemiological and biochemical support for a theory on the cause and prevention of breast cancer. Med Hypotheses 36(4):389–396
54. Brown NS, Jones A, Fujiyama C, Harris AL, Bicknell R (2000) Thymidine phosphorylase induces carcinoma cell oxidative stress and promotes secretion of angiogenic factors. Cancer Res 60(22):6298–6302
55. Arnold RS, He J, Remo A, Ritsick D, Yin-Goen Q, Lambeth JD, Datta MW, Young AN, Petros JA (2007) Nox1 expression determines cellular reactive oxygen and modulates c-fos-induced growth factor, interleukin-8, and Cav-1. Am J Pathol 171(6):2021–2032
56. Lim SD, Sun C, Lambeth JD, Marshall F, Amin M, Chung L, Petros JA, Arnold RS (2005) Increased Nox1 and hydrogen peroxide in prostate cancer. Prostate 62(2):200–207
57. Azad N, Rojanasakul Y, Vallyathan V (2008) Inflammation and lung cancer: roles of reactive oxygen/nitrogen species. J Toxicol Environ Health Part B 11(1):1–15
58. Atta EM, Mohamed NH, Silaev AAA (2017) Antioxidants: An overview on the natural and synthetic types. Eur Chem Bull 6(8):365–375
59. Moussa Z, Judeh Z, Ahmed SA (2019) Nonenzymatic exogenous and endogenous antioxidants. Free Rad Med Biol:1–22
60. Admassu S, Kebede M (2019) Application of antioxidants in food processing industry: Options to improve the extraction yields and market value of natural products. Adv Food Technol Nutr Sci 5:38–49
61. Bouayed J, Bohn T (2010) Exogenous antioxidants – double-edged swords in cellular redox state: Health beneficial effects at physiologic doses versus deleterious effects at high doses. Oxidative Med Cell Longev 3(4):228–237. https://doi.org/10.4161/oxim.3.4.12858
62. Shahidi F (2015) Handbook of antioxidants for food preservation. Woodhead Publishing
63. Tan BL, Norhaizan ME, Liew WP, Sulaiman Rahman H (2018) Antioxidant and oxidative stress: a mutual interplay in age-related diseases. Front Pharmacol 9:1162. https://doi.org/10.3389/fphar.2018.01162
64. Scalbert A, Johnson IT, Saltmarsh M (2005) Polyphenols: antioxidants and beyond. Am J Clin Nutr 81 (1 Suppl):215s–217s. https://doi.org/10.1093/ajcn/81.1.215S
65. Gammeren DV (2008) Vitamins and minerals. Essentials of sports nutrition and supplements. Springer, In, pp 313–328
66. Hands ES (1990) Food finder: food sources of vitamins and minerals. Esha Research

67. Xavier AA, Pérez-Gálvez A (2016) Carotenoids as a source of antioxidants in the diet. Subcell Biochem 79:359–375. https://doi.org/10.1007/978-3-319-39126-7_14
68. Petropoulos S, Di Gioia F, Ntatsi G (2017) Vegetable organosulfur compounds and their health promoting effects. Curr Pharm Des 23(19):2850–2875. https://doi.org/10.2174/1381612823666170111100531
69. Nicastro HL, Ross SA, Milner JA (2015) Garlic and onions: their cancer prevention properties. Cancer Prev Res (Phila) 8(3):181–189. https://doi.org/10.1158/1940-6207.Capr-14-0172
70. Netzel M, Strass G, Janssen M, Bitsch I, Bitsch R (2001) Bioactive anthocyanins detected in human urine after ingestion of blackcurrant juice. J Environ Pathol Toxicol Oncol 20(2):89–95
71. Arts IC, van De Putte B, Hollman PC (2000) Catechin contents of foods commonly consumed in The Netherlands. 2. Tea, wine, fruit juices, and chocolate milk. J Agric Food Chem 48(5):1752–1757. https://doi.org/10.1021/jf000026+
72. Panche AN, Diwan AD, Chandra SR (2016) Flavonoids: an overview. J Nutr Sci 5:e47. https://doi.org/10.1017/jns.2016.41
73. Bhattacharyya A, Chattopadhyay R, Mitra S, Crowe SE (2014) Oxidative stress: an essential factor in the pathogenesis of gastrointestinal mucosal diseases. Physiol Rev 94(2):329–354. https://doi.org/10.1152/physrev.00040.2012
74. Fridovich I (1989) Superoxide dismutases. An adaptation to a paramagnetic gas. J Biol Chem 264(14):7761–7764
75. Battino M, Ferreiro MS, Gallardo I, Newman HN, Bullon P (2002) The antioxidant capacity of saliva. J Clin Periodontol 29(3):189–194. https://doi.org/10.1034/j.1600-051x.2002.290301x.x
76. Farghaly TA, Al-Hussain SA, Zaki ME, Asghar BH, Muhammad ZA (2022) Synthesis of spiropyrazoles under organic and nonorganic catalysis. Curr Org Chem 26(9):834–856
77. Piatkivskyi A, Lau JK-C, Berden G, Oomens J, Hopkinson AC, Siu KM, Ryzhov V (2019) Hydrogen atom transfer in the radical cations of tryptophan-containing peptides AW and WA studied by mass spectrometry, infrared multiple-photon dissociation spectroscopy, and theoretical calculations. Eur J Mass Spectromet 25(1):112–121
78. Guo Y, Baschieri A, Mollica F, Valgimigli L, Cedrowski J, Litwinienko G, Amorati R (2021) Hydrogen atom transfer from HOO(.) to ortho-quinones explains the antioxidant activity of polydopamine. Angew Chem Int Ed Eng 60(28):15220–15224. https://doi.org/10.1002/anie.202101033
79. Galeotti M, Trasatti C, Sisti S, Salamone M, Bietti M (2022) Factors governing reactivity and selectivity in hydrogen atom transfer from C(sp(3))-H bonds of nitrogen-containing heterocycles to the cumyloxyl radical. J Org Chem 87(11):7456–7463. https://doi.org/10.1021/acs.joc.2c00955
80. Lesslie M, Lau JK, Lawler JT, Siu KW, Oomens J, Berden G, Hopkinson AC, Ryzhov V (2016) Alkali-metal-ion-assisted hydrogen atom transfer in the homocysteine radical. Chemistry 22(7):2243–2246. https://doi.org/10.1002/chem.201504631
81. León EI, Martín Á, Montes AS, Pérez-Martín I, Del Sol RM, Suárez E (2021) 1,5-hydrogen atom transfer/Surzur-Tanner rearrangement: a radical cascade approach for the synthesis of 1,6-dioxaspiro[4.5]decane and 6,8-dioxabicyclo[3.2.1]octane scaffolds in carbohydrate systems. J Org Chem 86(21):14508–14552. https://doi.org/10.1021/acs.joc.1c01376
82. Zheng YZ, Deng G, Zhang YC (2021) Multiple free radical scavenging reactions of aurones. Phytochemistry 190:112853. https://doi.org/10.1016/j.phytochem.2021.112853
83. Dalla Tiezza M, Hamlin TA, Bickelhaupt FM, Orian L (2021) Radical scavenging potential of the phenothiazine scaffold: a computational analysis. ChemMedChem 16(24):3763–3771. https://doi.org/10.1002/cmdc.202100546
84. Boulebd H (2021) Modeling the peroxyl radical scavenging behavior of carnosic acid: mechanism, kinetics, and effects of physiological environments. Phytochemistry 192:112950. https://doi.org/10.1016/j.phytochem.2021.112950

85. Vo QV, Hoa NT, Nam PC, Quang DT, Mechler A (2020) In silico evaluation of the radical scavenging mechanism of mactanamide. ACS Omega 5(37):24106–24110. https://doi.org/10.1021/acsomega.0c03646
86. Alisi IO, Uzairu A, Abechi SE (2020) Free radical scavenging mechanism of 1,3,4-oxadiazole derivatives: thermodynamics of O-H and N-H bond cleavage. Heliyon 6(3):e03683. https://doi.org/10.1016/j.heliyon.2020.e03683
87. Miro P, Marin ML, Miranda MA (2016) Radical-mediated dehydrogenation of bile acids by means of hydrogen atom transfer to triplet carbonyls. Org Biomol Chem 14(9):2679–2683. https://doi.org/10.1039/c5ob02561c
88. Herman A, Herman AP (2013) Caffeine's mechanisms of action and its cosmetic use. Skin Pharmacol Physiol 26(1):8–14. https://doi.org/10.1159/000343174
89. Ranganatha VL, Begum AB, Naveen P, Zameer F, Hegdekatte R, Khanum SA (2014) Synthesis, xanthine oxidase inhibition, and antioxidant screening of benzophenone tagged thiazolidinone analogs. Arch Pharm (Weinheim) 347(8):589–598. https://doi.org/10.1002/ardp.201400058
90. Mohamed Isa SSP, Ablat A, Mohamad J (2018) The antioxidant and xanthine oxidase inhibitory activity of plumeria rubra flowers. Molecules 23(2). https://doi.org/10.3390/molecules23020400
91. Simunkova M, Alwasel SH, Alhazza IM, Jomova K, Kollar V, Rusko M, Valko M (2019) Management of oxidative stress and other pathologies in Alzheimer's disease. Arch Toxicol 93(9):2491–2513. https://doi.org/10.1007/s00204-019-02538-y
92. Patrick L (2002) Mercury toxicity and antioxidants: Part 1: role of glutathione and alpha-lipoic acid in the treatment of mercury toxicity. Altern Med Rev 7(6):456–471
93. Parvin MS, Chlebek J, Hošťálková A, Catapano MC, Lomozová Z, Macáková K, Mladěnka P (2022) Interactions of isoquinoline alkaloids with transition metals iron and copper. Molecules 27(19). https://doi.org/10.3390/molecules27196429
94. Filipský T, Říha M, Macáková K, Anzenbacherová E, Karlíčková J, Mladěnka P (2015) Antioxidant effects of coumarins include direct radical scavenging, metal chelation and inhibition of ROS-producing enzymes. Curr Top Med Chem 15(5):415–431. https://doi.org/10.2174/1568026615666150206152233
95. Ajisaka K, Oyanagi Y, Miyazaki T, Suzuki Y (2016) Effect of the chelation of metal cation on the antioxidant activity of chondroitin sulfates. Biosci Biotechnol Biochem 80(6):1179–1185. https://doi.org/10.1080/09168451.2016.1141036
96. Lin TK, Zhong L, Santiago JL (2017) Anti-inflammatory and skin barrier repair effects of topical application of some plant oils. Int J Mol Sci 19(1). https://doi.org/10.3390/ijms19010070
97. de Oliveira SE, Batista R (2017) Ferulic acid and naturally occurring compounds bearing a feruloyl moiety: a review on their structures, occurrence, and potential health benefits. Compr Rev Food Sci Food Saf 16(4):580–616. https://doi.org/10.1111/1541-4337.12266
98. Deng L, Du C, Song P, Chen T, Rui S, Armstrong DG, Deng W (2021) The role of oxidative stress and antioxidants in diabetic wound healing. Oxidative Med Cell Longev 2021:8852759. https://doi.org/10.1155/2021/8852759
99. Xu Z, Han S, Gu Z, Wu J (2020) Advances and impact of antioxidant hydrogel in chronic wound healing. Adv Healthc Mater 9(5):e1901502. https://doi.org/10.1002/adhm.201901502
100. Dai J, Mumper RJ (2010) Plant phenolics: extraction, analysis and their antioxidant and anticancer properties. Molecules 15(10):7313–7352. https://doi.org/10.3390/molecules15107313
101. Pfister R, Heider K, Illgen B, Beglinger R (1990) Trichospirura leptostoma: a possible cause of wasting disease in the marmoset. Z Versuchstierkd 33(4):157–161
102. LT FMAAA, Anderson HR, Bhutta ZA, Biryukov S et al (2016) Global, regional, and national comparative risk assessment of 79 behavioural, environmental and occupational, and metabolic risks or clusters of risks, 1990–2015: a systematic analysis for the global burden of disease study 2015. Lancet 388:1659–1724

103. Colafella KMM, Denton KM (2018) Sex-specific differences in hypertension and associated cardiovascular disease. Nat Rev Nephrol 14(3):185–201. https://doi.org/10.1038/nrneph.2017.189
104. Di Raimondo D, Buscemi S, Musiari G, Rizzo G, Pirera E, Corleo D, Pinto A, Tuttolomondo A (2021) Ketogenic diet, physical activity, and hypertension – a narrative review. Nutrients 13(8). https://doi.org/10.3390/nu13082567
105. Aidietis A, Laucevicius A, Marinskis G (2007) Hypertension and cardiac arrhythmias. Curr Pharm Des 13(25):2545–2555. https://doi.org/10.2174/138161207781663037
106. Ferdinand DP, Nedunchezhian S, Ferdinand KC (2020) Hypertension in African Americans: Advances in community outreach and public health approaches. Prog Cardiovasc Dis 63(1):40–45. https://doi.org/10.1016/j.pcad.2019.12.005
107. Hermida RC, Crespo JJ, Domínguez-Sardiña M, Otero A, Moyá A, Ríos MT, Sineiro E, Castiñeira MC, Callejas PA, Pousa L, Salgado JL, Durán C, Sánchez JJ, Fernández JR, Mojón A, Ayala DE (2020) Bedtime hypertension treatment improves cardiovascular risk reduction: the hygia chronotherapy trial. Eur Heart J 41(48):4565–4576. https://doi.org/10.1093/eurheartj/ehz754
108. Ondimu DO, Kikuvi GM, Otieno WN (2019) Risk factors for hypertension among young adults (18-35) years attending in Tenwek Mission Hospital, Bomet County, Kenya in 2018. Pan Afr Med J 33:210. doi:https://doi.org/10.11604/pamj.2019.33.210.18407
109. Vaziri N (2000) Ni Z, Oveisi F, and Trnavsky-Hobbs DL. Effect of antioxidant therapy on blood pressure and NO synthase expression in hypertensive rats. Hypertension 36:957–964
110. Ahmad KA, Yuan Yuan D, Nawaz W, Ze H, Zhuo CX, Talal B, Taleb A, Mais E, Qilong D (2017) Antioxidant therapy for management of oxidative stress induced hypertension. Free Radic Res 51(4):428–438
111. Mazzali M, Hughes J, Kim YG, Jefferson JA, Kang DH, Gordon KL, Lan HY, Kivlighn S, Johnson RJ (2001) Elevated uric acid increases blood pressure in the rat by a novel crystal-independent mechanism. Hypertension 38:1101–1106
112. Greig JA, Shirley R, Graham D, Denby L, Dominiczak AF, Work LM, Baker AH (2010) Vascular-targeting antioxidant therapy in a model of hypertension and stroke. J Cardiovasc Pharmacol 56(6):642–650
113. Yao E-H, Fukuda N, Matsumoto T, Katakawa M, Yamamoto C, Han Y, Ueno T, Kobayashi N, Matsumoto K (2008) Effects of the antioxidative β-blocker celiprolol on endothelial progenitor cells in hypertensive rats. Am J Hypertens 21(9):1062–1068
114. Mak IT, Weglicki WB (1988) Protection by beta-blocking agents against free radical-mediated sarcolemmal lipid peroxidation. Circ Res 63(1):262–266
115. Kukin ML, Kalman J, Charney RH, Levy DK, Buchholz-Varley C, Ocampo ON, Eng C (1999) Prospective, randomized comparison of effect of long-term treatment with metoprolol or carvedilol on symptoms, exercise, ejection fraction, and oxidative stress in heart failure. Circulation 99(20):2645–2651
116. Pasini AF, Garbin U, Nava MC, Stranieri C, Davoli A, Sawamura T, Cascio VL, Cominacini L (2005) Nebivolol decreases oxidative stress in essential hypertensive patients and increases nitric oxide by reducing its oxidative inactivation. J Hypertens 23(3):589–596
117. Lin SJ, Austriaco N (2014) Aging and cell death in the other yeasts, Schizosaccharomyces pombe and Candida albicans. FEMS Yeast Res 14(1):119–135. https://doi.org/10.1111/1567-1364.12113
118. Książek K (2010) Let's stop overlooking bacterial aging. Biogerontology 11(6):717–723. https://doi.org/10.1007/s10522-010-9278-3
119. Liguori I, Russo G, Curcio F, Bulli G, Aran L, Della-Morte D, Gargiulo G, Testa G, Cacciatore F, Bonaduce D, Abete P (2018) Oxidative stress, aging, and diseases. Clin Interv Aging 13:757–772. https://doi.org/10.2147/cia.S158513
120. Liguori I, Russo G, Curcio F, Bulli G, Aran L, Della-Morte D, Gargiulo G, Testa G, Cacciatore F, Bonaduce D et al (2018) Oxidative stress, aging, and diseases. Clin Interv Aging 13:757–772

121. Sadgrove NJ, Simmonds MSJ (2021) Topical and nutricosmetic products for healthy hair and dermal antiaging using "dual-acting" (2 for 1) plant-based peptides, hormones, and cannabinoids. FASEB Bioadv 3(8):601–610. https://doi.org/10.1096/fba.2021-00022
122. Milosheska D, Roškar R (2022) Use of retinoids in topical antiaging treatments: a focused review of clinical evidence for conventional and nanoformulations. Adv Ther 39(12):5351–5375. https://doi.org/10.1007/s12325-022-02319-7
123. Berry K, Hallock K, Lam C (2022) Photoaging and topical rejuvenation. Facial Plast Surg Clin North Am 30(3):291–300. https://doi.org/10.1016/j.fsc.2022.03.003
124. Mumtaz S, Ali S, Tahir HM, Kazmi SAR, Shakir HA, Mughal TA, Mumtaz S, Summer M, Farooq MA (2021) Aging and its treatment with vitamin C: a comprehensive mechanistic review. Mol Biol Rep 48(12):8141–8153. https://doi.org/10.1007/s11033-021-06781-4
125. Tonnard P, Verpaele A, Carvas M (2020) Fat grafting for facial rejuvenation with nanofat grafts. Clin Plast Surg 47(1):53–62. https://doi.org/10.1016/j.cps.2019.08.006
126. Masaki H (2010) Role of antioxidants in the skin: anti-aging effects. J Dermatol Sci 58(2):85–90. https://doi.org/10.1016/j.jdermsci.2010.03.003
127. Boo YC (2022) Ascorbic acid (Vitamin C) as a cosmeceutical to increase dermal collagen for skin antiaging purposes: emerging combination therapies. Antioxidants (Basel) 11(9). https://doi.org/10.3390/antiox11091663
128. Kamel NS, Gammack J, Cepeda O, Flaherty JH (2006) Antioxidants and hormones as antiaging therapies: high hopes, disappointing results. Cleve Clin J Med 73 (12):1049-1056, 1058. https://doi.org/10.3949/ccjm.73.12.1049
129. Gao Y, Wei Y, Wang Y, Gao F, Chen Z (2017) Lycium Barbarum: a traditional Chinese herb and a promising anti-aging agent. Aging Dis 8(6):778–791. https://doi.org/10.14336/ad.2017.0725
130. Zou H, Ye H, Kamaraj R, Zhang T, Zhang J, Pavek P (2021) A review on pharmacological activities and synergistic effect of quercetin with small molecule agents. Phytomedicine 92:153736. https://doi.org/10.1016/j.phymed.2021.153736
131. Si H, Lai CQ, Liu D (2021) Dietary epicatechin, a novel anti-aging bioactive small molecule. Curr Med Chem 28(1):3–18. https://doi.org/10.2174/0929867327666191230104958
132. Ide K, Matsuoka N, Yamada H, Furushima D, Kawakami K (2018) Effects of tea catechins on Alzheimer's disease: recent updates and perspectives. Molecules 23(9). https://doi.org/10.3390/molecules23092357
133. Kim H, Eliuk S, Deshane J, Meleth S, Sanderson T, Pinner A, Robinson G, Wilson L, Kirk M, Barnes S (2007) 2D gel proteomics: an approach to study age-related differences in protein abundance or isoform complexity in biological samples. Methods Mol Biol 371:349–391. https://doi.org/10.1007/978-1-59745-361-5_24
134. Milani A, Basirnejad M, Shahbazi S, Bolhassani A (2017) Carotenoids: biochemistry, pharmacology and treatment. Br J Pharmacol 174(11):1290–1324. https://doi.org/10.1111/bph.13625
135. Sunder S (2019) Relevant topical skin care products for prevention and treatment of aging skin. Facial Plast Surg Clin North Am 27(3):413–418. https://doi.org/10.1016/j.fsc.2019.04.007
136. Pullar JM, Carr AC, Vissers MCM (2017) The roles of Vitamin C in skin health. Nutrients 9(8). https://doi.org/10.3390/nu9080866
137. Carr AC, Maggini S (2017) Vitamin C and immune function. Nutrients 9(11). https://doi.org/10.3390/nu9111211
138. Shenoy N, Creagan E, Witzig T, Levine M (2018) Ascorbic acid in cancer treatment: let the phoenix fly. Cancer Cell 34(5):700–706. https://doi.org/10.1016/j.ccell.2018.07.014
139. Milani GP, Macchi M, Guz-Mark A (2021) Vitamin C in the treatment of COVID-19. Nutrients 13(4). https://doi.org/10.3390/nu13041172
140. Moritz B, Schmitz AE, Rodrigues ALS, Dafre AL, Cunha MP (2020) The role of vitamin C in stress-related disorders. J Nutr Biochem 85:108459. https://doi.org/10.1016/j.jnutbio.2020.108459

141. Njus D, Kelley PM, Tu YJ, Schlegel HB (2020) Ascorbic acid: the chemistry underlying its antioxidant properties. Free Radic Biol Med 159:37–43. https://doi.org/10.1016/j.freeradbiomed.2020.07.013
142. Naveed M, Hejazi V, Abbas M, Kamboh AA, Khan GJ, Shumzaid M, Ahmad F, Babazadeh D, FangFang X, Modarresi-Ghazani F, WenHua L, XiaoHui Z (2018) Chlorogenic acid (CGA): a pharmacological review and call for further research. Biomed Pharmacother 97:67–74. https://doi.org/10.1016/j.biopha.2017.10.064
143. Mlcek J, Jurikova T, Skrovankova S, Sochor J (2016) Quercetin and its anti-allergic immune response. Molecules 21(5). https://doi.org/10.3390/molecules21050623
144. Saric S, Sivamani RK (2016) Polyphenols and sunburn. Int J Mol Sci 17(9). https://doi.org/10.3390/ijms17091521
145. Ritchie KJ, Walsh S, Sansom OJ, Henderson CJ, Wolf CR (2009) Markedly enhanced colon tumorigenesis in Apc(Min) mice lacking glutathione S-transferase Pi. Proc Natl Acad Sci U S A 106(49):20859–20864. https://doi.org/10.1073/pnas.0911351106
146. Henderson CJ, Ritchie KJ, McLaren A, Chakravarty P, Wolf CR (2011) Increased skin papilloma formation in mice lacking glutathione transferase GSTP. Cancer Res 71(22):7048–7060
147. Abel EL, Angel JM, Riggs PK, Langfield L, Lo H-H, Person MD, Awasthi YC, Wang L-E, Strom SS, Wei Q (2010) Evidence that Gsta4 modifies susceptibility to skin tumor development in mice and humans. J Natl Cancer Inst 102(21):1663–1675
148. Zhang Y, Ikeno Y, Qi W, Chaudhuri A, Li Y, Bokov A, Thorpe SR, Baynes JW, Epstein C, Richardson A, Van Remmen H (2009) Mice deficient in both Mn superoxide dismutase and glutathione peroxidase-1 have increased oxidative damage and a greater incidence of pathology but no reduction in longevity. J Gerontol A Biol Sci Med Sci 64(12):1212–1220. https://doi.org/10.1093/gerona/glp132
149. Van Remmen H, Ikeno Y, Hamilton M, Pahlavani M, Wolf N, Thorpe SR, Alderson NL, Baynes JW, Epstein CJ, Huang TT, Nelson J, Strong R, Richardson A (2003) Life-long reduction in MnSOD activity results in increased DNA damage and higher incidence of cancer but does not accelerate aging. Physiol Genomics 16(1):29–37. https://doi.org/10.1152/physiolgenomics.00122.2003
150. Shen G, Xu C, Hu R, Jain MR, Nair S, Lin W, Yang CS, Chan JY, Kong AN (2005) Comparison of (-)-epigallocatechin-3-gallate elicited liver and small intestine gene expression profiles between C57BL/6J mice and C57BL/6J/Nrf2 (-/-) mice. Pharm Res 22(11):1805–1820. https://doi.org/10.1007/s11095-005-7546-8
151. Saeidnia S, Abdollahi M (2013) Antioxidants: friends or foe in prevention or treatment of cancer: the debate of the century. Toxicol Appl Pharmacol 271(1):49–63. https://doi.org/10.1016/j.taap.2013.05.004

Chapter 4
Oxidative Stress as a Triggering Mechanism of Various Diseases

Haseeb Anwar, Laaraib Nawaz, Ayesha Saddiqa, Sanam Shaheen, Humaira Muzaffar, Muhammad Umar Ijaz, and Imran Mukhtaar

4.1 Introduction

Controlled combustion for energy provision is required for aerobic life and it is regulated and catalyzed by metabolic machinery. Uncontrolled free radicals can damage this controlled combustion which is associated with the production of energy. To regulate these reactions and replace or repair the impaired machinery, aerobic life has devolved into a complex system of antioxidants. Concomitantly, the systems of enzymes also devolved which is required for chemical defense, biosynthetic reactions, biological signaling, as well as detoxification. A simple definition of oxidative stress is precluded due to both beneficial and toxic consequences of reactive species [1]. It is a universally accepted fact that oxygen is the crucial factor that causes life to finite and it is considered one of the major elements of aerobic life. Although in various conditions, it can be toxic for many cells which is the main cause of necrosis and ultimately cell death, by generating reactive species [2].

While in the study of aging, "Harmna" first discovered oxidative stress and afterward in 1985, a scientist Sies conceptually explained oxidative stress. The study of oxidative stress has been carried out for above 30 years. Currently, it is accepted that the *accumulation of oxygen-reactive species in the human body occurs when it is beyond the scavenging capacity of the body which causes organ and tissue damage in particular parts of the body as oxidative stress* [3].

H. Anwar (✉) · L. Nawaz · A. Saddiqa · S. Shaheen · H. Muzaffar · I. Mukhtaar
Department of Physiology, Government College University Faisalabad, Punjab, Pakistan
e-mail: drhaseebanwar@gcuf.edu.pk

M. U. Ijaz
Department of Zoology, Wildlife and Fisheries, University of Agriculture, Faisalabad, Pakistan

A. Imran, G. Hussain (eds.), *The Role of Natural Antioxidants in Brain Disorders*, Food Bioactive Ingredients, https://doi.org/10.1007/978-3-031-41188-5_4

Pro-oxidants Free radicals such as singlet oxygen, Superoxide radicals (O^2 −), hydroxyl, and hydrogen peroxide (H_2O_2) are produced by biological systems as by-products of metabolism [4]. Because of various factors, for instance, actions of drugs and toxicity, aging, addiction, or/and inflammation [5] the homeostasis between anti-oxidants and pro-oxidants can be disturbed which is referred to as "Oxidative stress". In other words, insufficient removal or/and uncontrolled production of free radicals like reactive oxygen species as well as reactive nitrogen species is oxidative stress [6].

Oxygen has the potential to become a component of probably damaging and harmful particles because it is an extremely reactive molecule. Oxidative stress attacks the body's healthy cells which cause damage to their structure and function. This damage is because of a reduction in antioxidant levels and it imposes huge damage to proteins, DNA, and other different macromolecules which triggers the pathogenesis of different diseases particularly, heart and cancer disease. Thus, Oxidative stress is implicated in the pathogenesis of the above 50 diseases [5].

Antioxidants A substance whose accessibility, though in minute concentrations can delay or inhibits the oxidation of substrate can be labeled as "antioxidants". Antioxidants offer great defense either exogenously or endogenously and they may also serve as the biomarkers of oxidative stress. Antioxidants may be classified depending on their action mechanism, these antioxidants include preventive antioxidants and the other one is chain-breaking antioxidants [7]. Vitamin E and C, Glutathione (reduced/oxidized) and cysteine are types of biological antioxidants [8].

To protect from cellular damage induced by free radicals, cells use a defensive system of antioxidants that depends on components of enzymes (Fig. 4.1) especially, glutathione peroxidase, catalase, and superoxide dismutase (SOD) [9]. Different mechanisms in a body exogenous or endogenous may participate in the production of antioxidants, which will help in the protection of cells against the harmful effect of free radicals and elevate or neutralize the amount of reactive oxygen species and contributing towards the prevention of various diseases. Thus, these mechanisms help in scavenging the deleterious effects of these reactive oxygen species [10]. Small non-enzymatic molecules that are involved in detoxification are referred to as antioxidants. Glutathione is the most important antioxidant which is involved in various intracellular defense systems against the pathogenic effects of free radicals [11].

Different types of important processes including immunity, transcriptional factor activation, protein phosphorylation, apoptosis as well as differentiation utterly rely upon proper free radicals synthesis and its occurrence inside the cells [12]. If the production of free radicals increases, it causes toxic effects on lipids, nucleic acids, and proteins, which disrupt their functions and various cellular structures [13]. It leads to the progression or onset of different types of diseases like diabetes, cardiovascular, cancer, atherosclerosis, and endocrine disorders [14].

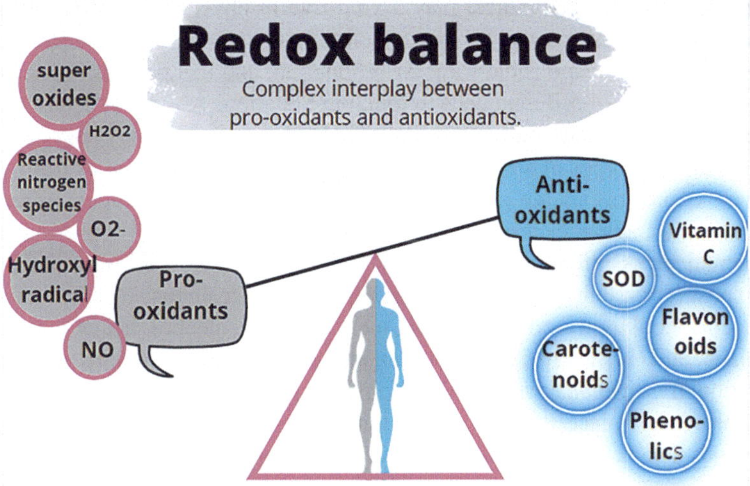

Fig. 4.1 represents redox balance maintained by pro-oxidants and antioxidants. Low levels of antioxidants and high levels of pro-oxidants causes imbalance in redox homeostasis leading towards oxidative stress

4.1.1 Oxidants and Free Radical Production

Reactive oxygen species that are short-lived molecules are involved in cell signaling and considered important in normal processes of migration and differentiation. However, when reactive oxygen species pass the electron (unpaired), it leads to the molecular and cell component's oxidation [15]. Reactive Nitrogen species and Reactive oxygen species are the classified as reactive species or free radicals [16].

Mitochondria is the main site of ROS production during both pathological as well as physiological conditions. Free radicals can be generated through cellular respiration by cyclooxygenases and lipo-oxygenases during arachidonic acid metabolism and also by inflammatory as well as endothelial cells [17]. It is important to note that despite having built-in scavenging capacity of free radicals in these organelles [18], it is not good enough to wash out the amount of free radicals being produced by mitochondria.

Non-enzymatic as well as enzymatic reactions are engaged in the production of free radicals. *Enzymatic reactions* which contribute to prostaglandin production, cytochrome P450 system, respiratory chain, and phagocytosis can generate oxidative stress [19]. Superoxide radical is produced by peroxidases and NADPH oxidase and xanthine oxidase. After the formation of these radicals, it is involved in a different types of reactions that generate hydroxyl radical, hypochlorous acid, peroxynitrite, hydrogen peroxide, etc. Among all the reactive oxygen molecules, the highly reactive species is *hydroxyl radical* which is produced by the reaction of H_2O_2 with O_2^-, with Cu^+ or Fe^{2+} (Fenton reaction) as a reaction catalyst [20]. Nitric oxide radical is synthesized by nitric oxide synthase, through the oxidation from arginine toward citrulline and this radical play a vital role in physiological functions [20]. For the production of reactive oxygen species, some *non-enzymatic reactions* may also be involved or responsible for their production, and this process is carried out when the exposure of different cells to ionizing radiations or the reaction of oxygen with other organic compounds takes place [21].

There are two types of sources exogenous and endogenous sources from which reactive oxygen species are generated. For the production of *endogenous* reactive oxygen species, aging, excessive exercise, ischemia, immune cell activation, infection, mental stress, inflammation, and cancer are responsible. Environmental pollutants exposure, certain drugs such as (gentamycin, tacrolimus, cyclosporine, and bleomycin), cooking (used oil, smoked meat, and fat), alcohol, heavy metals (Pb, Cd, As, Fe, and Hg), chemical solvents, cigarette smoke, and radiations all cause the production of *exogenous* reactive oxygen species [22]. The production of these molecules also takes place through non-enzymatic reactions of oxygen with organic compounds and as well as initiated by ionizing radiations [10].

Other sources may include chemicals, free radicals, macrophages and neutrophils production, industrial effluents, RNS, smoking of beedi, cigarettes, cigars, and radiations [23].

4.1.2 Biological Roles of Free Radicals

Due to its vital role in the evolution and the origin of life, reactive oxygen species are said a a necessary evil. Different pathways such as the extracellular-signal-regulated kinase, Mitogen-activated protein kinase, and pathways leading to alteration in gene expression are activated by reactive oxygen species. ROS also causes cell death in association with superoxide dismutase [24]. Similarly, neurons produce reactive nitrogen species which work as neurotransmitters; species that are produced by macrophages act as immunity mediators. For thrombosis, signal transduction, angiogenesis vascular tone, transcription of genes, and leukocyte adhesion these reactive species are involved [25].

Reactive oxygen species play several beneficial roles when these are species kept at moderate or low concentrations. For instance, these species are used to fight against pathogens by the defense system of the host. The storage and syncretization of reactive oxygen species in phagocytes enable them to release free radicals while destroying the pathogenic microbes by invading [16]. Patients with a disease like granulomatous demonstrates the vital role of free radicals in the immune system of the body. Because of the NADPH oxidase system's defect in this disease, the production of O_2^- is unable in these individuals, so they are prone to persistent infections [21].

In addition to regulating different pathways of cellular signaling [26] reactive oxygen species plays an important *regulatory role* in various types of cells like endothelial cells, cardiac myocytes, vascular smooth muscle cells, fibroblasts, and thyroid tissue, in cascades of intracellular signaling. Probably, *Nitric oxide* (NO) is

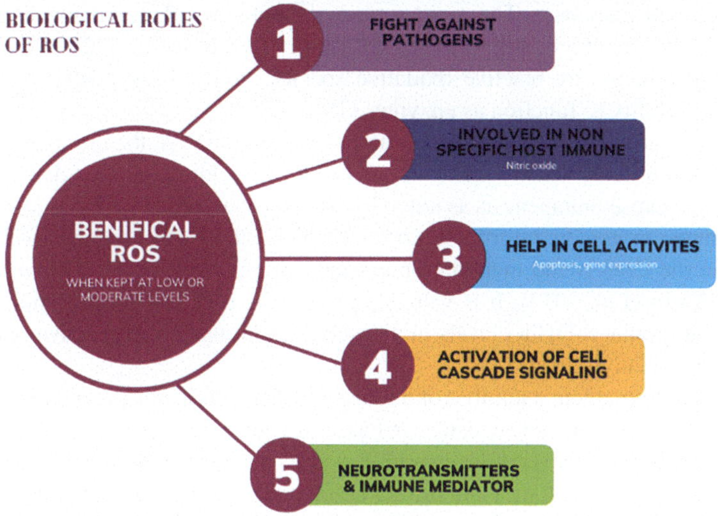

Fig. 4.2 Shows physiologically significant roles played by low or moderate levels of pro-oxidants in body

well known reactive oxygen species and it acts as a signaling molecule. Nitric oxide (NO) is crucial for the normal activity of neurons and involved in proper blood flow, is required for thrombosis, and plays an important role in cell-to-cell signaling [26]. Nitric oxide is required to eliminate tumor cells and intracellular pathogens because it is involved in the nonspecific defense system of the host. Mitogenic response induction is another physiological activity of reactive oxygen species [27]. However, because of the defense system in host cells, the phagocytic cells are responsible for free radical production. Various studies demonstrated that free radicals play a crucial role in apoptosis, gene expression, cell signaling, and the activation of cell cascade signaling [28]. Summarizing, when reactive oxygen species maintained at moderate or low levels, have pivotal importance to the health of humans (Fig. 4.2).

4.2 Detrimental Effects of Free Radicals on Human Health

As previously mentioned, oxidants and free radicals can cause oxidative stress when they are present in excess. This destructive process can disrupt various cellular structures, including deoxyribonucleic acid (DNA), lipids, proteins, lipoproteins, and membranes [63]. When there are instability between the production of reactive oxidant species and cells' capacity to get rid of them, oxidative stress results. For instance, an overabundance of peroxynitrite and hydroxyl radicals can lead to lipid peroxidation, which harms lipoproteins and cell membranes. Conjugated diene compounds and malondialdehyde (MDA), which are known to be mutagenic and cytotoxic, will then arise. The rapid spread of lipid peroxidation damages a huge number of lipidic atoms due to its radical chain reaction nature [22]. Proteins are also being affected by reactive oxidative species, going through alterations that affect their ability to function as enzymes [19].

Lesions caused by oxidative stress can occur even in DNA, the utmost typical of which is the 8-oxo-2′-deoxyguanosine (8-OHdG) development. This deadly DNA damage can cause mutagenesis as noted by Nishida et al. (2013). After that, the loss of epigenetic or genetic information occurs which can be a consequence of a deficiency in the CpG island methylation capacity of gene enhancers or transcription factors (Yasui et al. 2014). It is valuable to take the remarks of Valavanidis et al. (2013) stating that 8-OHdG levels in a tissue are a biomarker of reactive oxidative species or oxidative stress.

Cells can implement a number of defense mechanisms against deoxyribonucleic acid damage, such as antioxidants or base excision repair (BER; Willcox et al. 2004). Conclusively, several degenerative or chronic diseases as well as the body's accelerated acute pathologies and further aging process can be brought on by oxidative stress if it is not rigorously controlled (i.e., stroke and trauma). Different studies indicate that oxidative reactions may result in various disease processes and aging, including, pulmonary diseases [29], cancer [30], neurodegenerative diseases [31], and cardiovascular disease [32].

4.2.1 Cancer and Oxidative Stress

Cancer growth in human beings is a complicated process that undergoes both molecular and cellular changes, which are mediated by exogenous and endogenous stimuli. It is well established that one of the constituents contributing to the developing cancer is oxidative DNA damage [21]. Chromosomal aberrations and oncogene activators are determined by oxidants that cause the carcinogenic conditions. Deoxyribonucleic acid oxidation frequently develops hydrolyzed deoxyribonucleic acid bases that are regarded as the most important developments in carcinogenicity [21].

Normal cell growth is hampered by the creation of these adducts because they change the physiological transcriptome profile and result in gene alterations. Additionally, oxidative stress can alter DNA structure in a variety of ways, including sugar and base damage, Base-free regions, DNA-protein, and chain breakage crosslinks. As oxidative DNA damage can be caused by environmental toxins, tobacco use, and chronic inflammation, these factors may all have a role in the development of tumors [64]. The high link between dietary fat consumption (a constituent/factor that put lipid peroxidation at high risk in the body) and the death rates from various types of carcinogenic conditions suggested oxidative stress arising from lifestyle behaviors can also play a vital role in neoplasm (cancer formation) [63].

4.2.2 Cardiovascular Disease and Oxidative Stress

Low levels of reactive oxygen species generation are equivalent to their elimination under physiological parameters and are crucial for cellular signaling and function (Tsutsui et al. 2009). Redox signaling is a process that controls excitation-contraction coupling, gene expression, migration, cell growth, death, and differentiation by specifically and irreversibly altering cellular signaling components (Sack et al. 2017). Redox signaling is mediated by a number of kinases. H_2O_2, in particular, may activate $Ca^{2+/}$calmodulin-dependent kinase II (CAMKII), which leads toward p38 mitogen-activated protein kinase (p38 MAPK) or c-Jun N-terminal kinase (JNKs) and excitation-contraction coupling which results in the suppression of insulin signal transduction. Additionally translocated from the cytosol, cAMP-induced protein kinase A (PKA) is triggered by oxidation of its regulatory component R1. PKA controls the vasodilation in the arteries on the membrane and also cardiac excitation-contraction coupling in the heart (Burgoyne et al. 2012).

Under normal circumstances, NO acts as a cytoprotective and vasodilator molecule. NO protects against coronary failure and ischemia-reperfusion injury by preventing platelet and neutrophil activation and adhesion (Loyer et al. 2008). NO can activate protein kinase G (PKG) by interacting with the soluble guanylate cyclase and generating cyclic guanosine monophosphate (Hammond et al. 2012). It can also activate PKG by S-nitrosylation. The functions of various proteins, including, myosin heavy chain, pro-caspase 3, peroxiredoxins, ryanodine receptor, and tropomyosin, may be modulated by the RyR. RyR, a redox-regulated protein that causes Ca2+ release from the sarcoplasmic reticulum, is likewise activated by phosphorylation via PKA and CAMKII. Additionally, PKG is activated by oxidation independently of NO and controls the contraction or hypertrophy of cardiomyocytes and vascular tone (Burgoyne et al. 2012).

Major biological macromolecules can undergo oxidative alteration by ROS in pathogenic circumstances (such as DNA, lipids, or proteins). The intracellular organelles such as the mitochondria, sarcolemma, nucleus, and sarcoplasmic reticulum, undergo alterations as a result of this oxidation. For instance, ROS may influence contractility by *oxidizing actin and tropomyosin*, two proteins involved in contractility, as well as the Sarco or endoplasmic reticulum Calcium ATPase (SERCA 2A) (Lancel et al. 2010). This could lead to *contractile failure* (Steinberg 2013).

Radical species' toxicity is influenced by both their production site and half-life (the shorter the half-life, the highly unstable and poisonous ROS is. The ability of non-radical reactive oxygen species to produce radical species determines how hazardous they are. As a result, the hydroxyl radical, although the more unstable of the antioxidant radicals, also has the highest toxicity (Zorov et al. 2014). Numerous cardiac conditions, including myocardial fibrosis (Hermida et al. 2018), type 2 diabetes (Serpillon et al. 2009), metabolic syndrome (Jiménez-González et al. 2020), cardiomyopathy (Dai et al. 2011), heart failure (Dai et al. 2012), and myocardial infarction, have been linked to an increase in ROS production (Merabet et al. 2012).

The primary risk factor for heart failure, stroke, and myocardial infarction, is endothelial damage, which includes hypertension and atherosclerosis. Reduced availability and formation of NO, and instability between relaxing and contracting substances produced from the endothelium, are all signs of endothelial dysfunction, which is also known as a pro-inflammatory and prothrombotic state (Scioli et al. 2020). By interfering with the pathway of vasoprotective NO signaling, ROS can mediate endothelial damage and vascular abnormalities in the presence of oxidative stress causing NO synthase (NOS) uncoupling. The enzymes known as NOS are capable of producing superoxide anions in addition to synthesizing NO from L-arginine in the presence of cofactors and dioxygen. Type I or neuronal NOS (NOS1 or nNOS), type II or inducible NOS (NOS2 or iNOS), and type III or endothelial NOS have all been identified (NOS3 or eNOS). The heart has a calcium-dependent activity for the NOS1 and NOS3 isoforms, which are specifically expressed in striated muscle and endothelial cells. The NOS2 isoform functions without the need for calcium. Although it does not intrinsically express itself in a heart that is healthy, it does so under pathological situations like inflammation (Umar et al. 2010). Reduced NO bioavailability and vasoconstriction result from uncoupled NOS as they move from producing NO to peroxynitrite and superoxide anion ion (Santillo et al. 2015). Peroxynitrite inhibits vasorelaxation, reduces the positive effects of NO on vascular smooth muscle cell proliferation and platelet aggregation, and oxidizes DNA, all of which contribute to the progression of atherosclerosis (Cai and Harrison 2000). Additionally, aging increases peroxynitrite synthesis in the blood of aged mice, which increases oxidative stress in resistant arteries (Ma et al. 2014). However, mice exposed to smoking and modest levels of Angiotensin-II exhibit substantial endothelium dysfunction, particularly with regard to acetylcholine-dependent relaxation, which is accompanied by the inactivation of SOD2 activity (Dikalov et al. 2019). Additionally, from cardiomyopathic arteries, heart failure causes higher production of superoxide anions generated by NOX and NOS (Mollnau et al. 2005).

Cardiovascular diseases (CVDs) are clinical entities with a multifactorial etiology, typically accompanied by a greater number of harmful factors. The most widely acknowledged of these risk factors are hypercholesterolemia, diabetes, hypertension, smoking an irregular diet, stress, and a sedentary lifestyle (Ceriello 2008). Research findings over the past few years have indicated that reactive oxidants species should be regarded as either a major or secondary cause of many CVDs (Pacher et al. 2007).

Atherosclerosis is mostly brought on by oxidative stress. It is generally recognized that early inflammation of the thin membrane which lies on the inner side of blood vessels and the heart triggers the recruitment of macrophages to produce ROS, causing atheromatous plaque development. Reactive oxygen species subsequently oxidize the circulating LDL, causing foam cells to develop and fat buildup. An atherosclerotic plaque develops as a result of these circumstances. Experimentation both in-vivo and ex-vivo have shown that ROS plays a part in the development of atherosclerosis, cardiac hypertrophy, congestive heart failure, ischemia, cardiomyopathy, and hypertension (Chatterjee et al. 2007). The artery wall becomes thicker and harder as a result of the chronic illness of atherosclerosis. Currently, it is thought to be a chronic inflammatory condition that is accompanied by risk factors such as hyperlipidemia, hypertension, and diabetes. The pathophysiology of each of these diseases is thought to be influenced by excessive ROS generation. In response to oxidative stress, cells produce intracellular adhesion molecules that help smooth muscle cells, monocytes, and lymphocytes having "scavenger" receptors for oxidized low-density lipoprotein (LDL) invade the artery wall. A reaction that breaks down the extremely reactive anion of superoxide into peroxides of hydrogen and oxygen is catalyzed by the binding of oxidized LDL, which activates macrophages and monocytes and stimulates the production of SOD. Massive macrophage apoptosis (programmed cell death) can result from increased hydrogen peroxide generation, which can disrupt steady-state levels of ROS a process i.e. linked to the development of atherosclerotic plaques (Molavi and Mehta 2004).

4.2.3 Diabetes Mellitus and Oxidative Stress

Diabetes mellitus (DM) is one of the 4 major non-communicable illnesses that need immediate attention from all key stakeholders globally. Oxidative stress due to hyperglycemia, and inflammation, are all strongly correlated with the onset and progression of diabetes (Oguntibeju and pharmacology 2019).

Infiltrating *macrophages* secrete inflammatory cytokines as a result of hyperglycemia-induced ROS, which raises pro-inflammatory protein levels and causes both local and systemic inflammation (Wellen and Hotamisligil 2005). For the development of type 2 diabetes the increased tumor necrosis factor-alpha (TNF-alpha) release has been linked to insulin resistance related to obesity condition (Derosa et al. 2013). As ROS is a reported mechanism in the pathophysiology of diabetes (Giacco and Brownlee 2010), increases in oxidative pressure-induced DNA dysfunction markers like 8-hydroxy-2′-deoxyguanosine (8-OHdG) and 8-oxo-7, 8-dihydro-2′-deoxyguanosine, as well as protein oxidation products like carbonyl and nitrotyrosine levels, products of lipid peroxidation like thiobarbituric acid-reactive substances (TBARS) can all occur in a hyperglycemic state. Smooth muscle cells in arteries, pancreatic beta cells, cells in blood, and heart in cell culture also demonstrated an increase in reactive oxygen species generation in diabetes (Lee et al. 2010). Oxidative stress causes the *inhibition* of the insulin gene's promoter activity and mRNA expression in cell lines and isolated pancreatic islet cells, resulting in a reduction in insulin gene expression (Kawahito et al. 2009).

Chronic hyperglycemia-induced insulin resistance is also strongly suspected to be caused by oxidative damage (Eriksson 2007). This is because prolonged exposure of both human beings and animal tissues and cells to hyperglycemia results in non-enzymatic glycation of proteins, which results in the production of free radicals in the form of Schiff base and Amadori products (Hojs et al. 2016). In type 2 diabetes, chronic hyperglycemia is considered to play a major role in the development of microvascular and macrovascular complications. Chronic hyperglycemia also further causes damage to proteins, lipids, and DNA with the severity of damage being correlated with the amount of oxidative stress induced by hyperglycemia's production of free radicals (Butkowski and Jelinek 2017).

Chronic hyperglycemia, which characterizes diabetes, can cause a *pro-oxidative shift* in the glutathione redox balance in the blood via a number of processes. Superoxide and glycated proteins are both products of glucose auto-oxidation. Glycated proteins' interactions with cell surface receptors increase the production of ROS and reduce intracellular glutathione. Atherosclerotic lesions may eventually emerge in artery walls as a result of hyperglycemia's enhancement of cell-mediated 3 LDL peroxidation in endothelial cells (Dröge 2002).

4.2.4 Neurological Disease and Oxidative Stress

Various neurological conditions, including Alzheimer's disease (AD), amyotrophic lateral sclerosis (ALS), Parkinson's disease, depression, multiple sclerosis, and memory loss, have been associated with ROS (Allan 2002). Various experimental as well as clinical research on AD revealed that oxidative stress is responsible for the loss of neurons and the development of dementia (Christen 2000). It is well-recognised that the toxic peptide -amyloid, which is frequently present in the brains of AD patients, is at least partially to blame for the neurodegeneration seen during the start and progression of AD (Allan Butterfield 2002). Along with Parkinsonism, Alzheimer's dementia, ALS, and other neuropathologies like bovine spongiform encephalopathy, disruption of oxidative homeostasis has also been linked to various neurodegenerative diseases. The oxidative breakdown of *dopamine* in the central nervous system, which is associated with aging, produces ROS (Luo et al. 2000). Cerebral spinal fluid in Alzheimer's possessed typically higher levels of 4-hydroxynonenal, which is a by-product produced after lipid peroxidation, and lipid peroxidation levels in the brain. Additionally, ROS was discovered to be a mediator of ß-amyloid protein-induced neuronal damage (Multhaup et al. 1997).

4.2.5 Respiratory Disease and Oxidative Stress

Various lung diseases particularly chronic obstructive pulmonary disease and asthma after determined by local and systemic inflammation, are linked with oxidative stress [33]. Free radicals increase this inflammation by activating various kinases, focusing pathways, and transcriptional factors e.g. AP-1 and nuclear factor-kappa B [34].

Various studies have reported that a broad range of biologically significant molecules in the lungs is damaged in asthma due to a rise in reactive oxygen species. An elevation in both respiratory ethane [35] isoprostanes, [36], and urinary isoprostanes [37] predicts that oxidative stress is taking place both at endothelial cell membranes and epithelial membranes. Protein damage in the airway is confirmed by elevated chlorotyrosine [38] and nitrotyrosine [39]. Due to these modifications of proteins, their activity is diminished for instance inhibition of α1-protease [40]. Steroid therapy is helpful in the attenuation of nitrotyrosine, ethane formation, and hydrogen peroxide that is predictive of the correlation between oxidative stress and inflammation. These elevated free radical species overtake endogenously produced antioxidants. Although glutathione in the airway is elevated during asthma however there is also the elevation of the ratio of oxidized and reduced glutathione. This increased glutathione is suggestive of an adaptive response whereas the activity of other antioxidants including α-tocopherol, ascorbate [41], and SOD is lowered in asthma [42]. This association of oxidative stress and inflammation may initiate a positive feedback circuit that favors lung damage. During inflammation, various cytokines particularly heparin-bound epidermal growth factor, angiotensin II, fibroblast growth factor 2, TNF-α, thrombin, and serotonin are present in the lungs. They cause the activation of oxidases leading to a rise in free radicals [43]. Targets of these free radicals might involve phospholipids, receptor kinases, phosphatases, or mitogen-activated protein (MAP) kinases [44]. Nitric oxide is another target of these free radicals as it is dysregulated in asthmatic conditions. The function of nitric oxide is a little complicated as nitric oxide synthases (NOSs) have three sources.

1. NOS I (neuronal NOS or nNOS); is present at nerve terminals of smooth muscle (non-adrenergic) and is involved in bronchodilation.
2. NOS II (inducible NOS): Present in a broad variety of epithelial cells and inflammatory types
3. NOS III (extracellular): Primarily present on endothelium and controls vasodilation.

Extracellular NOS and nNOS are active however in asthma inducible NOS is mainly induced. Also, it is responsible for elevated concentrations of exhaled NO. The function of NO in asthma is still unexplored however due to lack of bronchorelaxation in asthma is predictive of the impairment of signaling pathways involved in the relaxation of bronchial smooth muscles. Thus, the concomitant elevation in oxidative stress in inflammation might be the source of impairment [45].

4.2.6 Rheumatoid Arthritis and Oxidative Stress

Rheumatoid arthritis is a chronic inflammatory disorder that primarily affects joints as well as tissues surrounding that joints. It is characterized by the filtration of activated T cells and macrophages [46]. Reactive oxygen species present at the inflammatory site play a significant role in the initial and progressive stages of this disorder. It is demonstrated by elevated prostaglandin and isoprostane levels in the synovial fluid of joints [47].

Rheumatoid arthritis is one of the diseases that induce reactive oxygen species as well as oxidative stress. This is demonstrated by the fivefold production of mitochondrial reactive oxygen species in monocytes and whole blood of RA individuals. This increase suggests that oxidative stress is a pathological hallmark in rheumatoid arthritis. ROS is *indirectly* implied in joint damage due to their role as second messengers during immunological cell response (Fig. 4.3) and inflammation. Exposure of T cells to oxidative stress becomes refractory to various stimuli, also including those required for death and growth. It eventually perpetuate the abnormal immunological response [48]. On the contrary, reactive oxygen species can also degrade the cartilage of the joint *directly* by targeting proteoglycans (Fig. 4.3) and halting their production [49]. In RA, protein oxidation resulting from oxidative stress causes carbonyl increment, lipoperoxidation products, oxidative damage of hyaluronic acid, and oxidation of low-density lipoproteins. Also, genotoxic phases resulting from ROS are related to p53 mutation in rheumatoid arthritis derived fibroblast-like synoviocytes [50]. Moreover, there is impairment of antioxidant systems in RA either non-enzymatic or enzymatic. Decreased concentrations of GSH [51], β-carotene, retinols, and tocopherols as well as reduced activity of superoxide dismutase and GR are also correlated [52]. Increased intra-articular pressure in joints during RA causes increased production of ROS during oxidative phosphorylation in cells. It then induces multiple cycles of hypoxia and re-oxygenation. The origin of this hypoxic event in RA is the result of the rapid proliferation of cells induced by inflammatory reaction [53].

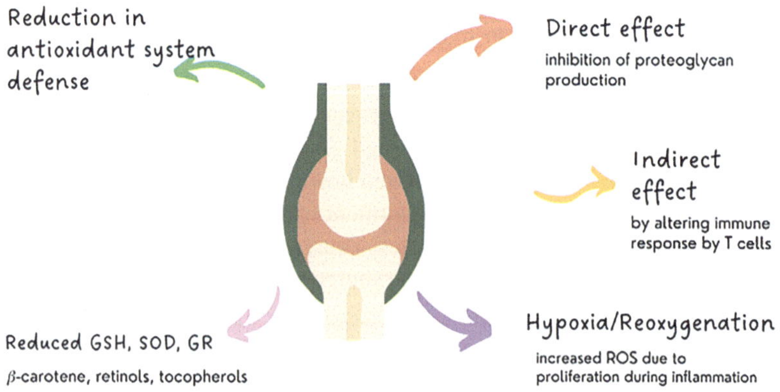

Fig. 4.3 Shows direct and indirect worsening effects of ROS on synovial joints during arthritic condition

4.2.7 Kidney Diseases and Oxidative Stress

The role of oxidative stress is extensively studied in the progression of kidney diseases and resulting functional loss [54]. Free radicals have a significant functions in the physiological control of kidney activities which ultimately makes kidneys more prone to oxidative stress and redox imbalances. Any alteration in the production or formation of free radicals may take place at the medulla and cortex, causing various effects from retention of sodium or fluid, renal blood circulation, proteinuria onset, fibrosis to inflammation [55]. Mitochondrial dysfunction and increased free radical generation in mitochondria are reported in chronic kidney diseases, particularly in diabetic nephropathy, functional and morphological disturbances associated with dysfunction of mitochondria [56]. Oxidative stress is involved in a broad range of disorders targeting renal apparatus like glomerulo- as well as tubular interstitial nephritis, uremia, renal failure, and proteinuria [57].

Kidneys are negatively influenced by oxidative stress. Free radicals induce the inflammatory cells recruitment and production of proinflammatory cytokines that further leads to the early inflammatory phase. During this initial phase, IL-1β and tumor necrosis factor-alpha play a significant role as proinflammatory mediators. Also transcriptional factor, nuclear factor- κB is needed to sustain the process of inflammation. At the later stage, there is elevated production of transforming growth factor beta which orchestrates the synthesis of the extracellular matrix. Thus, whenever in kidney tissue oxidative stress acts chronically, there will be an early phase of inflammation followed by a later phase of formation of excessive fibrotic tissue (Fig. 4.4) ultimately impairing the functional potential of the kidney i.e. renal failure. Various drugs like tacrolimus, bleomycin, cyclosporine, and gentamycin work

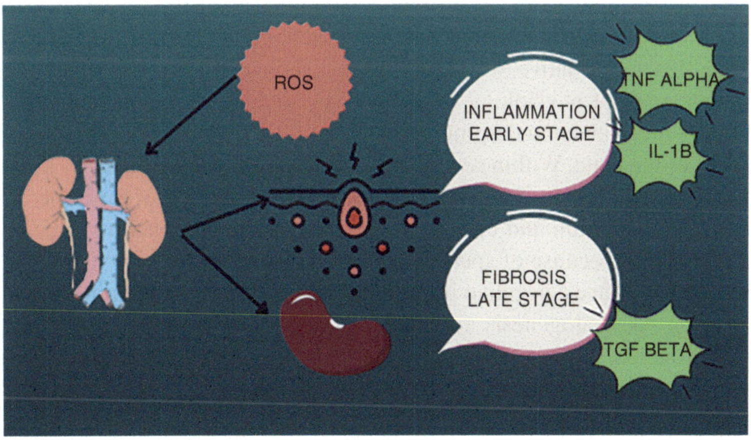

Fig. 4.4 Shows early inflammatory and late fibrotic stage of kidney damage due to oxidative stress

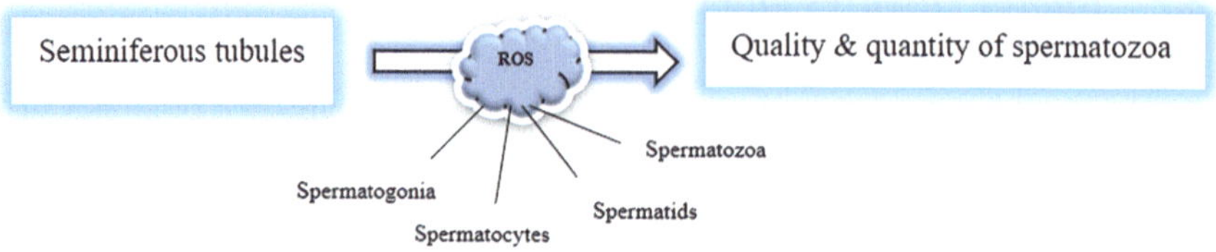

Fig. 4.5 Schematic flowchart of reactive oxygen species targeting reproductive system

as nephrotoxic because they elevate the synthesis of free radicals as well as oxidative stress by peroxidation of lipids [58]. There are also a few transitions (Cr, Co, Cu, and Fe) and heavy metals (As, Pb, Hg, and Cd) that may act as powerful inducers of oxidative stress. These are also responsible for nephropathy and various types of cancers [59].

4.2.8 Sexual Maturation and Oxidative Stress

It is stated that oxidative stress might be accountable for the delay in puberty onset and sexual maturation [60]. This seems to be a fact when prepubertal-aged children are exposed to metals like cadmium which is reportedly responsible for elevated reactive oxygen species and when women in the gestational period have exposure to the same element. Collectively, it is affirmed that reactive free species and oxidative stress are responsible for various pathological diseases targeting systems and tissues. Thus it is one of the most pervasive harms to health [58].

Male infertility and sperm dysfunction occur where *in the testis* there is an altered redox balance leading to the overproduction of reactive oxygen species and an impaired antioxidant system [61]. There is a need to clarify the reproductive toxicity of bisphenol A to oxidative stress. However, keeping in view the significance of redox balance in the physiology of sperm, oxidative stress might be the primary deteriorating mechanism underpinning the toxicity of bisphenol A in the reproductive system of the male. Within *seminiferous tubules*, Bisphenol A causes the reduction of antioxidant enzyme's activity and enhances the activity of myeloperoxidase supporting inflammation and causing testicular dysfunction lined to increase free radical production because of spermatogenic cell membrane's peroxidation. These coinciding events have a negative impact (Fig. 4.5) on meiotic and mitotic cell processes during spermatogenesis such as spermatozoa, spermatids, spermatocytes, and spermatogonia. It results in impairment in the quantity and quality of spermatozoa in seminiferous tubules [62].

4.3 Conclusion

Free radicals which are exogenously or endogenously produced are highly toxic and reactive metabolites. Their uncontrolled production leads to oxidative stress. These species target antioxidant defense system and cause DNA and protein damage which is responsible for triggering the onset and progression of various diseases.

References

1. Jones DP (2006) Redefining oxidative stress. Antioxid Redox Signal 8(9–10):1865–1879
2. Weseler AR, Bast A (2010) Oxidative stress and vascular function: implications for pharmacologic treatments. Curr Hypertens Rep 12(3):154–161
3. Qi J-H, Dong F-X (2021) The relevant targets of anti-oxidative stress: a review. Drug Target 29(7):677–686
4. Navarro-Yepes J, Zavala-Flores L, Anandhan A, Wang F, Skotak M, Chandra N, Li M, Pappa A, Martinez-Fong D, Del Razo LMJP (2014, Therapeutics) Antioxidant gene therapy against neuronal cell death. 142(2):206–230
5. Asmat U, Abad K, Ismail KJ (2016) Diabetes mellitus and oxidative stress – a concise review. Saudi Pharm J 24(5):547–553
6. Johansen JS, Harris AK, Rychly DJ, Ergul AJ (2005) Oxidative stress and the use of antioxidants in diabetes: linking basic science to clinical practice. Cardiovasc Diabetol 4 (1):1–11
7. Somogyi A, Rosta K, Pusztai P, Tulassay Z, Nagy G (2007) Antioxidant measurements. Physiol Meas 28(4):R41
8. Susan MD (2001) Thiol-based antioxidants. Curr Top Cell Regul 36:151–180
9. Deponte M (2013) Glutathione catalysis and the reaction mechanisms of glutathione-dependent enzymes. Biochim Biophys Acta 1830 (5):3217–3266
10. Pham-Huy LA, He H, Pham-Huy CJ (2008) Free radicals, antioxidants in disease and health. Int J Biomed Sci 4(2):89
11. Jones DP (2002) [11] Redox potential of GSH/GSSG couple: assay and biological significance. In: Methods in enzymology, vol 348. Elsevier, pp 93–112
12. Rajendran P, Nandakumar N, Rengarajan T, Palaniswami R, Gnanadhas EN, Lakshminarasaiah U, Gopas J, Nishigaki I (2014) Antioxidants and human diseases. Clin Chim Acta 436:332–347
13. Wu JQ, Kosten TR, Zhang XY (2013) Free radicals, antioxidant defense systems, and schizophrenia. Prog Neuropsychopharmacol Biol Psychiatr 46:200–206
14. Taniyama Y, Griendling KKJH (2003) Reactive oxygen species in the vasculature: molecular and cellular mechanisms. Hypertension 42(6):1075–1081
15. Bansal AK, Bilaspuri G (2011) Impacts of oxidative stress and antioxidants on semen functions. Vet Med Intl. https://doi.org/10.4061/2011/686137
16. Dröge W (2002) Free radicals in the physiological control of cell function. Physiol Rev 82(1):47–95
17. Al-Gubory KH, Garrel C, Faure P, Sugino N (2012) Roles of antioxidant enzymes in corpus luteum rescue from reactive oxygen species-induced oxidative stress. Reprod Biomed 25(6):551–560
18. Hansen JM, Go Y-M, Jones DP (2006) Nuclear and mitochondrial compartmentation of oxidative stress and redox signaling. Annu Rev Pharmacol Toxicol 46:215-234
19. Halliwell B, Gutteridge J (2007) Free radicals in biology and medicine, vol. 4th. Oxford
20. Kumar S, Pandey A (2015) Free radicals: health implications and their mitigation by herbals. Br J Med Med Res 7 (6):438–457

21. Valko M, Leibfritz D, Moncol J, Cronin MT, Mazur M, Telser J (2007) Free radicals and antioxidants in normal physiological functions and human disease. Int J Biochem Cell Biol 39 (1):44–84
22. Frei B (1997) Reactive oxygen species and antioxidant vitamins. Am J Med 97(3A):5S–13S
23. Sen S, Chakraborty R, Sridhar C, Reddy Y, De BJIjopsr (2010) Free radicals, antioxidants, diseases and phytomedicines: current status and future prospect. Intl J Pharm Sci Rev Res 3(1):91–100
24. Cho K-H, Wolkenhauer O (2003) Analysis and modelling of signal transduction pathways in systems biology. Portland Press Ltd
25. Fang Y-Z, Yang S, Wu G (2002) Free radicals, antioxidants, and nutrition. Nutrition 18(10):872–879
26. Pacher P, Beckman JS, Liaudet L (2007) Nitric oxide and peroxynitrite in health and disease. Phsiol Rev 87 (1):315–424
27. Genestra MJ (2007) Oxyl radicals, redox-sensitive signalling cascades and antioxidants. Cell Signal 19(9):1807–1819
28. Hancock J, Desikan R, Neill SJ (2001) Role of reactive oxygen species in cell signalling pathways. Biochem Soc Trans 29(2):345–349
29. Cantin AM (2004) Potential for antioxidant therapy of cystic fibrosis. Curr Opin Pulm Med 10(6):531–536
30. Berger MM (2005) Can oxidative damage be treated nutritionally? Clin Nutr 24(2):172–183
31. Viña J, Lloret A, Ortı R, Alonso D (2004) Molecular bases of the treatment of Alzheimer's disease with antioxidants: prevention of oxidative stress. Mol Aspects Med 25(1-2):117–123
32. Stocker R, Keaney JF (2004) Role of oxidative modifications in atherosclerosis. Physiol Rev 84:1381–1478
33. Caramori G, Papi A (2004) Oxidants and asthma. Thorax 59 (2):170–173
34. Hoshino Y, Mishima M (2008) Redox-based therapeutics for lung diseases. Antioxid Redox Signal 10(4):701–704
35. Lim S, Roche N, Oliver BG, Mattos W, Barnes PJ, Chung KF (2000) Balance of matrix metalloprotease-9 and tissue inhibitor of metalloprotease-1 from alveolar macrophages in cigarette smokers: regulation by interleukin-10. Am J Respir Crit Care Med 162(4):1355–1360
36. Montuschi P, Corradi M, Ciabattoni G, Nightingale J, Kharitonov SA, Barnes PJ (1999) Increased 8-isoprostane, a marker of oxidative stress, in exhaled condensate of asthma patients. Am J Respir Crit Care Med 160(1):216–220
37. Dworski R, Jackson Roberts Ii L, Murray J, Morrow J, Hartert T, Sheller JR (2001) Assessment of oxidant stress in allergic asthma by measurement of the major urinary metabolite of F2-isoprostane, 15-F2t-IsoP (8-iso-PGF2α). Clin Exp Allergy 31(3):387–390
38. Wu W, Samoszuk MK, Comhair SA, Thomassen MJ, Farver CF, Dweik RA, Kavuru MS, Erzurum SC, Hazen SL (2000) Eosinophils generate brominating oxidants in allergen-induced asthma. J Clin Invest 105(10):1455–1463
39. Hanazawa T, Kharitonov SA (2000) Increased nitrotyrosine in exhaled breath condensate of patients with asthma. Am J Respir Crit Care Med 162(4):1273–1276
40. Gaillard MC, Kilroe-Smith TA, Nogueira C, Dunn D, Jenkins T, Fine B, Kallenbach J (1992) Alpha-1-protease inhibitor in bronchial asthma: phenotypes and biochemical characteristics. Am Rev Resp Dis 145(6):1311–1315
41. Kelly FJ, Mudway I, Blomberg A, Frew A, Sandström T (1999) Altered lung antioxidant status in patients with mild asthma. Lancet 354 (9177):482–483
42. Smith LJ, Shamsuddin M, Sporn PH, Denenberg M, Anderson J (1997) Reduced superoxide dismutase in lung cells of patients with asthma. Free Rad Biol Med 22(7):1301–1307
43. Thannickal VJ, Fanburg BL (2000) Reactive oxygen species in cell signaling. Am J Physiol Lung Cell Mol Physiol 279(6):L1005–L1028
44. Abe MK, Chao T, Solway J, Rosner MR, Hershenson MB (1994) Hydrogen peroxide stimulates mitogen-activated protein kinase in bovine tracheal myocytes: implications for human airway disease. Am J Respir Cell Mol Biol 11(5):577–585

45. Bowler RP, Crapo JD (2002) Oxidative stress in allergic respiratory diseases. J Allergy Clin Immunol 110 (3):349–356
46. Walston J, Xue Q, Semba R, Ferrucci L, Cappola A, Ricks M, Guralnik J, Fried L (2006) Serum antioxidants, inflammation, and total mortality in older women. Am J Epidemiol 163(1):18–26
47. Mahajan A, Tandon VR (2004) Antioxidants and rheumatoid arthritis, Medicine, Biology 12:139–142
48. Hassan SZ, Gheita TA, Kenawy SA, Fahim AT, El-Sorougy IM, Abdou MS (2011) Oxidative stress in systemic lupus erythematosus and rheumatoid arthritis patients: relationship to disease manifestations and activity. Int J Rheum Dis 14(4):325–331
49. Hadjigogos K (2003) The role of free radicals in the pathogenesis of rheumatoid arthritis. Panminerva Med 45(1):7–13
50. Hitchon CA, El-Gabalawy HS (2004) Oxidation in rheumatoid arthritis. Arthritis Res Ther 6(6):1–14
51. Kalpakcioglu B, Şenel K (2008) The interrelation of glutathione reductase, catalase, glutathione peroxidase, superoxide dismutase, and glucose-6-phosphate in the pathogenesis of rheumatoid arthritis. Clin Rheumatol 27(2):141–145
52. Hassan M, Hadi R, Al-Rawi Z, Padron V, Stohs SJ (2001) The glutathione defense system in the pathogenesis of rheumatoid arthritis. J Appl Toxicol 21(1):69–73
53. Jeon C, Ahn J, Chai J, Kim H, Bae E, Park S, Cho E, Cha H, Ahn K, Koh E (2008) Hypoxia appears at pre-arthritic stage and shows co-localization with early synovial inflammation in collagen induced arthritis. Clin Exp Rheumatol 26 (4):646–648
54. Pérez JM, Alessi CJ (2018) Critical approach to the alternative treatment of chronic kidney disease in dogs and cats Slovenian Veter Res 55 (2):59–71
55. Daenen K, Andries A, Mekahli D, Van Schepdael A, Jouret F (2019) Oxidative stress in chronic kidney disease. Iran J Kidney Dis 34(6):975–991
56. Galvan DL, Green NH, Danesh FR (2017) The hallmarks of mitochondrial dysfunction in chronic kidney disease. Kidney Int 92(5):1051–1057
57. Galle J (2001) Oxidative stress in chronic renal failure. Nephrol Dial Transplant 16(11):2135–2137
58. Pizzino G, Irrera N, Cucinotta M, Pallio G, Mannino F, Arcoraci V, Squadrito F, Altavilla D, Bitto A (2017) Oxidative stress: harms and benefits for human health. Oxid Med Cell Longev 2017:8416763
59. Valko M, Rhodes C, Moncol J, Izakovic M, Mazur M (2006) Free radicals, metals and antioxidants in oxidative stress-induced cancer. Chemico-Biol Intrat 160(1):1–40
60. Samuel JB, Stanley JA, Princess RA, Shanthi P, Sebastian MS (2011) Gestational cadmium exposure-induced ovotoxicity delays puberty through oxidative stress and impaired steroid hormone levels. J Med Toxicol 7(3):195–204
61. Bisht S, Faiq M, Tolahunase M, Dada RJ (2017) Oxidative stress and male infertility. Nat Rev Urol 14(8):470–485
62. Meli R, Monnolo A, Annunziata C, Pirozzi C, Ferrante MC (2020) Oxidative stress and BPA toxicity: an antioxidant approach for male and female reproductive dysfunction. Antioxidants (Basel) 9(5):405
63. Young IS, Woodside JV (2001) Antioxidants in health and disease. J Clin Pathol 54(3):176–186. https://doi.org/10.1136/jcp.54.3.176
64. Pizzino G, Bitto A, Interdonato M, Galfo F, Irrera N, Mecchio A, Pallio G, Ramistella V, De Luca F, Minutoli L, Squadrito F (2014) Oxidative stress and DNA repair and detoxification gene expression in adolescents exposed to heavy metals living in the Milazzo-Valle del Mela area (Sicily, Italy). Redox Biol 2:686–693

Chapter 5
Chief Role of Neuroinflammation and Oxidative Stress in Brain Disorders

Shamaila Zafar

5.1 Introduction

Neuroinflammation and oxidative stress are interconnected pathological mechanisms in developing brain disorders [1]. Inflammation is a representative hallmark of infectious conditions and injury in the living body [2]. Being part of the innate immune response, it is supposed to provide protection against pathogenic bodies through the activation of immune cells, involvement of endothelial cells, and release of specific biochemical mediators. However, uncontrolled inflammatory condition if persists for a long, can lead to chronic inflammatory diseases [3].

The term "neuroinflammation" states those processes that trigger the brain's innate immune system as a result of an inflammatory challenge (injury, infection, hypoxia, exposure to a neurotoxin, aging, or neurodegenerative disease). It is characterized by the instigation of glial cells that include microglia and astrocytes, and the letting out of inflammatory mediators (interleukins and cytokines) that repair the brain [4]. In case, the stimulus persists for a long, the uncontrolled inflammatory condition shifts the system towards neuronal degeneration by dysregulating several cellular pathways that, under ideal physiological conditions, are involved in clearing the waste products/debris out of their cell bodies. Ultimately, the buildup of these waste molecules in neurons establishes oxidative stress and exaggerates the inflammatory condition leading to apoptosis and necrosis. The severity of neuroinflammation is defined by the context, duration, and progression of initial stimulation or insult [5]. Moreover, environmental factors, genetic background, age, and prior

S. Zafar (✉)
Department of Pathology and Genomic Medicine, Thomas Jefferson University, Philadelphia, PA, USA
e-mail: shamaila.zafar@jefferson.edu

A. Imran, G. Hussain (eds.), *The Role of Natural Antioxidants in Brain Disorders*, Food Bioactive Ingredients, https://doi.org/10.1007/978-3-031-41188-5_5

inflammatory experiences also contribute to determining the level of neuroinflammation [6].

Neuroinflammatory reactions in the brain suggestively affect its health and ailment state because of their chief role in amending the development, preservation, and provision of brain cells and their connections. Under physiological conditions, microglia (the brain's immune cells) protect the nervous system through the removal of damaged cellular remains and infectious pathogens and regulate both innate and adaptive brain's immune systems. Under pathological statuses like injury, infection, or ischemic condition, the activation of microglia induces the release of inflammatory compounds that further trigger the activation of astrocytes and immune cells that respond to the injury. Under disease conditions, these activated microglia facilitate neuronal and glial cell damage leading to their death [7–10].

5.2 Chief Players of Neuroinflammation

The neuronal inflammation is pictured by the activation of microglial cells and astrocytes along with the release of the inflammatory arbitrators, neurotrophic factors, and reactive oxygen species (ROS).

5.2.1 Microglia

Microglia are 10% of total brain cells and have been presented as the brain's garbage collector [11]. Alongside, they perform also as housekeepers who maintain brain homeostasis at all stages of life like development, adulthood, and aging [12]. These cells act in the same way as peripheral macrophages do. That is why they are also considered neuronal macrophages. In response to inflammatory challenge, the sleeping or latent microglial cells get activated, adopt the ameboid morphology, transfer to the affected site and start phagocytizing the cellular debris and/or infected cells. Activated microglia also release inflammatory mediators as well as neurotrophic factors and free radicals. These substances may then affect the surrounding tissues whether harmfully or beneficially, depending upon the nature and extent of stimuli [13]. At present, it is valued that microglia have not been considered passersby anymore that just respond to the brain's pathological insult, but in contrast, offer a more vigilant role in the pathophysiological progression of several nervous system disorders [14].

Activated microglia have been characterized traditionally into distinctive phenotypes depending upon their cellular appearance, surface markers expression, and physiological attributes [15]. In response to inflammatory insult, the markers, expressed on the surface of activated microglia are named immune-pattern-recognition receptors (PRRs). These include Toll-like receptors (TLRs), scavenger receptors (SRs), and nucleotide-binding oligomerization domain (NOD)-like receptors (NLRs) [16]. The goal of PRRs is to recognize the external pathogenic moieties which are identified as pathogen-associated molecular patterns or PAMPs, and/or endogenous host-derived molecules which are identified as damage-associated-molecular-patterns (DAMPs) [17, 18]. Both patterns have separate reactions; PAMPs persuade a microbicidal and an inflammatory reaction to counter the infection, whereas DAMPs instigate sterile inflammation. Sterile inflammation is a kind of inflammatory reaction elicited by the instigation of surface receptors that identify indicators/signals released from injured/broken cells in reaction to neuronal damage like hypoxia, trauma, and neurodegenerations [19]. The microglia's PRRs interaction with PAMP/DAMPs initiates the overflow of several intra-cellular processes including the release of certain kinases, and stimulation of downstream transcription factors, leading to the dismissal of inflammation mediating molecules and other cellular reactions [18, 19].

In a simple way, we can categorize the activation states of latent microglia into two types; M1 and M2, both have different roles in neuronal degeneration and repair. Here, latent microglia are denoted as M0. M1 phenotype is the classical activated form that is linked to pro-inflammatory and neurotoxic reactions; for instance, the stimulus having IFN-γ (Interferon-gamma) and LPS (lipopolysaccharide) endorses the conversion of latent microglia (M0) into phenotype M1. This influences the expression of pro-inflammatory cytokines like IL-1β, TNFα, IL-6, iNOS, and free radicals that contribute to impairing the brain repairing processes and lead to establishing the chronic type of neuroinflammation, uncontrolled oxidative stress, and overlong neurological damage. Whereas, the M2 phenotype intercedes anti-inflammatory and neuroprotective tasks; an inflammatory situation having higher levels of anti-inflammatory cytokine IL-4 or IL-13 pushes the state of microglia into the M2 phase that upregulates protein markers (CD206 (macrophage mannose receptor), CD163 (macrophage-specific membrane marker), FCγR (Fc gamma receptors), arginase 1, and TGFβ (Transforming growth factor beta)), discharge anti-inflammatory cytokines as well as neurotrophic factors along with proteases, that eventually increase the phagocytic activity, promote immunosuppression, control M1-mediated neuroinflammation, and contribute to neuronal repairment by modifying neurorestorative processes like neurogenesis, angiogenesis, oligodendrogenesis, and remyelination (Fig. 5.1) [20].

5.2.2 *Cytokines and Chemokines*

Cytokines are small protein molecules that, in very minute concentrations (nanomolar to picomolar), act as signaling mediators to control inflammation and modify cellular growth, survival, and differentiation [21]. Chemokines are a class of those chemogenic cytokines that are responsible to induce cell migration [22]. In the neuronal system, cytokines and chemokines play a neuromodulatory job by regulating neurodevelopment, synaptic transmission, and neuroinflammation. They are vital for neuronal immune activities as they serve to uphold immune inspection through

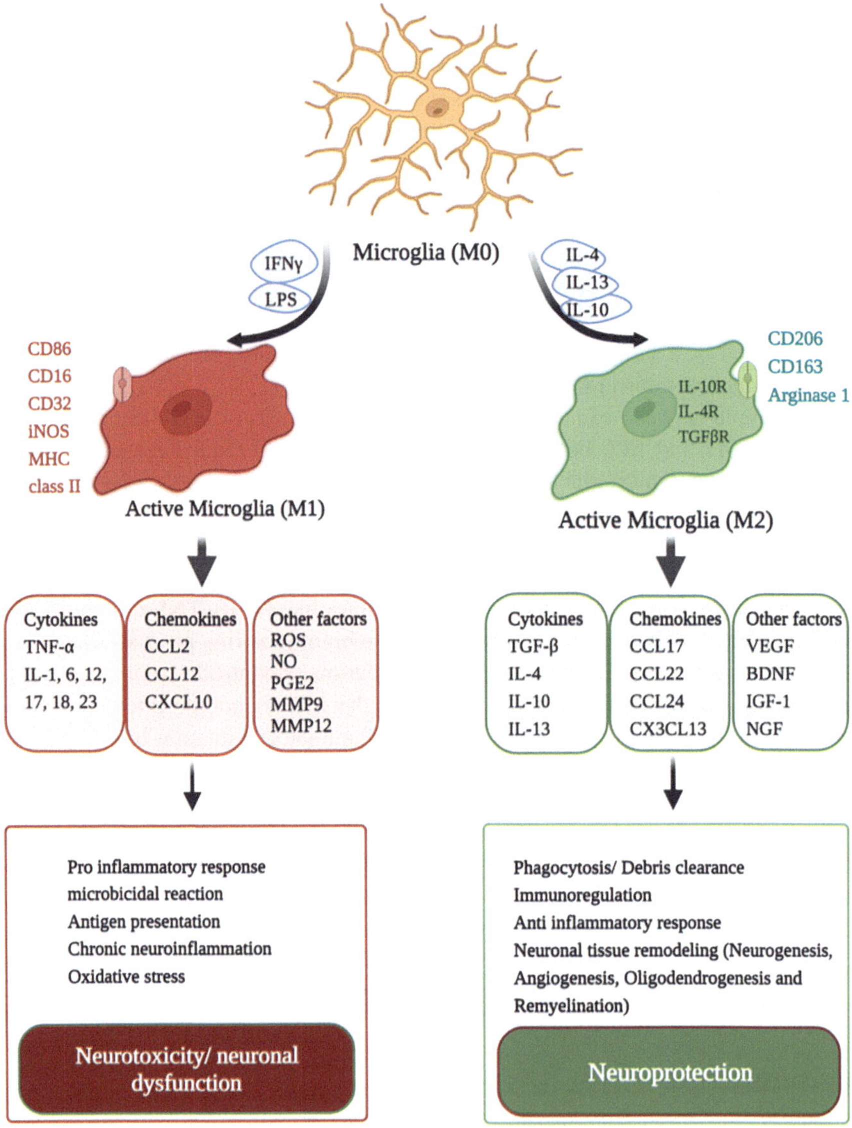

Fig. 5.1 Immunogenic response of neuronal tissues is displayed by activation of resting microglia (M0) into its activated phenotypes M1 and M2. The illustration represents the heterogeneity and functional diversity of both phenotypes and their distinctive roles in neural dysfunction and neuroprotection (figure created with BioRender.com)

migrating brain's immune cells like active microglia, astrocytes and oligodendrocytes (in the case of the central nervous system), and Schwann cells (in the case of the peripheral nervous system), and employee other inflammatory factors in reaction to invading pathogens or damaged cells [23]. Moreover, the endothelial cells of the nervous system's microvasculature and even the neurons can both discharge and respond to cytokines and chemokines because of having cytokine and chemokine receptors. The attachment of cytokines to their receptors results in the initiation of an overflow of signaling events that control cellular activities like adhesion, proliferation, survival, death, apoptosis, etc [24].

Some are proinflammatory cytokines while others are anti-inflammatory. Activated microglia (M1 phenotype) discharge proinflammatory cytokines (IL-1β, IL-6, and TNF-α) that are responsible for the up-regulation of inflammatory reactions including pathological pain. Some proinflammatory cytokines play both protective and neurodegenerative roles. For instance, IL-6 has a twin role in reaction to brain injury and ailment. It is released by active astrocytes as a reaction to neuronal injury, endorsing neuronal survival in a neurotrophin-like style, however, its abnormal levels also contribute equally to pathological characteristics of several brain diseases [25]. Ample evidence suggests that IL-1β and TNF persuade neurotoxicity by raising-up glutamate release that eventually causes neuronal excitotoxic death [26]. A major anti-inflammatory cytokine that is abundantly released by active microglia in response to PAMPs is IL-10 [27]. Other predominantly released anti-inflammatory cytokines are as follows; IL-1 receptor antagonist, ILs 4, 10, 11 & 13; Leukemia inhibitory factor (LIF), IF-α. IL-6 and TGF-β are characterized either as anti-inflammatory or proinflammatory, in various situations. Moreover, the receptors for IL-1, TNF-α, and IL-18 also play as a blocker for proinflammatory cytokines [28].

5.2.3 *Astrocytes*

Astrocytes are star-shaped glial cells and constitute 20–40% of the whole brain cells [29]. They provide support to all neurons and the rest of the glial cells. Being a component of the neurovascular unit, astrocytes act as one of the main pillars that are responsible to maintain the BBB permeability and function. They control the trafficking of immune cells and play a regulatory role in neuroinflammatory processes [30]. These regulatory functions are executed through specific signaling pathways such as NF-κB (nuclear factor kappa-B), TLRs, MAPK (mitogen-activated protein kinase), JAK/STAT3 (Janus kinase/ signal transducer and activator of transcription-3), SIPR (sphingosine 1-phosphate receptor) and PI3K/AKT (phosphatidyl inositol-3-kinases/ protein kinase-B) [31].

Like microglia, astrocytes' response to inflammatory stimuli can either be beneficial or damaging for neuronal repair, well again, it depends upon the nature of instigating stimuli. [32]. Active microglia induce resting astrocytes through molecular triggers such as TGF-β1 [33], leukemia inhibitory factor (LIF) [34], and ciliary neurotrophic factor [35] to convert them into their active forms either A1 state (neurotoxic) or A2 state (neuroprotective) [36] by upregulation of certain genes while downregulation of others [37, 38]. A1 astrocytes release proinflammatory molecules like TNF, IL-1β, and ROS [36]. They also instigate the release of some chemokines in response to IL-1. These include β chemokine-2 or CCL2, β chemokine-20 or CCL20, α chemokine-1 or CXCL1, α chemokine-10 or CXCL10, α chemokine-12 or CXCL12, and cell adhesion molecules [31, 39]. The A2 phase of astrocytes is associated with neuroprotection and involves the release of brain-derived neurotrophic factor (BDNF), vascular endothelial growth factor (VEGF), and nerve growth factor; all are known for their neuroprotective jobs [40, 41]. Moreover, the A2 morph of astrocytes acts as dominant cells in combating oxidative stress by increasing antioxidant enzymes level [42].

5.2.4 Blood-Brain Barrier

The blood-brain barrier, abbreviated as BBB, is an extremely selective semipermeable lining made by a monolayer of tightly packed endothelial cells that hinders the non-selective crossing of solutes from the systemic blood circulation into the surroundings of neuronal tissues in the brain. These endothelial cells have specialized tight junctions and are wrapped by a basal membrane, pericytes, and astroglial endfeet along with some macrophages, neurons, and interneurons to form a neurovascular unit (NVU). This multicellular unit is specialized to regulate cerebral blood flow and nutrients and oxygen transport to neuronal tissues [43]. BBB is a highly complex and dynamic barrier that acts as a protective interface for separating blood circulation from the brain parenchyma. It acts as an important contributor to establishing and maintaining the microenvironment of the CNS which is essential for normal neuronal activity [44, 45].

BBB was believed to be responsible for making the brain an organ having no access to the immune system as this barrier protects the entering of any immune cells from the systemic circulation into brain parenchyma. Current researches however negate this point of view. Inflammatory responses are now been reported in CNS in response to peripheral injury or systemic infection in the body (ref) and access to systemic infection to the nervous system happens through disrupted BBB. The breakdown of BBB happens through the disruption of NVU, a crucial characteristic of neuroinflammation [46]. Proinflammatory cytokines and chemokines released by activated microglia and astrocytes can affect the brain microvasculature and disrupt the BBB integrity. Peripheral macrophages, lymphocytes, and plasma proteins can then penetrate the brain parenchyma (Fig. 5.2) and exacerbate the neuroinflammatory condition and oxidative stress because peripherally infiltered immune cells and plasma proteins further trigger the conversion of resting glia into their active state, thereby worsening the inflammatory situation that fastens the neurodegeneration [47, 48].

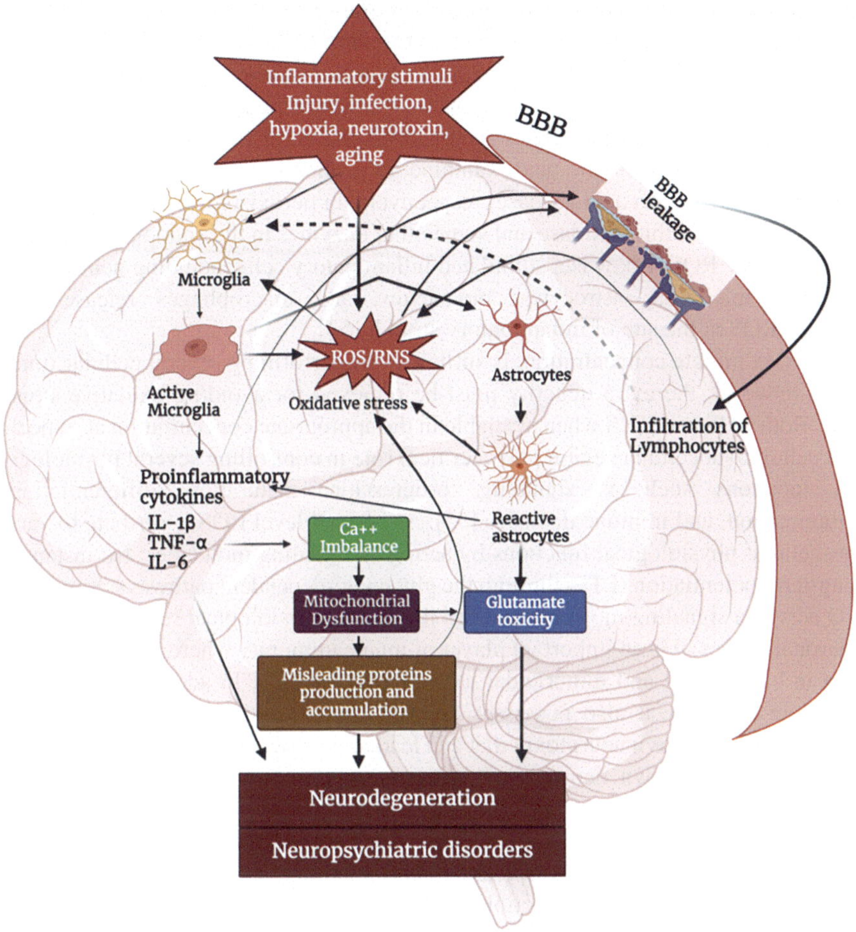

Fig. 5.2 Illustration represents the interlink between neuroinflammation and oxidative stress in brain disorders (figure created with BioRender.com)

5.3 Oxidative Stress; Trigger of Inflammation

Oxidation-reduction reactions or redox homeostasis is complimentary for all cellular physiological functions and disruption in this homeostasis establishes oxidative stress [49, 50]. Oxidative stress is a state of dissimilarity between the antioxidants and oxidants in the advantage of oxidants, leading to the bioavailability of free radicals that eventually cause tissue damage [51]. These free radicals are produced as byproducts under various cellular processes and include ROS e.g., hydroxyl radicals, superoxide, singlet oxygen, hydrogen peroxide, peroxides; reactive nitrogen species (RNS) e.g., nitric oxide (NO) and peroxynitrite (ONOO·); and some reactive chlorine species (RCS) e.g., hypochlorous acid (HOCl). The cellular organelle "mitochondria" is the center for cellular respiration, and is responsible for ATP production (a major source of energy for all bodily functions) through the way of oxidative phosphorylation that includes electron transport-chain reaction whereby ROS are released as a byproduct [52]. Moreover, several enzymatic reactions, happening in other cellular organelles such as β-oxidation of fatty acids by acyl CoA oxidase, metabolism of uric acid, xanthine, D-proline by urate oxidase, xanthine oxidase, and D-amino acid oxidase, respectively, in peroxisomes; and activities of cytochrome P450, thiol oxidase and some other enzymes in the endoplasmic reticulum, generate ROS. Likewise, stimulated inflammatory cells including neutrophils, activated microglia, astrocytes, monocytes, and macrophages release free radicals/ROS at the site of inflammatory stimuli [53].

ROS, in minute concentration, is sufficient for performing several cellular functions, however, the extra quantity must be removed for avoiding oxidative stress [54]. Both ROS and RNS when available in the appropriate concentration at a specified cellular/subcellular region play a critical role in controlling several physiological functions such as signaling, progression, maturation, differentiation, inflammation, and immune defenses [55]. At optimal level ROS contribute to various cellular physiological functions by acting as signaling molecules; for instance, long-term potentiation (LTP) through the glutamate-dependent pathways. Similarly, NO acts as a signaling molecule and has numerous physiological functions even in neuronal tissues. it is an important player of innate immunity where it offers its support in bactericidal and tumoricidal jobs of macrophages. In addition, to being a signaling molecule it also possesses neurotransmitter properties [56]. However, excessive NO acts as a neurotoxin that can lead to neurodegeneration for example it is a commonly known neurotoxin of dopaminergic neurons in Parkinson's disease [57].

Excessive ROS generation can cause protein aggregation, excitotoxicity, endoplasmic reticulum stress, and the activation of inflammatory pathways, all of which have been directly linked to the etiology of the disease. Being a consumer of a high amount of oxygen and having a higher content of lipids, neuronal tissues of the brain have a high susceptibility to being damaged. Neuronal cell membranes are enriched with polyunsaturated fatty acids that are highly prone to be affected by ROS because the primary mechanism of damage caused by ROS is lipid peroxidation [50].

There is always an oxidant scavenging system working in parallel, to keep ROS/RNS at an optimal level. In the brain, two types of such systems are working that oppose the threat of ROS; one system involves antioxidant enzymes, and the other one includes low molecular weight (LMW) antioxidants [42, 49, 58]. Superoxide dismutase (SOD), glyoxalase, catalase (CAT), glutathione reductase, and glutathione peroxidase are antioxidant enzymes that systemically scavenge the oxidants and thus prevent the establishment of oxidative stress [59, 60]. For instance, the SOD enzymes named Cu-Zn SOD and Mn-SOD, enable quick dismutation of superoxides [61]. This step produces hydrogen peroxide that is then replaced by glutathione peroxidase and CAT. The second system of scavenging oxidants includes LMW antioxidants such as uric acid, ascorbic acid (vitamin C), melatonin, and glutathione that scavenge the free radicals through the chelation of transition metals [52]. Glutathione protects the brain endothelial cells against oxidative stress by attenuating hydrogen peroxide-induced nitric oxide production, ROS, and 8-Oxo-2′-deoxyguanosine (8-OHdG). In this way, it is supposed to prevent hydrogen peroxide-induced decline in proteins of tight junctions. Glutathione also raises the Nrf2 (nuclear factor erythroid 2 related factors 2) and mediates its signaling cascades [62]. It indicated that during brain injury and stroke, glutathione could be a potential therapeutic target to cope the oxidative stress and protect the endothelial cells.

Disruption of the scavenging systems results in leveling up the ROS. High levels of ROS cause the mitochondrial permeability transition pore to open, which results in mitochondrial depolarization, a decline in ATP synthesis, and a rise in Ca^{2+} levels that eventually cause ferroptosis, necrotic cell death, and cellular apoptosis leading to neural degeneration [63, 64]. Ferroptosis is another morphologically and genetically unique type of cell death that has been recently discovered and is thought to be triggered by intracellular phospholipid peroxidation [65].

5.4 Role of Neuroinflammation and Oxidative Stress in Brain Disorders

Generally, brain disorders can be categorized into two major classes; neurodegenerative and neuropsychological. Both classes involve neuroinflammation and oxidative stress as major pathological perspectives in disease progression. In fact, neuroinflammation and oxidative stress are inducers of one another. As discussed above, active microglia release a variety of oxidants and several proteins and genes including proinflammatory cytokines [66]. Numerous cytokines and neurotoxic compounds are continuously released in persistent neuroinflammation, which prolongs the neurodegenerative process [67]. Oxidative stress-induced biochemical changes in biomolecular components of the neurons result in the development of neurodegenerative diseases [42] (Fig. 5.2).

5.5 Neurodegeneration

The condition when neurons lose their function and start dying is described as neurodegeneration. The specific area/group of neurons that undergo neurodegeneration results in a functional deficit of its respective organ/tissue. Neurotoxic accumulation of unwanted proteins in neuronal tissues is a major hallmark of neurodegenerative disorders like Parkinson's disorder (PD), Alzheimer's disorder(AD), and Huntington's disorder (HD) [68].

PD is characterized by a neurodegenerative loss of the dopaminergic neurons in the substantia nigra leading to motor function deficits and later-on cognitive function loss that happens as the disease progresses [69]. Abnormal accumulation of alpha-synuclein protein in dopaminergic neurons triggers an inflammatory cascade of reactions that result in neuronal cell death. Several pieces of evidence support the active role of neuroinflammation in disease inception and progression [70]. Briefly, activated microglia contribute as major players through the production of glutamate, proinflammatory cytokines, ROS, and quinolinic acid (neurotoxic metabolite) [71]. They mobilize the adaptive immune responses by releasing chemotactic cytokines that facilitate transendothelial migration of immunocytes across BBB and perpetuate neural damage [72]. The autopsy reports show the presence of reactive microglia (as discussed above; the main characteristic of neuroinflammatory response) in Substantia nigra (pars compacta) of PD brains, plus six times higher level of active microglia found in phagocytosing dopaminergic neurons and link with deposition of α-synuclein [73–75] suggest the active involvement of active microglia that over-express several inflammatory markers such as CD23, CD11a, CD54, ferritin, and release proinflammatory cytokines (IFN-γ, TNF-α, IL-β, iNOS, and cyclooxygenase I & II) [76–78]. Moreover, studies have shown the presence of significantly higher levels of M2 phenotype active microglia, not only in the SN region but also in its neighborhood tissues (putamen, transentorhinal cortex, hippocampus, temporal cortex, cingulate cortex, and) of the PD brain. The involvement of active microglia in the cell death of dopaminergic neurons suggests that it can be a sensitive biomarker for PD [76]. Furthermore, the epidemiological data reveals that NSAIDs (non-steroidal anti-inflammatory drugs) reduced PD risks [72, 79]. Likewise, various biochemical and histological pieces of evidence are there about the contribution of oxidative distress in PD pathology. oxidative stress endorses post-translational modification of α-synuclein and induces its accumulation in dopaminergic neuronal cells. That accumulated α-synuclein further produces intracellular ROS [76, 80]. Substantial nigra contain a remarkably high concentration of oxidized proteins as compared to that of the caudate, putamen, and frontal cortex, therefore, this region is more vulnerable to oxidative stress leading to selective neuronal degeneration [81]. Studies reported higher levels of lipid peroxidation, carbonyl and nitrotyrosine protein alterations, DNA impairment, and decline of glutathione plus ferritin in PD brains [82]. Excessive levels of NO exert its neurotoxic effects through the formation of peroxynitrile. It also binds with proteins having sulfur-attached cysteine residues, through the process called "S-nitrosylation" which may alter the structural confirmation of proteins and/or enzymes and ultimately their activities [57].

The upregulated level of NADPH oxidase (the main fabricator of ROS), as well as the high level of GP91 phlox protein (a transmembrane component of NADPH-oxidase) in substantia nigra pars compacta, suggest that active microglia's upregulated capacity for ROS production [83, 84]. Higher ROS and NO levels in healthy dopaminergic neurons are rigorously regulated by antioxidative processes involving glutathione (GSH) [85], DJ-1 [86], and superoxide dismutase [87]. Several studies have reported the disruption of these defensive agents against oxidative stress in the pathogenesis of PD. A greater knowledge of the complicated role that oxidative stress plays in the pathophysiology of Parkinson's disease may thus provide novel targets for therapeutic intervention and early clinical diagnostics [88].

Another common form of neurodegenerative disease is Alzheimer's disease (AD) which is estimated to contribute 60–70% of all cases of dementia around the globe [89]. According to the formerly known amyloid cascade hypothesis, abnormal formation and accumulation of misfolded amyloid beta (Ab) and tau-proteins in the brain is the instigating event in AD pathogenesis, although this hypothesis is insufficient to explain many pathological aspects. The identification of higher levels of inflammatory markers in patients with AD and the presence of AD risk genes linked with innate immune functions propose that neuroinflammation has a projecting role in AD pathology. [89]. Postmortem reports of AD brains provided evidence that Aβ plaques are surrounded by the presence of activated microglia and astrocytes, along with various inflammatory cytokines [90]. Aβ's direct binding to Toll-like receptors such as TLR4 and RAGE instigates important inflammatory pathways such as NF-kB [91]. In addition, Aβ is found to impair the integrity of BBB (Chan et al. 2018; Cuevas et al. 2019) thus facilitating the infiltration of neurotoxic substances in the brain parenchyma and worsening the neuroinflammatory reactions (Nelson et al. 2016) that mess up nerve terminals activity leading to dysfunctional or loss of synapses (a reason for memory deficit). Therefore, the foremost therapeutic focus should be to control the reactive microglia and astrocytes and to inhibit the excessive production of proinflammatory mediators and oxidants that could be valuable to stop or prevent neurodegeneration in dementia. Patients with AD have severe oxidative damage to their brains (Huang et al. 2016). As ROS overproduction is believed to play a crucial role in the accumulation and deposition of $A\beta$ in AD, ROS-induced oxidative stress is emerging as a significant element in the pathological condition of AD [92]. For instance, ROS induces lipid peroxidation, which leads to increased Ca +2 influx causing impaired membrane permeability and triggering excitotoxicity mechanisms. This is thought to significantly amend neurotransmission as well as cognitive functions. Plaques can cause cytosolic Ca2+ overload by reducing the amount of calcium ions (Ca^{2+}) stored in the endoplasmic reticulum (ER). Endogenous GSH levels are decreased and ROS might overaccumulation inside the cells in response to an increase in cytosolic Ca^{2+}. Mitochondrial malfunction can result in excitotoxicity, ATP production reduction, altered Ca^{2+} homeostasis, and ROS misregulation. All of these changes could be connected to the development of AD [93]. Moreover, by activating p38 mitogen-activated protein kinase, Aβ induced ROS increased production changes cellular signaling pathways and starts tau hyper-phosphorylation. NFT development within cells may result from an aberrant buildup of hyperphosphorylated tau enzymes [94]. For instance, p38, a member of the family of mitogen-activated protein kinases, is activated when Aβ-mediated oxidative stress occurs. In one of the many functions of p38, tau phosphorylation was seen to be induced in the primary neuronal model, that might be avoided by treatment with a p38 inhibitor or by vitamin E [95].

Another example of a neurodegenerative condition is Huntington's disease (HD). This is an autosomal dominant neurodegenerative ailment caused by a single mutation in the HTT gene that has instructions for the formation of Huntington protein [96]. HD is characterized by neuronal degeneration in the striatum, that propagates to the cerebral cortex and then to the thalamus as the disease progresses. Patients, at the start, manifest personality and mood changes. Later on, they show a cognitive decline leading to dementia, uncontrolled choreiform movements, bradykinesia, rigidity, and/or dystonia. Patients could survive only for 15–20 years after the disease onset [97].

Postmortem reports of HD patient brains showing higher levels of active microglia and astrocytes along with prominent signs of oxidative stress plus, upregulated levels of IL-6, IL-8, and TNF-α in the CSF of HD patients provide evidence about the existence of active inflammatory processes [98, 99]. Oxidative stress markers reported in HD brains specimen include a decline in several mitochondrial enzymes taking part in respiration [100], lack of aconitase activity in brain regions; caudate and putamen [101], higher levels of Protein carbonyls representing overproduction of ROS, [102]. Additionally, higher levels of 3-nitrotyrosine (representative of RNS) have been reported in the striatum and cortex of HD [103]. Plus accumulated levels of iron in HD brains were seen that indicate mitochondrial disruption and ROS over-production [104].

Amyotrophic Lateral Sclerosis (ALS) involves the neurodegeneration of large motor neurons found in the motor cortex (Pyramid Betz cells), cranial motor nuclei of the brain stem, and large anterior horn of the spinal cord [51, 105]. The motor neurons' degeneration is instigated by abnormal intracellular accumulation of waste products in form of distinct inclusion bodies. Ubiquitinated inclusions, hyaline conglomerate inclusions, and Bunina bodies are a few examples. Ubiquitinated inclusions are largely found in lower motor neurons of the brainstem and spinal cord. It affects motor neurons and is deadly within two to five years of diagnosis. There is currently no effective treatment for ALS. Data suggests that an excessive synthesis of ROS coupled with an ineffective antioxidant defense system represents a key pathological aspect of ALS [106]. Already published postmortem reports of ALS brains have shown elevated levels of oxidative damage to proteins, lipids, and DNA as well as higher ROS quantification were reported in cerebrospinal fluid, urine, and serum samples of ALS patients. Researchers found the mutant form of SOD1in ALS patients. SOD1 is a crucial player in clearing the excessive ROS, however, its mutant form (mSOD1) is unable to control the ROS [107]. Moreover, Glutathione level was changed in ALS patients, with raised quantities of GSSG and reduced amounts of GSH, resulting in a considerably higher GSSG / GSH ratio. Increased levels of 8-Hydrxy-2-deoxyguanosine and Malondialdehyde [108], 3-nitrotyrosine, hydroxyguanine, and protein carbonyl [109] in ALS patients are considered to be linked with disease progression. Excessive oxidants activate the microglia and initiate the cascade of inflammatory reactions [107]. An elevated level of several inflammatory markers has been reported such as higher levels of COX-2, iNOS, and neuronal NOS (all are expressed by reactive astrocytes), and IL1, TNF, and C1q (expressed by activated microglia) in ALS patients [110, 111] have found high levels of cytokines such as G-CSF, IL2, 15 & 17, MCP-1, MIP-1, TNF, and VEGF, in the CSF of ALS patients. All of these observations support the theory that neuroinflammatory processes contribute to the development of ALS [112, 113].

5.6 Neuropsychiatric Diseases

Neuropsychiatric diseases include certain mood disorders (bipolar disorder, mania, disruptive mood dysregulation disorder, cyclothymia, premenstrual dysphoric disorder) and Schizophrenia. These are multifaceted and diverse nature diseases that not even destroy the quality of life but also negatively disturb behavior and cognitive tasks [114, 115]. Genetic predisposition, circadian disruptions, monoamine deficiency, hormonal imbalancement (such as hypercortisolemia), and inflammation are remarkable pathological aspects of these diseases [115, 116]

Neuroinflammation plays its role through the oversecretion of proinflammatory cytokines that trigger the hypothalamus pituitary adrenal axis, raise-up the glucocorticoid resistance, intricate the serotonin synthesis and its metabolism, and thus interrupt the neurogenesis, neuronal cell apoptosis, and neuroplasticity [117, 118]. Astrocytes are supporting glial cells and play a vital role in neuronal synapse activity. However, under influence of inflammatory reactions and oxidative stress astrocytes become dysfunctional. Dysfunctional astrocytes are unable to uptake glutamate from the synaptic cleft which results in prolonged synaptic activation leading to excitotoxicity, a pathological phenomenon reported in many psychiatric illnesses like depression, bipolar disorder, and schizophrenia [115, 119].

Meanwhile, oxidative stress establishment is also a complimentary part of these pathological processes (Fig. 5.2). Psychological stress itself is a disruptor of the oxidant-antioxidant balance in the brain. Inflammation induced-Ca++ imbalance, glutamate toxicity, and mitochondrial dysfunction exaggerate the oxidative stress, triggering biochemical distress in the brain that disrupts neurocircuitry and deteriorates hippocampal, amygdalar, and cortical connections, eventually causing behavioral and cognitive deficits [120].

Several researchers in the early years are favoring that oxidative stress is an inducer of depression. This theory is further supported by the fact that some psychotropic drugs prevent depression symptoms by combating oxidative stress [121]. The positive effects of antidepressants against depression are thought to be due to their capability to suppress proinflammatory cytokine control ROS/RNS production and enhance antioxidant defense [122].

Researchers have reported the presence of elevated levels of oxidative stress markers in anxiety, depression, Autism, and Schizophrenia [120, 123–125]. Postpartum reports of depressed individuals' brain samples have shown higher levels of ROS, RNS, and altered levels of antioxidant GSH [120, 126–129]. Moreover, the polymorphic changes in genes coding SOD, CAT, and GPX4 (Glutathione peroxidase 4) are considered to have a link with depression [130, 131]. A recent finding has suggested that a single nucleotide polymorphic change in Vall6A1a-SOD2 (SOD-Mn dependent) gene causes superoxide and hydrogen peroxide-based oxidative imbalance that increases the risk of depression and psychological stress with increasing age [132]. Glutathione depletion raises oxidative stress and studies have reported the involvement of scant levels of Glutathione peroxidase in developing brain abnormalities leading to neuropsychiatric issues such as bipolar disorder [133]. Antipsychotic drugs that are prescribed to a patient with schizophrenia and bipolar disorders are found to increase the systemic SOD level in patients [133, 134].

In a nutshell, it looks reasonable to recommend that managing oxidative stress either by quenching the free radicals or by promoting the antioxidant enzyme activity could be a fruitful way to prevent or stop the prognosis of neuropsychiatric issues.

References

1. Parsons ALM, Bucknor EMV, Castroflorio E et al (2022) The interconnected mechanisms of oxidative stress and neuroinflammation in epilepsy. Antioxidants 11:157
2. Matsuda M, Huh Y, Ji RR (2018) Roles of inflammation, neurogenic inflammation, and neuroinflammation in pain. J Anesth 2018 331(33):131–139. https://doi.org/10.1007/S00540-018-2579-4
3. Furman D, Campisi J, Verdin E, et al (2019) Chronic inflammation in the etiology of disease across the life span. Nat Med 25:1822–1832. https://doi.org/10.1038/s41591-019-0675-0
4. Gilhus NE, Deuschl G (2019) Neuroinflammation – a common thread in neurological disorders. Nat Rev Neurol 15:429–430. https://doi.org/10.1038/S41582-019-0227-8
5. DiSabato DJ, Quan N, Godbout JP (2016) Neuroinflammation: the devil is in the details. J Neurochem 139:136–153. https://doi.org/10.1111/JNC.13607
6. Shabab T, Khanabdali R, Moghadamtousi SZ et al (2017) Neuroinflammation pathways: a general review. Int J Neurosci 127:624–633. https://doi.org/10.1080/00207454.2016.1212854
7. Nilupul Perera M, Ma HK, Arakawa S et al (2006) Inflammation following stroke. J Clin Neurosci 13:1–8. https://doi.org/10.1016/J.JOCN.2005.07.005
8. De Chiara G, Marcocci ME, Sgarbanti R et al (2012) Infectious agents and neurodegeneration. Mol Neurobiol 46:614–638. https://doi.org/10.1007/S12035-012-8320-7
9. Vilhardt F (2005) Microglia: phagocyte and glia cell. Int J Biochem Cell Biol 37:17–21. https://doi.org/10.1016/J.BIOCEL.2004.06.010
10. Gendelman HE (2002) Neural immunity: friend or foe? J Neurovirol 8:474–479. https://doi.org/10.1080/13550280290168631
11. Badimon A, Strasburger HJ, Ayata P et al (2020) Negative feedback control of neuronal activity by microglia. Nature 586:417–423. https://doi.org/10.1038/S41586-020-2777-8
12. Wendimu MY, Hooks SB (2022) Microglia phenotypes in aging and neurodegenerative diseases. Cells 11. https://doi.org/10.3390/CELLS11132091
13. Woodburn SC, Bollinger JL, Wohleb ES (2021) The semantics of microglia activation: neuroinflammation, homeostasis, and stress. J Neuroinflam 18:1–16. https://doi.org/10.1186/S12974-021-02309-6/FIGURES/2
14. Wright-Jin EC, Gutmann DH (2019) Microglia as dynamic cellular mediators of brain function. Trends Mol Med 25:967. https://doi.org/10.1016/J.MOLMED.2019.08.013
15. Prinz M, Jung S, Priller J (2019) Microglia biology: one century of evolving concepts. Cell 179:292–311. https://doi.org/10.1016/J.CELL.2019.08.053
16. Doens D, Fernández PL (2014) Microglia receptors and their implications in the response to amyloid β for Alzheimer's disease pathogenesis. J Neuroinflammation 11:48. https://doi.org/10.1186/1742-2094-11-48

17. Zindel J, Kubes P (2020) DAMPs, PAMPs, and LAMPs in immunity and sterile inflammation. Annu Rev Pathol 15:493–518. https://doi.org/10.1146/ANNUREV-PATHMECHDIS-012419-032847
18. Ransohoff RM, Brown MA (2012) Innate immunity in the central nervous system. J Clin Invest 122:1164. https://doi.org/10.1172/JCI58644
19. Broggi A, Granucci F (2015) Microbe- and danger-induced inflammation. Mol Immunol 63:127–133. https://doi.org/10.1016/J.MOLIMM.2014.06.037
20. Shao F, Wang X, Wu H, et al (2022) Microglia and neuroinflammation: crucial pathological mechanisms in traumatic brain injury-induced neurodegeneration. Front Aging Neurosci:14. https://doi.org/10.3389/FNAGI.2022.825086
21. Giansanti F, Giardi MF, Botti D (2007) Avian cytokines – an overview. Curr Pharm Des 12:3083–3099. https://doi.org/10.2174/138161206777947542
22. Moser B, Loetscher P (2001) Lymphocyte traffic control by chemokines. Nat Immunol 2001 22 2:123–128. https://doi.org/10.1038/84219
23. Takeshita Y, Ransohoff RM (2012) Inflammatory cell trafficking across the blood–brain barrier: chemokine regulation and in vitro models. Immunol Rev 248:228–239. https://doi.org/10.1111/J.1600-065X.2012.01127.X
24. Devi LA (2000) G-protein-coupled receptor dimers in the lime light. Trends Pharmacol Sci 21:324–326. https://doi.org/10.1016/S0165-6147(00)01519-4
25. Erta M, Quintana A, Hidalgo J (2012) Interleukin-6, a major cytokine in the central nervous system. Int J Biol Sci 8:1254. https://doi.org/10.7150/IJBS.4679
26. Ye L, Huang Y, Zhao L et al (2013) IL-1β and TNF-α induce neurotoxicity through glutamate production: a potential role for neuronal glutaminase. J Neurochem 125:897–908. https://doi.org/10.1111/JNC.12263
27. Wang YJ, Monteagudo A, Downey MA et al (2021) Cromolyn inhibits the secretion of inflammatory cytokines by human microglia (HMC3). Sci Reports 11:1–11. https://doi.org/10.1038/s41598-021-85702-8
28. Opal SM, DePalo VA (2000) Anti-inflammatory cytokines. Chest 117:1162–1172. https://doi.org/10.1378/CHEST.117.4.1162
29. Khakh BS, Sofroniew MV (2015) Diversity of astrocyte functions and phenotypes in neural circuits. Nat Neurosci 18:942. https://doi.org/10.1038/NN.4043
30. Colombo E, Farina C (2016) Astrocytes: key regulators of neuroinflammation. Trends Immunol 37:608–620. https://doi.org/10.1016/j.it.2016.06.006
31. Ding Z-B, Song L-J, Wang Q et al (2021) Astrocytes: a double-edged sword in neurodegenerative diseases. Neural Regen Res 16:1702–1710. https://doi.org/10.4103/1673-5374.306064
32. Bayraktar OA, Fuentealba LC, Alvarez-Buylla A, Rowitch DH (2015) Astrocyte development and heterogeneity. Cold Spring Harb Perspect Biol 7:a020362. https://doi.org/10.1101/CSHPERSPECT.A020362
33. Ongali B, Nicolakakis N, Tong X-K et al (2018) Transforming growth factor-β1 induces cerebrovascular dysfunction and astrogliosis through angiotensin II type 1 receptor-mediated signaling pathways. Can J Physiol Pharmacol 96:527–534. https://doi.org/10.1139/cjpp-2017-0640
34. Zhou H-J, Yang X, Cui H-J et al (2017) Leukemia inhibitory factor contributes to reactive astrogliosis via activation of signal transducer and activator of transcription 3 signaling after intracerebral hemorrhage in rats. J Neurotrauma 34:1658–1665. https://doi.org/10.1089/neu.2016.4711
35. Wang J, Sui R-X, Miao Q et al (2020) Effect of Fasudil on remyelination following cuprizone-induced demyelination. CNS Neurosci Ther 26:76–89. https://doi.org/10.1111/cns.13154
36. Liddelow SA, Barres BA (2017) Reactive astrocytes: production, function, and therapeutic potential. Immunity 46:957–967. https://doi.org/10.1016/J.IMMUNI.2017.06.006
37. Zamanian JL, Xu L, Foo LC et al (2012) Genomic analysis of reactive astrogliosis. J Neurosci Off J Soc Neurosci 32:6391–6410. https://doi.org/10.1523/JNEUROSCI.6221-11.2012

38. Rossi D (2015) Astrocyte physiopathology: At the crossroads of intercellular networking, inflammation and cell death. Prog Neurobiol 130:86–120. https://doi.org/10.1016/j.pneurobio.2015.04.003
39. Liddelow S, Hoyer D (2016) Astrocytes: adhesion molecules and immunomodulation. Curr Drug Targets 17:1871–1881. https://doi.org/10.2174/1389450117666160101120703
40. Ludwin SK, Rao VT, Moore CS, Antel JP (2016) Astrocytes in multiple sclerosis. Mult Scler 22:1114–1124. https://doi.org/10.1177/1352458516643396
41. Rizor A, Pajarillo E, Johnson J, et al (2019) Astrocytic oxidative/nitrosative stress contributes to Parkinson's disease pathogenesis: the dual role of reactive astrocytes. Antioxidants (Basel, Switzerland):8. https://doi.org/10.3390/antiox8080265
42. Yi W, Schlüter D, Wang X (2019) Astrocytes in multiple sclerosis and experimental autoimmune encephalomyelitis: star-shaped cells illuminating the darkness of CNS autoimmunity. Brain Behav Immun 80:10–24. https://doi.org/10.1016/j.bbi.2019.05.029
43. Prola Netto J, Iliff J, Stanimirovic D et al (2018) Neurovascular unit: basic and clinical imaging with emphasis on advantages of ferumoxytol. Neurosurgery 82:770–780. https://doi.org/10.1093/neuros/nyx357
44. Abbott NJ (2013) Blood-brain barrier structure and function and the challenges for CNS drug delivery. J Inherit Metab Dis 36:437–449
45. Liebner S, Dijkhuizen RM, Reiss Y et al (2018) Functional morphology of the blood–brain barrier in health and disease. Acta Neuropathol 135:311–336
46. Keaney J, Campbell M (2015) The dynamic blood–brain barrier. FEBS J 282:4067–4079. https://doi.org/10.1111/FEBS.13412
47. Laurent C, Dorothée G, Hunot S et al (2017) Hippocampal T cell infiltration promotes neuroinflammation and cognitive decline in a mouse model of tauopathy. Brain 140:184–200. https://doi.org/10.1093/BRAIN/AWW270
48. Mayne K, White JA, McMurran CE et al (2020) Aging and neurodegenerative disease: is the adaptive immune system a friend or foe? Front Aging Neurosci 12:305. https://doi.org/10.3389/FNAGI.2020.572090/BIBTEX
49. Kohen R, Nyska A (2002) Oxidation of biological systems: oxidative stress phenomena, antioxidants, redox reactions, and methods for their quantification. Toxicol Pathol 30:620–650. https://doi.org/10.1080/01926230290166724
50. Jelinek M, Jurajda M, Duris K (2021) Oxidative stress in the brain: basic concepts and treatment strategies in stroke. Antioxidants 10:1886. https://doi.org/10.3390/ANTIOX10121886
51. Obrador E, Salvador-Palmer R, López-Blanch R et al (2021) The link between oxidative stress, redox status, bioenergetics and mitochondria in the pathophysiology of ALS. Int J Mol Sci 22. https://doi.org/10.3390/IJMS22126352
52. Salim S (2017) Oxidative stress and the central nervous system. J Pharmacol Exp Ther 360:201. https://doi.org/10.1124/JPET.116.237503
53. D'Ambrosi N, Cozzolino M, Carrì MT (2018) Neuroinflammation in amyotrophic lateral sclerosis: role of redox (dys)Regulation. Antioxid Redox Signal 29:15–36. https://doi.org/10.1089/ARS.2017.7271
54. Pero RW, Roush GC, Markowitz MM, Miller DG (1990) Oxidative stress, DNA repair, and cancer susceptibility. Cancer Detect Prev 14:555–561
55. Montezano AC, Dulak-Lis M, Tsiropoulou S et al (2015) Oxidative stress and human hypertension: vascular mechanisms, biomarkers, and novel therapies. Can J Cardiol 31:631–641. https://doi.org/10.1016/J.CJCA.2015.02.008
56. Garthwaite J (2019) NO as a multimodal transmitter in the brain: discovery and current status. Br J Pharmacol 176:197–211. https://doi.org/10.1111/bph.14532
57. Bashkatova V (2021) Metabotropic glutamate receptors and nitric oxide in dopaminergic neurotoxicity. World J Psychiatr 11:830–840. https://doi.org/10.5498/wjp.v11.i10.830
58. Kohen R, Beit-Yannai E, Berry EM, Tirosh O (1999) Overall low molecular weight antioxidant activity of biological fluids and tissues by cyclic voltammetry. Methods Enzymol 300:285–296. https://doi.org/10.1016/S0076-6879(99)00135-4

59. Gusti AMT, Qusti SY, Alshammari EM, et al (2021) Antioxidants-Related Superoxide Dismutase (SOD), Catalase (CAT), Glutathione Peroxidase (GPX), Glutathione-S-Transferase (GST), and Nitric Oxide Synthase (NOS) gene variants analysis in an obese population: a preliminary case-control study. Antioxidants (Basel, Switzerland):10. https://doi.org/10.3390/antiox10040595
60. Griendling KK, Sorescu D, Lassègue B, Ushio-Fukai M (2000) Modulation of protein kinase activity and gene expression by reactive oxygen species and their role in vascular physiology and pathophysiology. Arterioscler Thromb Vasc Biol 20:2175–2183. https://doi.org/10.1161/01.ATV.20.10.2175
61. Saso L, Firuzi O (2014) Pharmacological applications of antioxidants: lights and shadows. Curr Drug Targets 15:1177–1199. https://doi.org/10.2174/1389450115666141024113925
62. Song J, Kang SM, Lee WT et al (2014) Glutathione protects brain endothelial cells from hydrogen peroxide-induced oxidative stress by increasing nrf2 expression. Exp Neurobiol 23:93–103. https://doi.org/10.5607/en.2014.23.1.93
63. Masaldan S, Bush AI, Devos D et al (2019) Striking while the iron is hot: Iron metabolism and ferroptosis in neurodegeneration. Free Radic Biol Med 133:221–233. https://doi.org/10.1016/J.FREERADBIOMED.2018.09.033
64. Kalani K, Yan SF, Du Yan SS (2018) Mitochondrial permeability transition pore: a potential drug target for neurodegeneration. Drug Discov Today 23:1983–1989. https://doi.org/10.1016/J.DRUDIS.2018.08.001
65. Sun Y, Chen P, Zhai B et al (2020) The emerging role of ferroptosis in inflammation. Biomed Pharmacother 127:110108. https://doi.org/10.1016/J.BIOPHA.2020.110108
66. Lyman M, Lloyd DG, Ji X et al (2014) Neuroinflammation: The role and consequences. Neurosci Res 79:1–12. https://doi.org/10.1016/J.NEURES.2013.10.004
67. Kaur N, Chugh H, Sakharkar MK et al (2020) Neuroinflammation mechanisms and phytotherapeutic intervention: a systematic review. ACS Chem Neurosci 11:3707–3731. https://doi.org/10.1021/ACSCHEMNEURO.0C00427/ASSET/IMAGES/MEDIUM/CN0C00427_0010.GIF
68. Bazan NG, Halabi A, Ertel M, Petasis NA (2011) Neuroinflammation. Basic Neurochem Princ Mol Cell Med Neurobiol Eighth Ed:610–620. https://doi.org/10.1016/B978-0-12-374947-5.00034-1
69. Schneider JS (2022) A critical role for GM1 ganglioside in the pathophysiology and potential treatment of Parkinson's disease. Glycoconj J 39:13–26. https://doi.org/10.1007/s10719-021-10002-2
70. Joers V, Tansey MG, Mulas G, Carta AR (2017) Microglial phenotypes in Parkinson's disease and animal models of the disease. Prog Neurobiol 155:57–75. https://doi.org/10.1016/j.pneurobio.2016.04.006
71. Espey MG, Chernyshev ON, Reinhard JFJ et al (1997) Activated human microglia produce the excitotoxin quinolinic acid. Neuroreport 8:431–434. https://doi.org/10.1097/00001756-199701200-00011
72. Mosley RL, Benner EJ, Kadiu I et al (2006) Neuroinflammation, oxidative stress and the pathogenesis of Parkinson's disease. Clin Neurosci Res 6:261–281. https://doi.org/10.1016/j.cnr.2006.09.006
73. McGeer PL, Itagaki S, Akiyama H, McGeer EG (1988) Rate of cell death in parkinsonism indicates active neuropathological process. Ann Neurol 24:574–576. https://doi.org/10.1002/ana.410240415
74. Butler CA, Popescu AS, Kitchener EJA et al (2021) Microglial phagocytosis of neurons in neurodegeneration, and its regulation. J Neurochem 158:621–639. https://doi.org/10.1111/jnc.15327
75. Croisier E, Moran LB, Dexter DT et al (2005) Microglial inflammation in the parkinsonian substantia nigra: relationship to alpha-synuclein deposition. J Neuroinflammation 2. https://doi.org/10.1186/1742-2094-2-14

76. Ferreira SA, Romero-Ramos M (2018) Microglia response during Parkinson's disease: alpha-synuclein intervention. Front Cell Neurosci 12. https://doi.org/10.3389/fncel.2018.00247
77. Imamura K, Hishikawa N, Sawada M et al (2003) Distribution of major histocompatibility complex class II-positive microglia and cytokine profile of Parkinson's disease brains. Acta Neuropathol 106:518–526. https://doi.org/10.1007/s00401-003-0766-2
78. Hunot S, Dugas N, Faucheux B et al (1999) FcepsilonRII/CD23 is expressed in Parkinson's disease and induces, in vitro, production of nitric oxide and tumor necrosis factor-alpha in glial cells. J Neurosci Off J Soc Neurosci 19:3440–3447. https://doi.org/10.1523/JNEUROSCI.19-09-03440.1999
79. Chen H, Zhang SM, Hernán MA et al (2003) Nonsteroidal anti-inflammatory drugs and the risk of Parkinson disease. Arch Neurol 60:1059–1064. https://doi.org/10.1001/archneur.60.8.1059
80. Xiang W, Schlachetzki JCM, Helling S et al (2013) Oxidative stress-induced posttranslational modifications of alpha-synuclein: specific modification of alpha-synuclein by 4-hydroxy-2-nonenal increases dopaminergic toxicity. Mol Cell Neurosci 54:71–83. https://doi.org/10.1016/J.MCN.2013.01.004
81. Puspita L, Chung SY, Shim JW (2017) Oxidative stress and cellular pathologies in Parkinson's disease. Mol Brain 101:1–12. https://doi.org/10.1186/S13041-017-0340-9
82. Hald A, Lotharius J (2005) Oxidative stress and inflammation in Parkinson's disease: is there a causal link? Exp Neurol 193:279–290. https://doi.org/10.1016/j.expneurol.2005.01.013
83. Wu D-C, Teismann P, Tieu K et al (2003) NADPH oxidase mediates oxidative stress in the 1-methyl-4-phenyl-1,2,3,6-tetrahydropyridine model of Parkinson's disease. Proc Natl Acad Sci U S A 100:6145–6150. https://doi.org/10.1073/pnas.0937239100
84. Babior BM (1999) NADPH oxidase: an update. Blood 93:1464–1476
85. Bjørklund G, Peana M, Maes M et al (2021) The glutathione system in Parkinson's disease and its progression. Neurosci Biobehav Rev 120:470–478. https://doi.org/10.1016/j.neubiorev.2020.10.004
86. Mita Y, Kataoka Y, Saito Y et al (2018) Distribution of oxidized DJ-1 in Parkinson's disease-related sites in the brain and in the peripheral tissues: effects of aging and a neurotoxin. Sci Rep 8:12056. https://doi.org/10.1038/s41598-018-30561-z
87. Bostantjopoulou S, Kyriazis G, Katsarou Z et al (1997) Superoxide dismutase activity in early and advanced Parkinson's disease. Funct Neurol 12:63–68
88. Trist BG, Hare DJ, Double KL (2019) Oxidative stress in the aging substantia nigra and the etiology of Parkinson's disease. Aging Cell 18:e13031. https://doi.org/10.1111/ACEL.13031
89. Leng F, Edison P (2021) Neuroinflammation and microglial activation in Alzheimer disease: where do we go from here? Nat Rev Neurol 17:157–172. https://doi.org/10.1038/s41582-020-00435-y
90. Frost GR, Li YM (2017) The role of astrocytes in amyloid production and Alzheimer's disease. Open Biol 7. https://doi.org/10.1098/RSOB.170228
91. Fu Y, Hsiao JHT, Paxinos G et al (2016) ABCA7 mediates phagocytic clearance of amyloid-β in the brain. J Alzheimers Dis 54:569–584. https://doi.org/10.3233/JAD-160456
92. Liu Z, Zhou T, Ziegler AC, et al (2017) Oxidative stress in neurodegenerative diseases: from molecular mechanisms to clinical applications. Oxid Med Cell Longev:2017. https://doi.org/10.1155/2017/2525967
93. Niedzielska E, Smaga I, Gawlik M, et al (2015) Oxidative stress in neurodegenerative diseases. Mol Neurobiol. 536(53):4094–4125. https://doi.org/10.1007/S12035-015-9337-5
94. Giraldo E, Lloret A, Fuchsberger T, Viña J (2014) Aβ and tau toxicities in Alzheimer's are linked via oxidative stress-induced p38 activation: protective role of vitamin E. Redox Biol 2:873–877. https://doi.org/10.1016/J.REDOX.2014.03.002
95. Cioffi F, Adam RHI, Broersen K (2019) Molecular mechanisms and genetics of oxidative stress in Alzheimer's disease. J Alzheimer's Dis 72:981–1017. https://doi.org/10.3233/JAD-190863

96. Kumar A, Ratan RR (2016) Oxidative stress and Huntington's disease: the good, the bad, and the ugly. J Huntingtons Dis 5:217–237. https://doi.org/10.3233/JHD-160205
97. Ross CA, Tabrizi SJ (2011) Huntington's disease: from molecular pathogenesis to clinical treatment. Lancet Neurol 10:83–98. https://doi.org/10.1016/S1474-4422(10)70245-3
98. Rosas HD, Lee SY, Bender AC et al (2010) Altered white matter microstructure in the corpus callosum in Huntington's disease: implications for cortical "disconnection". Neuroimage 49:2995–3004. https://doi.org/10.1016/j.neuroimage.2009.10.015
99. Björkqvist M, Wild EJ, Thiele J et al (2008) A novel pathogenic pathway of immune activation detectable before clinical onset in Huntington's disease. J Exp Med 205:1869–1877. https://doi.org/10.1084/jem.20080178
100. Brennan WAJ, Bird ED, Aprille JR (1985) Regional mitochondrial respiratory activity in Huntington's disease brain. J Neurochem 44:1948–1950. https://doi.org/10.1111/j.1471-4159.1985.tb07192.x
101. Tabrizi SJ, Cleeter MW, Xuereb J et al (1999) Biochemical abnormalities and excitotoxicity in Huntington's disease brain. Ann Neurol 45:25–32. https://doi.org/10.1002/1531-8249(199901)45:1<25::aid-art6>3.0.co;2-e
102. Sorolla MA, Reverter-Branchat G, Tamarit J et al (2008) Proteomic and oxidative stress analysis in human brain samples of Huntington disease. Free Radic Biol Med 45:667–678. https://doi.org/10.1016/j.freeradbiomed.2008.05.014
103. Browne SE, Bowling AC, MacGarvey U et al (1997) Oxidative damage and metabolic dysfunction in Huntington's disease: selective vulnerability of the basal ganglia. Ann Neurol 41:646–653. https://doi.org/10.1002/ana.410410514
104. van den Bogaard SJA, Dumas EM, Roos RAC (2013) The role of iron imaging in Huntington's disease. Int Rev Neurobiol 110:241–250. https://doi.org/10.1016/B978-0-12-410502-7.00011-9
105. Boillée S, Vande Velde C, Cleveland DW (2006) ALS: a disease of motor neurons and their nonneuronal neighbors. Neuron 52:39–59. https://doi.org/10.1016/j.neuron.2006.09.018
106. Cunha-Oliveira T, Montezinho L, Mendes C, et al (2020) Oxidative stress in amyotrophic lateral sclerosis: pathophysiology and opportunities for pharmacological intervention. Oxid Med Cell Longev:2020. https://doi.org/10.1155/2020/5021694
107. Liu J, Wang F (2017) Role of neuroinflammation in amyotrophic lateral sclerosis: cellular mechanisms and therapeutic implications. Front Immunol 8:1005. https://doi.org/10.3389/FIMMU.2017.01005/BIBTEX
108. Blasco H, Garcon G, Patin F et al (2017) Panel of oxidative stress and inflammatory biomarkers in ALS: a pilot study. Can J Neurol Sci 44:90–95. https://doi.org/10.1017/CJN.2016.284
109. Nagase M, Yamamoto Y, Miyazaki Y, Yoshino H (2016) Increased oxidative stress in patients with amyotrophic lateral sclerosis and the effect of edaravone administration. 21:104–112. https://doi.org/10.1179/1351000215Y.0000000026
110. Liddelow SA, Guttenplan KA, Clarke LE, et al (2017) Neurotoxic reactive astrocytes are induced by activated microglia. Nature 541:481–487. https://doi.org/10.1038/nature21029
111. Chen X, Hu Y, Cao Z et al (2018) Cerebrospinal fluid inflammatory cytokine aberrations in Alzheimer's disease, Parkinson's disease and amyotrophic lateral sclerosis: a systematic review and meta-analysis. Front Immunol 9:2122. https://doi.org/10.3389/FIMMU.2018.02122/BIBTEX
112. Oeckl P, Weydt P, Steinacker P et al (2019) Different neuroinflammatory profile in amyotrophic lateral sclerosis and frontotemporal dementia is linked to the clinical phase. J Neurol Neurosurg Psychiatr 90:4–10. https://doi.org/10.1136/JNNP-2018-318868
113. Chen Y, Xia K, Chen L, Fan D (2019) Increased interleukin-6 levels in the astrocyte-derived exosomes of sporadic amyotrophic lateral sclerosis patients. Front Neurosci 13:574. https://doi.org/10.3389/FNINS.2019.00574/BIBTEX
114. Post RM (1992) Transduction of psychosocial stress into the neurobiology of recurrent affective disorder. Am J Psychiatry 149:999–1010. https://doi.org/10.1176/ajp.149.8.999

115. Bauer ME, Teixeira AL (2019) Inflammation in psychiatric disorders: what comes first? Ann N Y Acad Sci 1437:57–67. https://doi.org/10.1111/nyas.13712
116. Belmaker RH, Agam G (2008) Major depressive disorder. N Engl J Med 358:55–68. https://doi.org/10.1056/NEJMra073096
117. Jeon SW, Kim Y-K (2018) The role of neuroinflammation and neurovascular dysfunction in major depressive disorder. J Inflamm Res 11:179–192. https://doi.org/10.2147/JIR.S141033
118. Abdoli A, Taghipour A, Pirestani M, et al (2020) Infections, inflammation, and risk of neuropsychiatric disorders: the neglected role of "co-infection." Heliyon 6:e05645. https://doi.org/10.1016/j.heliyon.2020.e05645
119. Afridi R, Suk K (2021) Neuroinflammatory basis of depression: learning from experimental models. Front Cell Neurosci 15:691067. https://doi.org/10.3389/fncel.2021.691067
120. das Gracas Fedoce A, Ferreira F, Bota RG et al (2018) The role of oxidative stress in anxiety disorder: cause or consequence? Free Radic Res 52:737–750. https://doi.org/10.1080/10715762.2018.1475733
121. Ribaudo G, Bortoli M, Pavan C, et al (2020) Antioxidant potential of psychotropic drugs: from clinical evidence to in vitro and in vivo assessment and toward a new challenge for in silico molecular design. Antioxidants (Basel, Switzerland):9. https://doi.org/10.3390/antiox9080714
122. Behr GA, Moreira JCF, Frey BN (2012) Preclinical and clinical evidence of antioxidant effects of antidepressant agents: implications for the pathophysiology of major depressive disorder. Oxid Med Cell Longev 2012:609421. https://doi.org/10.1155/2012/609421
123. Bhatt S, Nagappa AN, Patil CR (2020) Role of oxidative stress in depression. Drug Discov Today 25:1270–1276. https://doi.org/10.1016/j.drudis.2020.05.001
124. Emiliani FE, Sedlak TW, Sawa A (2014) Oxidative stress and schizophrenia: recent breakthroughs from an old story. Curr Opin Psychiatr 27:185–190. https://doi.org/10.1097/YCO.0000000000000054
125. Smaga I, Niedzielska E, Gawlik M et al (2015) Oxidative stress as an etiological factor and a potential treatment target of psychiatric disorders. Part 2. Depression, anxiety, schizophrenia and autism. Pharmacol Reports 67:569–580. https://doi.org/10.1016/j.pharep.2014.12.015
126. Goh XX, Tang PY, Tee SF (2021) 8-hydroxy-2'-deoxyguanosine and reactive oxygen species as biomarkers of oxidative stress in mental illnesses: a meta-analysis. Psychiatry Investig 18:603–618. https://doi.org/10.30773/PI.2020.0417
127. Ng F, Berk M, Dean O, Bush AI (2008) Oxidative stress in psychiatric disorders: evidence base and therapeutic implications. Int J Neuropsychopharmacol 11:851–876. https://doi.org/10.1017/S1461145707008401
128. Gawryluk JW, Wang J-F, Andreazza AC et al (2011) Decreased levels of glutathione, the major brain antioxidant, in post-mortem prefrontal cortex from patients with psychiatric disorders. Int J Neuropsychopharmacol 14:123–130. https://doi.org/10.1017/S1461145710000805
129. Salim S (2014) Oxidative stress and psychological disorders. Curr Neuropharmacol 12:140–147. https://doi.org/10.2174/1570159X11666131120230309
130. Wigner P, Czarny P, Synowiec E et al (2018) Variation of genes involved in oxidative and nitrosative stresses in depression. Eur Psychiatr 48:38–48. https://doi.org/10.1016/j.eurpsy.2017.10.012
131. Maes M, Galecki P, Chang YS, Berk M (2011) A review on the oxidative and nitrosative stress (O&NS) pathways in major depression and their possible contribution to the (neuro)degenerative processes in that illness. Prog Neuropsychopharmacol Biol Psychiatr 35:676–692. https://doi.org/10.1016/j.pnpbp.2010.05.004
132. da Cruz Jung IE, da Cruz IBM, Barbisan F et al (2020) Superoxide imbalance triggered by Val16Ala-SOD2 polymorphism increases the risk of depression and self-reported psychological stress in free-living elderly people. Mol Genet genomic Med 8:e1080. https://doi.org/10.1002/mgg3.1080

133. Lv Q, Hu Q, Zhang W et al (2020) Disturbance of oxidative stress parameters in treatment-resistant bipolar disorder and their association with electroconvulsive therapy response. Int J Neuropsychopharmacol 23:207–216. https://doi.org/10.1093/ijnp/pyaa003
134. Ranjekar PK, Hinge A, Hegde MV et al (2003) Decreased antioxidant enzymes and membrane essential polyunsaturated fatty acids in schizophrenic and bipolar mood disorder patients. Psychiatry Res 121:109–122. https://doi.org/10.1016/s0165-1781(03)00220-8

Chapter 6
Available Treatment Modules for Brain Disorders

Tehreem Iman, Rabia Akram, Muhammad Shahid Javed, Azhar Rasul, Faiqa Sajid, Ammara Tehreem, Sania Waris, and Ghulam Hussain

6.1 Introduction

The primary neurological conditions that affect the central nervous system (CNS), including Alzheimer's disease, Parkinson's disease, epilepsy, headache, neurotrauma, sleep disorders, brain cancer, autism, pain, multiple sclerosis, and Huntington's disease, have had severe impact on health of humans. These problems are the most prevalent medical burdens in the modern world compared to other major diseases since early diagnosis remains inadequate and no effective and affordable treatment has been found yet. One of the most intriguing and therapeutically significant objectives of research in neurology and psychiatry is effective neuroprotection against neurodegenerative insults. We give an overview of clinical

T. Iman · F. Sajid · A. Tehreem · S. Waris (✉)
Neurochemicalbiology and Genetics Laboratory (NGL), Department of Physiology, Faculty of Life Sciences, Government College University, Faisalabad, Pakistan

G. Hussain
Department of Physiology, Government College University Faisalabad, Faisalabad, Pakistan
e-mail: ghulamhussain@gcuf.edu.pk

R. Akram
Neurochemicalbiology and Genetics Laboratory (NGL), Department of Physiology, Faculty of Life Sciences, Government College University, Faisalabad, Pakistan

Department of Neuromuscular Diseases, UCL Institute of Neurology, University College London, London, United Kingdom

M. S. Javed
Department of Physiology, Sargodha Medical College, Sargodha, Pakistan

A. Rasul
Department of Zoology, Faculty of Life Sciences, Government College University, Faisalabad, Pakistan

A. Imran, G. Hussain (eds.), *The Role of Natural Antioxidants in Brain Disorders*, Food Bioactive Ingredients, https://doi.org/10.1007/978-3-031-41188-5_6

treatment applications in mental and neurological illnesses in this chapter. We enlist the various therapeutic modalities applied to cure neurological and mental conditions.

6.2 Neurological Disorders

6.2.1 Dementia

The loss of cognitive abilities—thinking, remembering, and reasoning—to the point where it affects a person's day-to-day functioning is known as dementia. Some dementia patients have emotional instability and personality changes. The intensity of dementia varies from the mildest stage, when it is just starting to interfere with a person's ability to function, to the most severe level, when the individual must fully rely on others for fundamental daily activities. The resulting mental problems affect memory, judgment, language, and understanding ability. About 36 million people all over the world are estimated to be living with dementia [1].

6.2.1.1 Alzheimer's Disease (AD)

The most common type of dementia is Alzheimer's disease (AD). AD, in people aged 65 and older, accounts for at least two-thirds of cases of dementia. It is a progressive and gradual neurodegenerative disease caused by neuronal cell death. It has an impact on behavior and cognitive functions which comprise judgment, understanding, attention, memory, thinking, and language and has a deceptive beginning [2, 3]. Among the pathologic signs of AD are the buildup of amyloid plaques, neurofibrillary tau tangles, and acetylcholine depletion. [4]. There is now no treatment for Alzheimer's disease, however, there has been substantial progress in this area in recent years. The U.S. Food and Drug Administration has approved several medications to treat Alzheimer's disease.

6.2.1.2 Vascular Dementia (VaD)

Vascular dementia (VaD) is directly associated with vascular injury to the brain. It signifies clinically important cognitive impairment and it is a neurocognitive disorder [5]. VaD is well-defined as the deficient or reduced flow of blood to the brain that causes dementia. Neurons of critical nutrients will deprive by reduced or blocked blood flow to the brain which causes VaD. The brain tissue starts to shrink because neurons die, which eventually cause by this deprivation [6]. The Elderly (age range between 65 years and older) are affected by VaD which is one of the most typical form of dementia, second only to AD. It has uncertain disease progression with a flexible presentation. Physical examination such as a measure of cognitive performance and a detailed history is helpful to attain the diagnosis of VaD (Uwagba 2022) [7].

6.2.2 *Huntington's Disease (HD)*

Clinically HD is described by neuropsychiatric, motor, cognitive and behavioral symptoms. It is a disturbing autosomal dominant disease, 35–44 years is the mean age at onset (AO) [8, 9]. In 1983, a correlation off chromosome 4 was placed, afterwards, in 1993, the huntington's disease (HD) gene was recognized [10]. With the definite loss of efferent medium spiny neurons (MSNs), HD is described by the deterioration of the striatum (putamen and caudate nucleus) and a general shrinkage of the brain. In patients with HD, a regionally-specific weakening of the cortical ribbon has been found, it seems that the most affected region of the brain is the striatum [11]. HD is amongst the most treatable neurodegenerative disorders because it has comprehensive penetrance and single-gene source.

6.2.3 *Parkinson's Disease (PD)*

PD is the second most common neurodegenerative disease after Alzheimer's disease and it was first described by Dr. James Parkinson in 1817 as a "shaking palsy." It is a chronic, progressive neurodegenerative disease characterized by both motor and non-motor features [12]. Unintentional or involuntary movements like shaking, stiffness, and issues with balance and coordination result from it. Typically, symptoms start mildly and get worse over time. People could experience difficulties speaking and walking as the illness worsens. Additionally, they could experience behavioral and mental changes, sleep issues, depression, memory issues, and exhaustion. [13]. Although there is no cure for Parkinson's disease, some symptoms can frequently be managed with medication, surgery, and other therapy.

6.2.4 *Restless Legs Syndrome (RLS)*

Restless legs syndrome is a neurological disease attributed to an uncontrollable desire to move during a resting state. RLS is also known as Willis-Ekbom disease. People dealing with this disorder also feel a burning or prickling sensation in the lower legs [14]. Restless leg syndrome also called WED (Willis-Ekbom disease) is a persistent, escalating basic primary sensory-motor disorder that usually remains underdiagnosed. RLS is specified by the irresistible desire of the patient to move the legs linked to intolerable and at times aching sensations following a diurnal pattern. Critical sleep disruption occurs in RLS and requires further consultation. However, the pathophysiology of RLS is not known, gene sequence changes along with dysregulation of brain dopamine and iron levels might play a crucial role [15]. Restless leg syndrome develops more frequently in women than men. The central nervous system occupying glutamate and dopamine transmission alterations and specific brain regions with iron insufficiency are all involved in probable pathophysiological mechanisms of RLS [16]. Patients on hemodialysis suffer from restless leg syndrome and their quality of life is compromised. Unluckily, RLS is a poorly treated and underdiagnosed disease [17].

6.2.5 Dystonia

Dystonia, the third most prevalent movement disorder, is generally known as an irregular posturing of the whole of the body or a part of the body. It is represented by unusual, tedious actions or postures due to continuous or scattered contractions of muscles. The movement disorder is categorized as an excess of movements (hyperkinetic) and a lack of movements (hypokinetic) [18]. Twisted postures that get more critical upon walking are due to muscle spasms. [19]. These movements are generally based on body distribution and are generally patterned; these may be observed in two or more non-contiguous body parts (multifocal), two or more adjoining body parts (segmental), and a particular body part (focal). In children, the commoner type of dystonia is generalized dystonia (Oppenheim's or torsion dystonia). On other hand in adults, focal hand dystonia (writer's cramp) and cervical dystonia (cervical torticollis) are the most common clinical appearances [18]. Since dystonia cannot be cured, treatment focuses on symptom management. The three tiers of treatment for dystonia include botulinum toxin (botox) injections, various medications, and surgery.

6.2.6 Multiple Sclerosis (MS)

Young adults are affected by the most prevalent non-traumatic debilitating disease known as multiple sclerosis (MS) [20]. Historically, it has been categorized as an autoimmune condition caused by T-cells, which target certain organs [21]. Multiple sclerosis is a long-term disorder caused by lymphocytes and monocytes undesirable infiltration into the central nervous system leading to axon demyelination. As a result of this impairment, physical and psychological disabilities might occur [22]. Inflammatory demyelination of axonal transection in the central nervous system supports the fact that multiple sclerosis is an autoimmune neurological disorder [23]. Contributing factors that affect the epidemiology of multiple sclerosis are smoking early age obesity, lower values of vitamin D in serum, etc. Certain viral infections like Epstein-Barr virus infection also play a role in MS development [24, 25]. The symptoms of MS can be controlled and eased with a range of medications, but there is currently no known cure.

6.2.7 Amyotrophic Lateral Sclerosis (ALS)

Amyotrophic lateral sclerosis (ALS), known colloquially as Lou Gehrig's disease is an idiopathic, fatal neurodegenerative disease of the human motor system. It is characterized by progressive muscular paralysis reflecting degeneration of motor neurons in the primary motor cortex, corticospinal tracts, brainstem, and spinal cord. The main symptoms of ALS include dysphagia, dysarthria, respiratory distress, pain, and psychological disorders. The management of ALS is supportive, palliative, and multidisciplinary.

6.2.8 Myasthenia Gravis (MG)

Myasthenia gravis is an immunological disease that infects the neuromuscular junction. Even though the treatment strategy of MG is very efficacious, noteworthy morbidity and mortality rates are observed. This can be prevented by timely identification and treatment of disease [26]. Muscle weakness that gets worse after periods of exertion and gets better after periods of rest is the defining feature of myasthenia gravis. The disorder frequently (but not always) involves specific muscles, such as those that regulate facial expression, chewing, talking, and swallowing. [27]. Myasthenia gravis is mostly under control today. Several treatments are available to assist lessen and treat muscle weakness.

6.2.9 Epilepsy

Epilepsy is a long-term disorder of the brain specified by continuous vulnerability to produce seizures. The symptoms of epilepsy include an ongoing (or persistent) propensity to have seizures that are not immediately triggered by an injury to the central nervous system, as well as the neurobiological, cognitive, psychosocial, and social effects of seizure recurrences. [28]. Seizures are a particular indication of epilepsy according to literature, for centuries they occur at random [29]. Postictal paresis, postictal automatisms, delirium, psychosis, postictal migraine, and headache are the frequent postictal symptoms. Gene therapy, ketogenic diet, antiepileptic drugs, and surgery are emergency treatments and are the available management for epilepsy [30].

6.2.10 Schizophrenia

Schizophrenia is a severe mental illness in which reality is perceived by sufferers strangely. It may include hallucinations, delusions, and severely irrational thinking and behavior, which can make it difficult to go about daily activities and be incapacitating. Schizophrenia patients need lifetime care. [31]. It is critical to identify the signs of schizophrenia and get care as soon as one can. Schizophrenia is typically detected in patients between the ages of 16 and 30. Current schizophrenia treatments put a strong emphasis on assisting patients in managing their symptoms, enhancing day-to-day functioning, and achieving their life objectives, such as completing their education, establishing a profession, and developing satisfying relationships.

6.3 Treatment Strategies for Neurological Disorders

With the type of brain disease, treatment differs. Some disorders cannot have a cure, but can be treated for example; the use of medicine to stop seizures and to regulate signs of Parkinson's disease. other must be assisted for example; to handle several balance disorders, a walker or cane can be used.

The treatments used for brain- disorders may include:

- Stress management
- Speech, physical or occupational therapy
- Cognitive behavior therapy and counseling
- Medications
- Surgery
- Diet and exercise
- Rest (Fig. 6.1)

6.3.1 Alzheimer's Disease

6.3.1.1 Pharmacological Treatment

The current pharmacological approach for AD treatment is based on vascular prevention and symptomatic therapy with two major classes of drugs recommended for AD treatment which are cholinesterase inhibitors (naturally-derived synthetic and hybrid analogs) and NMDA receptor antagonists [32]. Donepezil, rivastigmine, and galantamine are FDA-approved cholinesterase inhibitors, while Memantine works by blocking NMDA receptors. Galantamine exhibited a lower mortality rate when compared to placebo [33]. The therapeutic efficacy of rivastigmine is vast as it inhibits both acetylcholine esterase and butyrylcholine esterase [34]. Rivastigmine

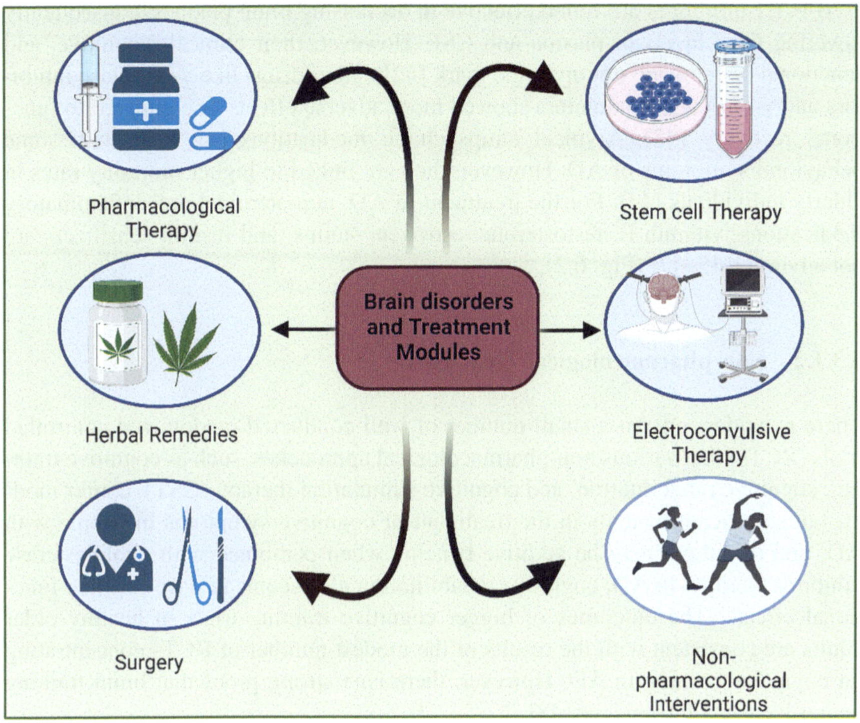

Fig. 6.1 Treatment modules for brain disorders

and Galantamine are used for mild to moderate phases of AD. The drug approved for the treatment of all stages of AD is Donepizel. Memantine is an orally active non-competitive N-methyl-D-aspartate receptor antagonist. It is recommended for moderate to severe AD management [35]. Memantine shows effects in combination with ChEI, and also as monotherapy [36].

Combination therapy is more beneficial than monotherapy. During AD treatment, the implication of early combination therapy leads to the gradual decline of cognitive impairment [37]. In the case of donepezil hydrochloride, alone therapy is less effective when compared to combination therapy of rivastigmine hydrogen tartrate with donepezil hydrochloride. Patients dealing with combination therapy disembarrass from symptoms of AD and feels betterment in their quality of life [38]. Likewise, donepezil, another standard care treatment is a combination therapy of galantamine and memantine. Combination therapy is considered one treatment contributing factor in improving AD symptoms [39]. Similarly, the drug's capability of targeting multiple pathways for clinical efficacy may also be needed during therapy [40]. An efficacious example of multi-target-directed ligands (MTDLs) strategy is rationally designed and synthesized novel apigenin-rivastigmine hybrids. The reason behind apigenin use is its strong antioxidant, anti-inflammatory capacity, and Neuroprotective effects [41].

BACE1 inhibitors are found effective in decreasing brain plaques, consequently lowering their levels in plasma and CSF. However, their clinical, cognitive, and functional benefits are not up to the mark [42]. Other drugs like γ-secretase inhibitors and γ-secretase modulators showed more adverse effects as compared to functional recovery [43]. Atypical antipsychotic medications can help with some behavioral symptoms of AD. However, they are linked to higher mortality rates in elderly individuals [44]. For the treatment of AD, non-steroidal anti-inflammatory medications, vitamin E, testosterone, estrogen, statins, and insulin sensitizers are not advised. [45–47] (Fig. 6.2).

6.3.1.2 Non-pharmacological Treatment

There is evidence from a small number of well-conducted randomized controlled trials (RCTs) that various non-pharmacological approaches, such as cognitive training, cognitive rehabilitation, and cognitive stimulation therapy (CST), confer modest but significant benefits in the treatment of cognitive symptoms in people with AD, and that there may be additive benefits when combined with cholinesterase inhibitor therapy. In AD, cognitive rehabilitation also seems to have positive functional effects. The outcomes of bigger cognitive training trials in healthy older adults are consistent with the results of the modest number of RCTs concentrating on cognitive training in AD. However, there isn't strong proof that brain training games have any advantages [48].

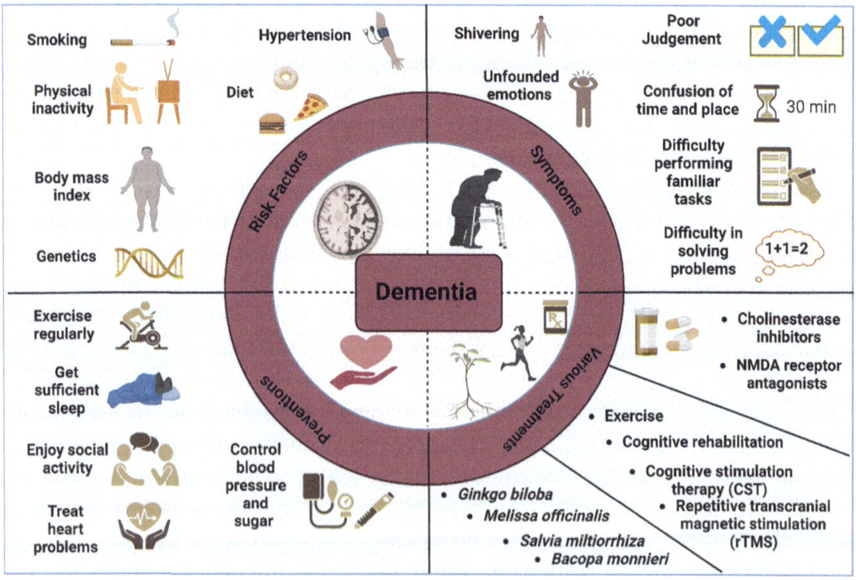

Fig. 6.2 Preventions and treatments for Dementia

The probability of AD in individuals is reduced by adopting healthy activities, mostly exercise [49]. Various clinical studies have confirmed the powerful correlation between exercise and enhanced cognitive response and it is affirmed that healthy physical activity decreases the risk of AD [50]. There are three types of cognitive intervention. These are cognitive stimulation, training, and rehabilitation. One most important factor contributing to increased cognitive refinement is improved neural plasticity [51]. Management of AD is directly linked to psychological health and cognitive ability. Music therapy can be a strong substitute for AD-related features [52] and it strengthens the reminiscence capacity of persons by different unconventional mechanisms. These mechanisms involve the activation of the sympathetic system, dopaminergic pathways, and various brain regions [53]. Another major asymptomatic and pain-free method for AD management is Repetitive transcranial magnetic stimulation (rTMS). It deals with magnetic stimulation of the cortical area [54]. As a result, expression of brain-derived neurotrophic factor is enhanced in the brain hippocampus. It causes enhanced neuronal excitability leading to remarkable neuroprotective effects [55]. Traditional Chinese therapies consist of a distinctive non-pharmacological treatment called acupuncture therapy. This procedure is immensely effective in treating degenerative disorders by both inhibiting degenerations of neurons and stimulating axonal regeneration [56]. The result of all these factors leads to recovery of cognitive function in AD patients.

6.3.1.3 Herbal Treatment

Although some substances, such as *Ginkgo biloba* extracts, have demonstrated some benefit in clinical studies, the strength of the data supporting their effects is insufficient to guide therapeutic practice. *G. biloba* shows improvement in the early phases of AD. It ameliorates cognitive impairment and slows disease progression. The mechanism of *G. biloba* involves the restoration of acetylcholine receptors in the hippocampus. The stimulation of neurotransmitter activity results in regain of memory and learning ability [57]. Beneficial effects of *Melissa officinalis* are observed in patients with dementia. It causes behavioral and motivational improvement [58]. Saffron is considered a foremost herb for repressing Aβ plaque formation in AD [59], through its anti-amyloidogenic activity [60]. It also shows neuroprotective capacity by suppressing glutamate-induced toxicity in AD. The accumulation of β-amyloid protein in AD is reduced by *Salvia miltiorrhiza, it* can reduce oxidative stress and boosts the cholinergic energy system [61].

6.3.2 *Vascular Dementia*

6.3.2.1 Pharmacological Treatment

Most of the drugs approved for the management of AD by the FDA are considered effective for VaD patients. NMDA receptor antagonists, memantine, and cholinesterase inhibitors are recommended for moderate effects [62]. Moxibustion therapy alone and in combination with piracetam shows relieving effects in VaD [63]. Statins and fibrates are known for their lipid-lowering effects. Some studies have appreciated their anti-oxidant, anti-inflammatory activities. These medications also show the potential to augment neuroplasticity and must be considered for VaD treatment [64]. Tetramethylpyrazine nitrone (TBN) has strong antioxidant potential. In a research conducted on VaD rats, this compound has shown strong neuroprotective properties [65]. Stem cell therapy is exclusively used nowadays in various neurodegenerative diseases. In VaD treatment variety of stem cells are used for transplantation. Examples are iPSCs (induced pluripotent stem cells), BMSCs (bone marrow mesenchymal stem cells), ESCs (embryonic stem cells), HUC-MSCs (human umbilical cord-derived mesenchymal stem cells), BMMNCs (bone marrow mononuclear cells, and NSCs (brain-derived neural stem cells) [66].

6.3.2.2 Non-pharmacological Treatment

Exercise has a positive effect on people with dementia regarding their lifestyle. People perform their daily chores in a better way. Therefore, their perceptive ability, anxiety, and neurological symptoms are not cured completely [67]. Regular mental activities to improve a person's ability to perform cognitive tasks are included in cognitive training and rehabilitation. Retraining of previously known skills is performed in dementia and AD [68]. These compensatory strategies are beneficial in lowering depression symptoms in people with dementia [69].

6.3.2.3 Herbal Treatment

According to preliminary research, herbal treatments for dementia, whether taken alone or in combination with other herbs, may have therapeutic advantages. Herbal remedies have long been used to improve memory and cognitive abilities as well as to treat the behavioral and emotional symptoms of dementia/VaD. *Ginkgo biloba, Huperzia serrata, Curcuma longa, Panax ginseng, Panax notoginseng, Bacopa monnieri, Salvia miltiorrhiza, Crocus sativus, Camellia sinensis*, and many others are among the herbs that are most frequently used and researched [70, 71]. Another study's findings showed that Chinese herbal medicine enhance daily activities and cognitive performance. Furthermore, CHMs were typically well tolerated and safe in VaD patients. Therefore, the systematic review's findings at least partially support the recommendation that CHM be used regularly to treat VaD [72].

6.3.3 Huntington's Disease

6.3.3.1 Pharmacological Treatment

Dopamine-depleting medications, dopamine antagonists, benzodiazepines, glutamate antagonists, acetylcholinesterase inhibitors, dopamine agonists, anti-seizure medications, cannabinoids, lithium, deep brain stimulation, and foetal cell transplantation have all been tested in HD for their effectiveness in reducing chorea [73]. The first FDA-approved drug for HD is Tetrabenazine (TBZ), named Xenazine and another one is Deutetrabenazine (DTBZ) which the Food and Drug Administration has expressly licensed to stop the jerking and writhing motions (chorea) linked to HD [74]. Deutetrabenazine was approved in April 2017 for HD while Tardive dyskinesia treatment, was approved in August 2017 [75]. Xenazine is specified for its capability of lowering the amount of those chemicals in the body that causes overexcitation. Huntington's chorea is treated by Xenazine with consistency. Severe adverse effects are experienced with Xenazine. It can increase depressive symptoms leading to suicidality [76]. However, the disease's development is unaffected by these medications, they may have negative effects. Except for suicidal tendencies, others adverse symptoms are anxiety, nervousness, and sleep disorders like insomnia [77].

6.3.3.2 Non-pharmacological Treatment

Numerous non-pharmacological therapies are now receiving more thorough evaluations. Most of the interventions are looking at the impact on gait and balance, and some of them employ music therapy, exercise, dancing, or video games. They may be thought of as an addition to drug therapy. There is currently a dearth of evidence supporting non-pharmacological therapies in HD. External assistance that helps to make up for a person's cognitive limitations is referred to as assistive technology for cognition (ATC). Additionally, managing people with HD requires that family members be educated [78].

6.3.3.3 Herbal Treatment

Many of the most promising anti-HD possibilities, including plants like *Bacopa monnieri, Centella asiatica, Cannabis sativa, Gastrodia elata, Ginkgo biloba, Panax ginseng, Withania somnifera*, etc., are well-known CNS-active medications. Several of the anti-HD substances are well-known as neuroprotectants, such as curcumin, epigallocatechin-gallate, ginsenosides, kaempferol, naringin, resveratrol, and S-allylcysteine [79].

Tripterygium wilfordi also called, Thunder of God vine is a Chinese herb that has anti-inflammatory properties. Celastrol is obtained from the root extract of *wilforidi* and *Celastrus regelii* and it is considered a strong neuroprotective agent for treating HD in recent clinical data [80]. In another study, the neuroprotective effects of Hesperidin and naringin have been analyzed. Muscle grip strength and locomotor activity were markedly enhanced after treating rats with Hesperidin and naringin [81]. In a research based on a yeast model of HD, Epigallocatechin gallate has been evaluated. This plant compound decreased the cytotoxicity and polyQ-mediated protein aggregation in the model. However, misfolding of huntingtin protein has been reported during the early phases of the study [82]. To assess the treatment efficacy of both well-known and innovative herbal extracts and chemicals in HD models, more study is still required.

6.3.4 Parkinson's Disease

6.3.4.1 Pharmacological Treatment

Dopaminergic medications are the typical treatment method for motor symptoms in PD. L-DOPA, the most potent anti-parkinsonian drug currently available, remains the "gold standard" for PD therapy. Dopamine does not easily pass through the blood-brain barrier, but its precursor, L-DOPA can. Strong evidence supports using levodopa and dopamine agonists for motor symptoms at all stages of PD. Dopamine agonists (Bromocriptine, Ropinirole, Pramipexole, Apomorphine) are effective for motor fluctuations and clozapine is effective for hallucinations. Cholinesterase inhibitors may improve symptoms of dementia and the antidepressant pramipexole may improve depression [83].

The striatum contains isoenzymes (MAO-A and MAO-B), which are responsible for the majority of the brain's oxidative metabolism of dopamine. Inactive monoamines of intestinal origin are also found there. Selegiline and Rasagiline, two selective monoamine oxidase-B inhibitors, have the potential to permanently inhibit the enzyme [84]. Investigations have shown that the levels and activation of Abelson non-receptor tyrosine kinase (c-Abl) were upregulated in the brain tissue of patients with PD. Nilotinib reduced c-Abl activation, levels of Parkin substrate, and neuronal cell death. [85]. Safinamide is a new option for treating motor fluctuations in mid- and late-stage PD; it is a reversible inhibitor of Monoamine oxidase B (MAO-B), with additional properties including blockade of voltage-dependent sodium channels, modulating calcium channels and inhibiting glutamate release [86]. Antidepressant Medications, specifically selective serotonin reuptake inhibitors (SSRIs) significantly improved depression among PD patients [87].

6.3.4.2 Non-pharmacological Treatment

The potential and restrictions of non-pharmacological therapies including exercise, deep brain stimulation, transcranial magnetic stimulation, and cell replacement therapy are highlighted by recent therapeutic advancements to treat PD. Other non-pharmacological methods, like cognitive behavioral treatment (CBT), instruction in good sleep habits, and bright light therapy (BLT), demonstrate appreciable improvements in PD patients [88].

Exercise has been shown to not only improve motor performance, but also improve learning, memory, and depression, facilitate synaptogenesis, induce neurotrophic factors, enhance neuroplasticity, and reverse some neurochemical deficits (e.g., increases D2 receptors) [89]. Additionally, physiotherapy, occupational therapy, and speech therapy (for speech and swallowing) are useful. Therapy interventions can help maintain or improve motor symptoms, balance, gait, and function and provide strategies for addressing hypophonia and dysphagia. Referrals for interdisciplinary therapy consultations are an important component of quality care for PD [90, 91].

On PD patients, it has been reported that mindfulness, massage [92], the Alexander Technique [93], vibration therapy [94], music therapy [95], and whole-body vibration all have positive effects. The Alexander Technique is a technique for improving posture and movement through mental concentration and immediate feedback from a professional.

6.3.4.3 Herbal Treatment

Extracts of herbal medicines (such Acanthopanax, Alpinia, and Astragalus) have been shown in current pharmacological studies to have consistent and significant effects on the models of PD [96, 97]. Levodopa content in Mucuna pruriens ranges from 4 to 6 percent, and its extracts are also said to have neuroprotective properties, such as restoring endogenous dopamine production and lowering oxidative stress in the substantia nigra [98]. Another Ayurvedic remedy, Banisteriopsis caapi, contains harmine and harmaline, two substances that have been demonstrated to have inhibitory effects on monoamine oxidase. However, the medicine has not been well-researched in humans [99].

The active chemical compounds present in each herb that are effective in the treatment of PD include Baicalei, *Erythrina velutin*, resveratrol, *Peganum Harmal*, *Curcuma longa* (Zingiberaceae), *Car-thamus tinctorius L.* (Safflower), *Pueraria lobate, Juglandis Semen* (Walnut), *Tianma Gouteng Yin* (TGY), *Lycium barbarum* L fruit, *Mucuna pruriens* (Velvet bean), Chunghyuldan (CHD), *Paeoniae Alba Radix* [100].

6.3.5 Restless Leg's Syndrome

6.3.5.1 Pharmacological Treatment

Various drugs are recommended by European guidelines for RLS treatment. Short term RLS treatment includes rotigotine, ropinirole, pramipexole, gabapentin, gabapentin encarbil and pregabalin. Rotigotine, ropinirole, and pramipexole are approved as first-line treatments by FDA and European Medicine Agency. Rotigotine is found effective in long-term therapy of RLS. However, gabapentin encarbil might show more efficacy than ropinirole, pramipexole, and gabapentin [101]. Some undesirable effects of gabapentin enacarbil are drowsiness and dizziness [102]. Patients dealing with RLS also face urinary problems. Vitamins C and E are recommended for treatment at doses of 200mg, and 400mg respectively [103], and both are effective for uremic RLS treatment. Vitamin E shows better efficacy [104]. Benzodiazepines are effective for augmentation caused by RLS prior therapy [105]. Benzodiazepines contain sedative and hypnotic agents. These drugs decrease RLS-linked depression [106]. American Academy of Sleep Medicine Practice guidelines does not recommend benzodiazepines as first-line management of RLS [107].

6.3.5.2 Non-pharmacological Treatment

Some patients with RLS may not respond well to conventional pharmacologic treatment. Up to 65% of RLS sufferers consistently use complementary therapies to relieve their symptoms. We reviewed the available clinical evidence and proposed physiologic underpinnings for some non-pharmacologic lifestyle interventions for RLS, such as pneumatic compression devices (PCDs), light therapy, cognitive-behavioral therapy (CBT), and neutraceuticals. Mind-body interventions for RLS include conventional exercise, yoga, and acupuncture (vitamins, valerian, and Chinese herbs) [108]. Six months of intradialytic stretching exercise has numerous beneficial effects on the mind and body. It enhances sleep quality, cures anxiety, restores functional ability, and minimizes RLS symptoms [109]. rTMS is repetitive transcranial magnetic stimulation which at high frequency relieves motor symptoms of RLS like depression and sleep problems [110].

6.3.5.3 Herbal Treatment

Chinese herbs have reportedly been shown to be secure and have a great deal of potential to be an efficient therapy option for RLS, however, the strength of this proof is constrained by the caliber of these investigations. Just over 150 years before Sir Thomas Willis originally described RLS in England, the ancient text Neike Zhaiyao (Internal summary) contains the first description of RLS that includes three of the four main current criteria. Chinese herbs that are anti-RLS need to be the subject of more in-depth research to be used in clinical trials as a valuable adjunct to Western pharmacology [111].

6.3.6 Dystonia

6.3.6.1 Pharmacological Treatment

Symptoms of dystonia are neutralized by alleviating the effects of trihexyphenidyl on the smooth muscles. It is a synthetic anticholinergic drug and shows antagonistic effects on muscarinic receptors against acetylcholine [112]. Muscle stiffness and abnormal involuntary movements in primary dystonia are treated by baclofen, an oral taking drug [113]. Cervical dystonia is treated by Botulinum toxin injection therapy. It is observed that this treatment gives symptomatic ease in 70 to 90% of people with a single dose. As a result of BoNT/A injection therapy, most symptoms are cured in CD (Hefter et al. 2020). Tetrabenazine, metyrosine, and other dopamine-depleting drugs, as well as anticonvulsants and mexiletine, have been shown to improve it.

In the case of static dystonic, urgent treatment is compulsory, this emergency can be handled by a deep brain stimulation strategy. This treatment remains for over 10 years in patients with generalized dystonia [114]. Patients with segmental dystonia also show improvement after this therapy [115]. When non-invasive measures do not give the required results, surgical treatment can be opted for cervical dystonia. Remarkable improvement in symptoms is noticed in most of the patients. A serious drawback of this procedure is that there is a considerable chance of re-innervation [116] (Fig. 6.3).

6.3.6.2 Non-pharmacological Treatment

Any medication or surgical procedure for the treatment of dystonia should be accompanied by physical and occupational therapy, immobilization, and constraint-induced therapy. Important objectives include enhancing strength and mobility and avoiding contractures. Numerous devices have been put into use under the theory that sensory trickery or geste antagonists temporarily improve dystonia. There have been descriptions of hand splints, dental prostheses, and head and neck braces [117]. Constraint-induced movement therapy has been tested as a treatment for musicians with focal hand dystonia. Focused hand dystonia has been improved with sensory retraining techniques that involve studying and using braille [118]. These procedures employ noninvasive methods that might complement other treatments, even though long-term benefits might not be realized.

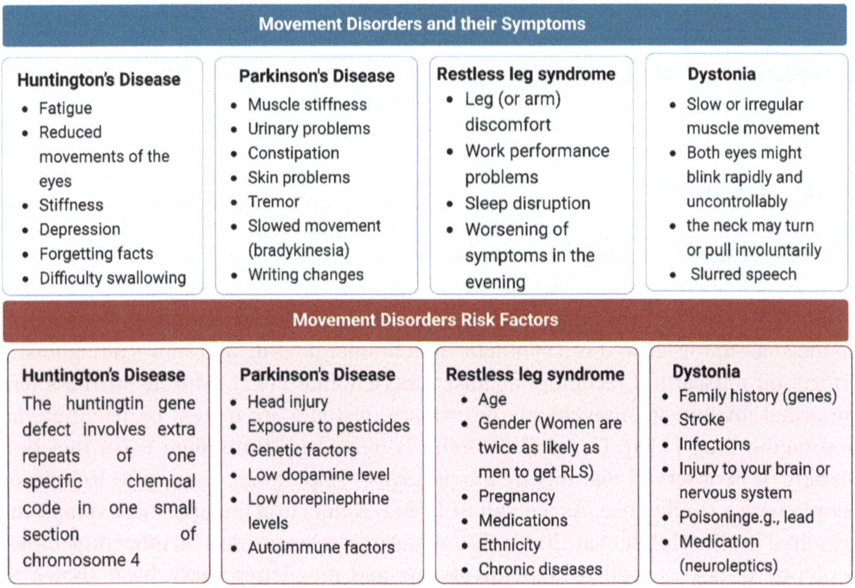

Fig. 6.3 Symptoms and risk factors of movement disorders

6.3.6.3 Herbal Treatment

Dystonia may respond favorably to traditional Korean medicine. [119]. Yokukan-san was added to the patient's regular treatment in a complementary manner in each case, and it had a clear but limited impact. The tardive dystonia was improved by 80% thanks to the addition of orengedoku-to to yokukan-san, and the patient's violent impulses were controlled. These effects were comparable to those of aripiprazole. [120]. There are no alternative natural remedies for dystonia in the literature.

6.3.7 Multiple Sclerosis

6.3.7.1 Pharmacological Treatment

Currently, the US FDA has only approved a few drugs to treat relapse forms of MS. Contrarily, neither a pharmaceutical agent nor a treatment for primary-progressive MS is effective in patients with secondary-progressive MS without relapses. These medications include mitoxantrone (Novantrone®), natalizumab (Tysabri®), IFN—1b (Betaseron®), and Copaxone [121]. With the availability of the first interferon in 1993, the management period for MS started [122]. To improve MS disorder, subcutaneous IFNβ-1a (SC IFNb-1a, Rebif®), intramuscular interferon beta-1a (IM IFNβ-1a, Avonex®), and subcutaneous IFNβ-1b (SC

IFN-beta-1b, Betaferon®, Extavia®) seems to be effective. The majority of these medicines work by inhibiting the immune system to prevent the auto-reactive immune cells attack on myelin sheath and protect neurons. The accessible three formulas of interferon beta such as Betaseron, Avonex, and Rebif are the most often prescribed medication for patients suffering from MS [123], especially for those patients of MS who have fewer physical disabilities and suffer more from mental illness, beta interferons are strongly advised [124]. Another first-line disease-modifying agent used is GA-glatiramer acetate, it affects different levels of the adaptive and innate immune response. It leads to deviation from the anti-inflammatory pathway from the pro-inflammatory pathway (Copaxone®/Glatopa®) [123, 125].

The second-line MS medications fingolimod (Gilenia®, Gilenya®), natalizumab (Tysabri®), and alemtuzumab (Campath®, MabCampath®, Lemtrada®) are more effective but less safe [126]. Mitoxantrone is also considered the second line of medication for MS patients which has immunomodulatory and immunosuppressive effects [127]. But MX was lesser successful in lowering the risk of MS development and the severity of relapsing-remitting MS (RRMS), progressive relapsing MS (PRMS), and secondary progressive MS (SPMS) during short-term follow-up (up to two years) [128, 129].

Fingolimod (Gylenia®) was approved as the first drug with oral administration to treat highly-active RRMS, and monoclonal antibody natalizumab [130] was approved by the US Food and Drug Administration (FDA) based on two major randomized, double-blinded, placebo-controlled Phase III studies [131]. Natalizumab works primarily by inhibiting adhesion molecules to stop lymphocytes from crossing the blood-brain barrier. While fingolimod is effective in reducing the relapse rate when compared to a placebo, it could not be demonstrated that it would stop the advancement of the impairment.

Inflammation in MS has been linked to decreased blood flow to the brain and spinal cord, which results in the cells receiving less oxygen. Early findings had, for the most part, been in favor of HBO_2T's capacity to stop development and even lessen handicaps in a variety of individuals. [132, 133].

6.3.7.2 Non-pharmacological Treatment

Numerous methods including aerobic training such as cognitive enhancers (CE), yoga, novel techniques; Whole Body cryostimulation (WBC) brain stimulation such as Transcranial Direct Current Stimulation (tDCS) and Transcranial Magnetic Stimulation (TMS) and invasive ones, including Deep Brain Stimulation (DBS), neuropsychological rehabilitation, and Computer Based Training Programs are useful non-pharmacological methods for treating MS. Meditation, physical exercise, better sleep, music, and several additional therapies (functional foods, supplements, and nutraceuticals) can also be helpful in this regard [134]. One of the numerous devastating adverse effects of MS development is muscle stiffness (spasticity). Non-drug approaches are being examined for this issue, however, there is little to no evidence that they can improve spasticity in MS patients. These approaches include physiotherapy, magnetic stimulation, electromagnetic therapy, and vibration therapy [135]. For the prevention of osteoporosis, cardiovascular problems, obesity, deconditioning, and many other MS-specific symptoms aerobic training is highly recommended [136, 137].

Transcutaneous electrical nerve stimulation (TENS), psychotherapy (telephone self-management, hypnosis, and electroencephalogram (EEG) biofeedback), transcranial random noise stimulation (tRNS), transcranial direct stimulation (tDCS), hydrotherapy (Ai Chi), and reflexology were some of the non-pharmacological interventions used to treat chronic pain in MS patients. For the use of these non-pharmacological therapies, there is very scant data. [138].

6.3.7.3 Herbal Treatment

For MS disease, herbal remedies have shown a hopeful treatment. According to a review of the literature, herbal remedies may help cure MS and its associated symptoms by promoting remyelination, lowering demyelination, and reducing CNS inflammation. These plants are *Crocus sativu, Curcuma longa, Vitis vinifera, Valeriana officinalis, Ginkgo biloba, Gastrodia elata, Boswellia papyrifera, Zingiber officinale, Nigella sativa, Camellia sinensis, Hypericum perforatum, Panax ginseng, Vaccinium macrocarpon, Cannabis sativa, Oenothera biennis and Piper methysticum* [139]. These plants improved the MS patients' condition by lowering fatigue, anti-inflammatory and platelet-activating factor (PAF)-inhibiting, anti-oxidant characteristics, increasing remyelination, and antidepressant properties [139–147]

6.3.8 Amyotrophic Lateral Sclerosis

6.3.8.1 Pharmacological Treatment

At the current time, there are no treatments that can reverse or arrest disease progression in ALS. The only two drugs, Riluzole and Edaravone, have been approved by Food and Drugs. In 1994, Riluzole was identified as the first and only disease-specific treatment. The drug now approved for clinical usage, which is an inhibitor of glutamate release, only significantly prolongs life for a few months. Riluzole is usually well tolerated, with the most common side effects being gastrointestinal disorders, the elevation of liver enzymes, dizziness, and asthenia. Edaravone was discovered and developed as a potential free radical scavenger to reduce oxidative stress [148]. Regarding the side effects, the drug is well tolerated, and adverse events (that include constipation, insomnia, headache, and transient leukopenia) are uncommon [149]. Another drug Nuedexta (dextromethorphan HBr and quinidine sulfate), received approval from FDA to target symptoms of pseudobulbar affect, a condition characterized by sudden and unpredictable episodes of crying or laughing seen in people with ALS and other neurological conditions [150]. In individuals with a typical disease advancement, additional therapy being analyzed is masitinib. To decline ALSFRS-R in patients, an oral tyrosine kinase inhibitor was used as an add-on therapy to riluzole which show an optimistic effect [151].

The cornerstone of disease management for ALS patients remains multidisciplinary care which has a positive effect on patient satisfaction and outcome. Several discomforting symptoms of ALS can be managed by symptomatic treatment options, including pharmacological and non-pharmacological interventions. For instance, spasticity can be treated with baclofen, tizanidine, cannabinoids, and muscle stretching, and sialorrhea can be treated with anticholinergic medications (amitriptyline, glycopyrronium bromide, and oxybutynin) and botulin toxin injections into the salivatory glands. Muscle cramps may respond to magnesium supplements, quinine sulfate, gabapentin, or carbamazepine. A selective serotonin reuptake inhibitor, amitriptyline, benzodiazepines, and dextromethorphan hydrobromide/quinidine sulfate, can be used in the case of emotional lability [152].

6.3.8.2 Non-pharmacological Treatment

Currently, the only pharmacological treatments for ALS are those that aim to lessen disease progression and symptoms. Multidisciplinary care, which includes symptom management and support for respiratory and nutritional needs, continues to be the cornerstone of treatment. Dietary changes can help to improve nutrition and a gastrostomy tube is an option if the caloric intake is insufficient or when swallowing becomes hazardous.

Creatine supplementation may improve or at least maintain muscular strength and stave off weariness. [153]. There are more non-pharmacological treatments that can be utilized to reduce weariness. For this goal, studies have examined the possible advantages of supported treadmill ambulation, physical activity, and repetitive transcranial magnetic stimulation (rTMS). There are more non-pharmacological treatments that can be utilized to reduce weariness. For this goal, studies have examined the possible advantages of supported treadmill ambulation, physical activity, and repetitive transcranial magnetic stimulation (rTMS) [154–157]. Speech therapy is frequently necessary and assisted communication (customized software) can also be used. Non-invasive ventilation is the preferred life-prolonging treatment for respiratory insufficiency. At all disease stages, the patient's wishes should be taken into account and advance care planning should be initiated early [158].

6.3.8.3 Herbal Treatment

Herbs are therapeutically effective against oxidative stress, excitatory amino acid toxicity, neuroinflammation, and calcium cytotoxicity in the case of ALS, according to studies conducted in vitro and in vivo. Certain compounds including

Madecassoside; *Centella asiatica* (L) Urban, *Epigallocatechin gallate*: Green Tea, Diallyl trisulfide; *Liliaceae allium*, Ampelopsin; *Ampelopsis grossedentata*, Morroniside; *Cornus officinalis*, Picroside-II; *Picrorhizae Rhizoma*, Astragaloside IV; *Radix Astragali* were effective against oxidative stress in ALS. Other compounds such as β-Asarone; Acorus tatarinowii Schott, Huperzine-A; *Huperzia serrata* (Thunb) Trev., Catalpol; Rehmannia glutinosa Libosch, Ferulic acid; Angelica, *Szechwan lovage rhizome*, *herba taraxaci*, Selaginellin; *Saussurea pulvinata Maxim*, Cryptotanshinone; *Saussurea* works through Antiexcitatory amino acids (EAAs).

Additionally, a few other compounds like *pulvinata Maxim*, Celastrol; *Tripterygium wilfordii*, Resveratrol; *Polygoni cuspidatus*, *Curcumin*; *Rhizome curcumae longa*, Obovatol; *Magnolia officinalis*, Wogonin; *Scutellaria root*, IRN; *Uncaria rhynchophylla*, Paeonol; *Paeonia suffruticosa*; *Cynanchum paniculatum*, were active as Antineuroinflammatory. Paeoniflorin; Peaoniae radix, Gastrodin; *Gastrodia elata Blume*, Muscone; Natural musk, Ligustrazine; Rhizoma Chuanxiong shows effect by Anticalcium cytotoxicity [159].

6.3.9 Myasthenia Gravis

6.3.9.1 Pharmacological Treatment

There are four fundamental treatments for MG: Acetylcholinesterase inhibitors for symptomatic treatment, plasmapheresis and intravenous immunoglobulin for quick short-term immunomodulation, glucocorticoids and other immunosuppressive medications for chronic long-term immunomodulation, and surgery for treatment are the other options. Immediate immunosuppressive drug treatment must be started to control MG severity [160]. Immunosuppressant drugs used for MS are IVIG (intravenous immunoglobin) and SCIG (subcutaneous immunoglobulin). Immunoglobulins have the safest potency for long-term MG therapy [161].

Pyridostigmine is a synthetic acetylcholinesterase inhibitor and the most frequently used drug for the symptomatic treatment of MG. Pyridostigmine gives immediate and prominent relief in muscle fatigue during early and moderate MG phases. It shows efficacy for longer period of time but higher doses may cause adverse effects. Initial most widely recommended immunosuppressive treatment for MG was corticosteroid treatment. Initially, oral prednisolone is administered, high dose of prednisone (100mg/day or every other day) was reported highly efficacious in treating MG. The adverse effects of prednisolone should be monitored continuously. However, in case of dependency, intolerability, or due to alarming side effects occurrence by prolonged use of the drug, alternative second-line therapies should be started. Azathioprine and mycophenolate are second-line drugs for immunosuppression [27, 162]. Azathioprine is a drug of choice for MG treatment not only as an immunosuppressive agent but also for limiting the use of steroids [163]. Except for azathioprine and mycophenolate, other effective non-steroidal immunosuppressive drugs are cyclosporine, methotrexate, tacrolimus, rituximab, eculizumab, and cyclophosphamide. For chronic management of MG, recurrent exposure to plasma exchange infusions of SCIG or IVIG is also recommended [164]. Rituximab is a monoclonal antibody and possesses better acceptance against MG as compared to standard immunotherapeutics. A recent retrospective cohort study proposed that rituximab has shown remarkable efficacy soon after the detection of MG [165]. The primary complement inhibitor approved for prescription is Eculizumab [165]. Quick and sustained minimal manifestation in patients with refractory MG (non-responsiveness to acetylcholinesterase) has been attained after Eculizumab treatment [166]. Another notable human anti-CD20 antibody recommended for MG treatment is ofatumumab [167].

Patients with thymomas must have a thymectomy, while nonthymomatous patients with widespread MG are advised to consider it as a treatment option. Several surgical approaches have been suggested, but the maximal approach, or extended trans-sternal thymectomy, is generally accepted. Thymectomy has the potential to reduce or, in some situations, abolish the need for continued chronic medical treatments in the long run [168].

6.3.9.2 Non-pharmacological Treatment

Combining a rehabilitation program with various medical treatments can help MG patients feel better and function better. The main objective is to increase the person's strength to make it easier for them to return to work and daily activities. The disease stage and general state of health affect the exercise's intensity and progression. It is advised to have an interdisciplinary strategy that combines respiratory treatment, physical medicine, and neuromuscular medicine. Restoring muscle strength over the long term is facilitated by physical therapy. Gradually harder exercises aid in keeping the person as functional as possible. Utilizing energy-saving and compensatory approaches, occupational therapy assists the patient in adapting to new ways of carrying out daily activities. After a tracheostomy, speech therapy is available to train esophageal speech. If the demands of the current employees cannot be satisfied, vocational counseling may be required. Psychological treatments to deal with the condition may be required. [169].

6.3.9.3 Herbal Treatment

Chinese herbal medicines (CHMs) para therapy can be advised for regular use in the treatment of MG, at least to some extent, according to the evidence from the few studies that have been conducted. Evidence supporting the likelihood that further conventional treatment will also help to improve MG symptoms is of moderate to low certainty. Giving CHMs in addition to conventional medicine might be safer than doing so [170].

6.3.10 Epilepsy

6.3.10.1 Pharmacological Treatment

Benzodiazepines are recommended for the treatment of epilepsy and sudden-onset seizures. They rank among the best anti-epileptic medications (AEDs) for treating patients with acute repetitive seizures or status epilepticus. They have minimal levels of toxicity, high efficacy, and fast action, all of which have clinical advantages. Because they can be delivered via a variety of methods and in a variety of formulations, benzodiazepines are useful in a wide range of clinical settings. BZDs, however, have drawbacks. BZDs may become ineffective for use in long-term epilepsy therapy as tolerance may set in over time. Clobazam is used for First-line adjunctive treatment for treatment-resistant partial and generalized seizures, intermittent therapy and non-convulsive status epilepticus. Clobazam sustains its therapeutic potential along with good tolerance capability as compared to other anti-epileptic drugs [171, 172]. Lennox-Gastaut syndrome, early status epilepticus, and partial and generalized (especially absence and myoclonic) seizures are all treated with clonazepam as a second-line adjunctive medication. It is a well-known anti-anxiety and anti-epileptic medication with a notable therapeutic effect on various medical disorders that are worsened together with epileptic symptoms as well as on the control of seizures [173]. The adjuvant medication for partial seizures is clorazepate. First-line treatment for status epilepticus in the early stages, second-line therapy for status epilepticus in the advanced stages, treatment for non-convulsive status epilepticus, intermittent preventive medication for febrile seizures, and at-home treatment for ARS all involve the use of diazepam. First-line treatment for early status epilepticus and status epilepticus that occurs outside of a hospital is lorazepam. Second-line treatment for early-status epilepticus is midazolam [174]. Among other medications, mostly prescribed medication is Carbamazepine [175]. CBZ minimizes seizure recurrence by different mechanical pathways. The collective decrease in recurrent seizure frequency by CBZ is caused by either converting convulsive seizures to non-convulsive seizures or by inhibiting a few non-convulsive seizures [176]. CBZ has been reported to decrease calcium and vitamin D during treatment. It has been suggested that patients suffering from osteoporosis and skeletal disorders should avoid long-term CBZ treatment. Regular bone health monitoring should be conducted in both epileptic adults and children with skeletal problems [177] (Fig. 6.4).

Eslicarbazepine acetate is usually recommended both as single agent and as add-on therapy for the treatment of partial-onset epilepsy. Treatment-refractory patients respond to ESL resulting in better life quality. ESL has good tolerability in patients with common adverse effects like headache, Drowsiness, and diplopia [178, 179]. Gabapentin is mostly prescribed by physicians to inhibit partial seizures. This is a symptomatic drug for psychological disorders like seizures and restless leg

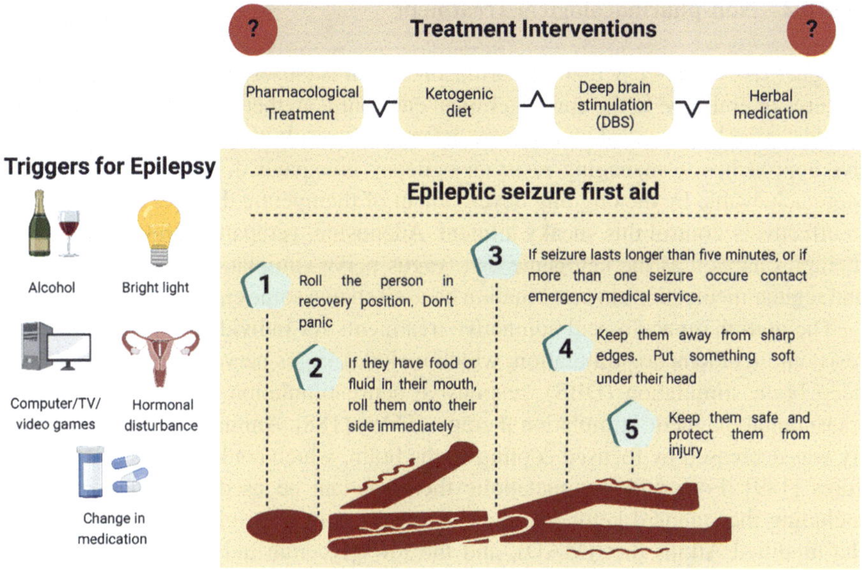

Fig. 6.4 Epilepsy, its triggers and treatment interventions

syndrome. In children, Gabapentin use can cause serious adverse effects. Gabapentin should be used with caution in children as it can provoke behavioral problems. Anxiety, aggression, poor efficiency in school, restlessness, hyperactivity, altering mood swings, and depressive condition are also observed in children after taking gabapentin as an epileptic treatment [180].

Lacosamide is a novel therapeutic agent used as supportive therapy in patients with partial-onset seizures [181]. LCM has shown proven improvement in behavioral problems linked with epilepsy, like stress-oriented conditions, depression, and anxiety. Patients feel better in their life quality as the drug possesses good efficacy and safety [182, 183]. Lamotrigine is an anti-epileptic drug with mood-enhancing properties. It is recommended not only for treating epilepsy but also for the majority of mental disorders. Even though lamotrignine is a well-tolerated drug, an overdose of the drug might lead to life-threatening problems [184].

Perampanel is a newly developed drug in AEDs with an innovative mechanism of action and exhibits maximum therapeutic efficacy in adult epileptic patients. It has shown remarkable effects in reducing various types of seizure symptoms, especially in patients of 12 years or having more age dealing with partial-onset seizures and primary-generalized tonic-clonic seizures [185]. Along with a higher efficacy rate, perampanel possesses a good tolerability profile with very few mild adverse effects. The most prominent side effects are dizziness, somnolence, and fatigue [186, 187].

6.3.10.2 Non-pharmacological Treatment

A higher risk of epilepsy that is pharmacoresistant is closely linked to uncontrolled seizures or epilepsy. In the current clinical environment, there are no potential treatments available to control pharmacoresistant. As a result, the usefulness of alternative approaches to managing resistant epilepsy is expanded, which is made even more achievable by the ongoing development of therapeutic therapies that promise to effectively control this sneaky ailment. Adenosine, verapamil, and other add-on therapies as well as the ketogenic diet, vagus nerve stimulation, targeted cooling, and regular medications in combination have all shown some encouraging outcomes.

The non-pharmacological adjunctive treatments for individuals with pharmacoresistant include neurostimulation, which includes vagus nerve stimulation (VNS), deep brain stimulation (DBS), responsive neurostimulation therapy (RNS), and transcranial magnetic stimulation therapy (TMS) [188]. Animal epileptiform activity was decreased by focused cooling of the brain, which could also stop focal seizures [189]. Four different metabolic therapies can be used to reduce seizures, including the standard ketogenic diet, the medium-chain triglyceride (MCT) diet, the modified Atkins diet (MAD), and the low glycemic index treatment (LGIT) [190]. In the treatment of pharmacoresistant epilepsies, complementary therapies including biofeedback and music therapy have shown encouraging outcomes [191].

6.3.10.3 Herbal Treatment

Numerous chemical compounds produced from plants have also been demonstrated in a variety of in vivo and in vitro studies to have anticonvulsant properties. *Bacopa monnieri* is a traditional medicine and a nervine tonic obtained from it has been reported to cure epilepsy. Another efficient metabolite of *B.monnieri* is bacoside which has been found potent in lowering seizures by directly affecting the peripheral nervous system [192]. In the case of pilocarpine-induced epilepsy, down-regulation of NMDAR1 gene expression and glutamate receptor function has been restored after treatment with *B.monnieri*. In another study, *B.monnieri* ameliorated the epileptic symptoms by antagonizing the up-regulation of 5-hydroxytryptamine-2C in the hippocampus of drug-induced epileptic rat model [193]. Withanolide A is the most active constituent of *Withania somnifera* and carbamazepine has been used in the pilocarpine-induced temporal lobe epilepsy model (TLE). This treatment reversed the motor dysfunctioning in rats that were observed after epilepsy induction [194]. *Nardostachys jatamansi*, usually called jatamansi is used for reducing seizure symptoms [195]. *Scutellaria,* commonly known as skullcap, belongs to the family Lamiaceae and also possesses anticonvulsant properties [196]. *Uncaria rhynchophylla* is a Chinese traditional herb known as wild Asian liana, it is used for curing seizure disorders for centuries but collectively, pharmacological and toxicological evaluation data concerning its epileptic efficiency is insufficient [197]. *Viscum album*, commonly called mistletoe is highly effective in treating epilepsy, blood pressure, and other neurological disorders [198]. Studies were conducted on

Viscum album leaf extract for its epileptic activity in mice and rat models. Both aqueous and aqueous-ethanolic extract of Viscum album leaves from maple exhibited strong anti-convulsant potential [199].

6.3.11 Schizophrenia

6.3.11.1 Pharmacological Treatment

Antipsychotic drugs are generally prescribed for schizophrenia patients. They are further divided into two categories, typical and atypical drugs, based on their specific mechanism of action. Typical drugs such as Chlorpromazine (Thorazine), Fluphenazine (Prolixin), Haloperidol (Haldol), Perphenazine (Trilafon), Thioridazine (Mellaril), Thiothixene (Navane), regulate the dopaminergic system, while serotonin, norepinephrine, and histamine neurotransmitters are modulated by atypical psychotics i.e., Aripiprazole, Asenapine, Clozapine, Iloperidone, Lurasidone, Olanzapine, Paliperidone, Quetiapine, and Risperidone [200, 201] (Fig. 6.5).

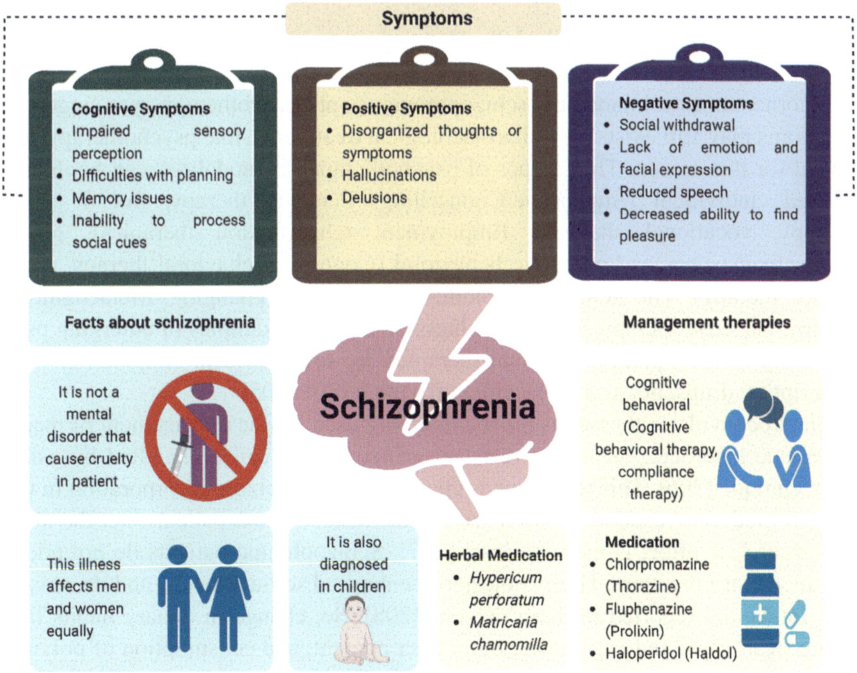

Fig. 6.5 Symptoms, facts, and management therapies for schizophrenia

The first atypical antipsychotic drugs with maximum efficacy against positive symptoms of schizophrenia were phenothiazines. They did not exhibit much improvement in negative and perceptive features. Many adverse effects were also reported. Butyrophenones had shown few undesirable effects along with higher responses against positive symptoms. Another potent therapeutic agent among second-generation antipsychotic drugs is Olanzapine. Approval of Olanzapine by the FDA was done in September 1996. Even though recorded efficacy was shown by olanzapine in schizophrenia patients, changes in metabolic functions and weight gain are the most expected adverse effects. That's why Olanzapine is the second choice drug for treating this disease [202]. The third-generation approved antipsychotic drug is Aripiprazole for schizophrenia. This drug possesses many advantageous features as compared to above mention drugs. Aripiprazole use is obvious because of its minimum adverse effects and well-tolerated capacity [203]. Schizophrenia is also managed by an FDA-approved novel antipsychotic agent called Asenapine. Due to asenapine's complete metabolization by liver enzymes, the oral route of this drug is not recommended. The sublingual route is preferred over the oral route, to attain sustained drug concentration in general circulation [204]. Taking drugs for schizophrenia may have several unrelated side effects. Therefore, when creating a thorough treatment plan, clinicians must take the possibility of non-adherence and treatment-related side effects into account.

6.3.11.2 Non-pharmacological Treatment

The cornerstone of managing schizophrenia is pharmacotherapy, yet lingering symptoms may still exist. Non-pharmacological treatments, like psychotherapy, are crucial for this reason. Three types of psychotherapeutic modalities can be distinguished: individual (Supportive/Counselling, Personal therapy, Social Skills Therapy, vocational sheltered Employment rehabilitation therapies), group (Interactive/social), and cognitive behavioral (Cognitive behavioral therapy, compliance therapy). The field of psychotherapy is always changing. Metacognitive therapy, narrative therapy, and mindfulness therapy are examples of emerging psychotherapies. Non-pharmacological therapies are to be used in conjunction with prescription drugs, not as a replacement for them [201, 205].

Electroconvulsive therapy is most commonly used for the management of major depressive disorders. Treatment-resistant schizophrenia can be handled by using this technique [206]. This technique is characterized by seizure incorporation in the body. A moderate response is observed in treatment-resistant patients, but the most dangerous side effect is memory loss. [207]. Schizophrenic patients do not adopt healthy dietary practices. High cholesterol, enhanced sodium levels, and decreased fiber intake are reported in these persons [208]. So, change in dietary habits like taking fresh fruits and vegetables, more fiber content, and consumption of polyunsaturated fats have useful effects on the health of schizophrenic patients [209].

6.3.11.3 Herbal Treatment

The use of herbal medicine in a modern medicine context has been evaluated in certain trials, although these are limited due to their sample size and study length. The results of these six trials suggest that using herbal medicine as monotherapy for psychiatric disorders may not be effective, but if used adjuvant therapy with the first or second generation of antipsychotic drugs, they may be useful.

Ginkgo biloba is getting a lot of attention within scientific communities for its therapeutic potential to treat schizophrenia. According to a 2008 Canadian research study, *P. ginseng* also helps reduce some of the behavioral traits associated with schizophrenia, including the inability to show emotions [210, 211]. *Hypericum perforatum*, is one of the most frequently used herbal remedies for the treatment of psychiatric disorders and this medicinal plant might be effective in treating cognitive dysfunction in patients with schizophrenia [212]. The herb *glycyrrhiza glabra* is very effective at reducing tension and anxiety. Schizophrenia could be successfully treated with *G. glabra's* therapeutic properties [213]. *O. basilicum* leaves can boost brain functionality and consequently assist in controlling the symptoms associated with schizophrenia. *Matricaria chamomilla* has compounds and crucial minerals along with relaxing and soothing properties which help a great deal with issues of the brain and nervous system. *M. chamomilla* may be utilized as a mood enhancer, or to encourage restful sleep for those suffering from schizophrenia [210, 211].

6.4 Conclusion

The umbrella term "Brain disorders" contains all dysfunctions and diseases of the peripheral or central nervous system. It is well documented that, on a global scale BD effects thousands of millions of individuals. We all know that in our society, these disorders cause an economic burden. A better understanding of BD and treatment modules is provided in this chapter. It is also determined that in all aspects, BD can impact the lives of individuals. According to the advancement and nature of BD, therapeutical methodologies are adopted. These methodologies include interdisciplinary treatments. To understand the basis and foundations for the development of respective diseases, additional research is required because numerous neurological and mental diseases have deficits in treatment. Furthermore, to refine symptoms, more study is required for advanced treatments which transform disease development.

References

1. Kindell J, Keady J, Sage K, Wilkinson R (2017) Everyday conversation in dementia: a review of the literature to inform research and practice. Int J Lang Commun Disord 52:392–406
2. Kumar A, Tsao JW (2018) Alzheimer Disease: REVUE. StatPearls:1–27
3. Nicolas G, Acuña-Hidalgo R, Keogh MJ et al (2018) Somatic variants in autosomal dominant genes are a rare cause of sporadic Alzheimer's disease. Alzheimer's Dement 14:1632–1639. https://doi.org/10.1016/j.jalz.2018.06.3056

4. McGirr S, Venegas C, Swaminathan A (2020) Alzheimers disease: a brief review. J Exp Neurol 1:89–98. https://doi.org/10.33696/neurol.1.015
5. Bir SC, Khan MW, Javalkar V et al (2021) Emerging concepts in vascular dementia: a review. J Stroke Cerebrovasc Dis 30:105864
6. Akhter F, Persaud A, Zaokari Y et al (2021) Vascular dementia and underlying sex differences. Front Aging Neurosci 13:583
7. Vascular Dementia – StatPearls – NCBI Bookshelf. https://www.ncbi.nlm.nih.gov/books/NBK430817/. Accessed 28 Mar 2023
8. Tabrizi SJ, Leavitt BR, Landwehrmeyer GB et al (2019) Targeting Huntingtin Expression in Patients with Huntington's Disease. N Engl J Med 380:2307–2316. https://doi.org/10.1056/nejmoa1900907
9. Ghosh R, Tabrizi SJ (2018) Clinical features of huntington's disease. In: Advances in experimental medicine and biology. Springer, New York LLC, pp 1–28
10. Andhale R, Shrivastava D (2022) Huntington's disease: a clinical review. Cureus 14. https://doi.org/10.7759/cureus.28484
11. Gatto EM, Rojas NG, Persi G et al (2020) Huntington disease: advances in the understanding of its mechanisms. Clin Park Relat Disord 3:100056
12. DeMaagd G, Philip A (2015) Parkinson's disease and its management: Part 1: disease entity, risk factors, pathophysiology, clinical presentation, and diagnosis. Pharm Ther 40:504
13. Duarte-Jurado AP, Gopar-Cuevas Y, Saucedo-Cardenas O, et al (2021) Antioxidant therapeutics in Parkinson's disease: current challenges and opportunities. Antioxidants 10:453. https://doi.org/10.3390/ANTIOX10030453
14. Lv Q, Wang X, Asakawa T, Wang XP (2020) Pharmacologic treatment of restless legs syndrome. Curr Neuropharmacol 19:372–382. https://doi.org/10.2174/1570159x19666201230150127
15. Memon MD, Faiz S, Zaveri MP et al (2020) Unraveling the mysteries of restless leg syndrome. Cureus 12. https://doi.org/10.7759/CUREUS.10951
16. Seeman MV (2020) Why are women prone to restless legs syndrome? Int J Environ Res Public Heal 17:368. https://doi.org/10.3390/IJERPH17010368
17. Salib M, Memon AN, Gowda AS et al (2020) Dialysis patients with restless leg syndrome: can we relieve their suffering? Cureus 12. https://doi.org/10.7759/CUREUS.10053
18. Batla A (2018) Dystonia: a review. Neurol India 66:S48–S58
19. Grütz K, Klein C (2021) Dystonia updates: definition, nomenclature, clinical classification, and etiology. J Neural Transm 128:395–404
20. Kobelt G, Thompson A, Berg J et al (2017) New insights into the burden and costs of multiple sclerosis in Europe. Mult Scler 23:1123–1136. https://doi.org/10.1177/1352458517694432
21. Greenfield AL, Hauser SL (2018) B-cell therapy for multiple sclerosis: entering an era. Ann Neurol 83:13–26. https://doi.org/10.1002/ANA.25119
22. Ramaglia V, Rojas O, Naouar I, Gommerman JL (2021) The ins and outs of central nervous system inflammation—lessons learned from Multiple Sclerosis 39:199–226. https://doi.org/10.1146/ANNUREV-IMMUNOL-093019-124155
23. McGinley MP, Goldschmidt CH, Rae-Grant AD (2021) Diagnosis and treatment of multiple sclerosis: a review. JAMA 325:765–779. https://doi.org/10.1001/JAMA.2020.26858
24. Dobson R, Giovannoni G (2019) Multiple sclerosis – a review. Eur J Neurol 26:27–40
25. Silveira C, Guedes R, Maia D et al (2019) Neuropsychiatric symptoms of multiple sclerosis: state of the art. Psychiatr Investig 16:877. https://doi.org/10.30773/PI.2019.0106
26. Dresser L, Wlodarski R, Rezania K, Soliven B (2021) Myasthenia gravis: epidemiology, pathophysiology and clinical manifestations. J Clin Med 10. https://doi.org/10.3390/jcm10112235
27. Farrugia ME, Goodfellow JA (2020) A practical approach to managing patients with myasthenia gravis—opinions and a review of the literature. Front Neurol 11:604
28. Beghi E (2020) The Epidemiology of Epilepsy. Neuroepidemiology 54:185–191. https://doi.org/10.1159/000503831

29. Karoly PJ, Rao VR, Gregg NM et al (2021) Cycles in epilepsy. Nat Rev Neurol 175(17):267–284. https://doi.org/10.1038/s41582-021-00464-1
30. Anwar H, Khan QU, Nadeem N, et al (2020) Epileptic seizures. Discoveries 8:e128. https://doi.org/10.15190/D.2020.7
31. Hu ML, Zong XF, Mann JJ et al (2017) A review of the functional and anatomical default mode network in Schizophrenia. Neurosci Bull 33:73–84. https://doi.org/10.1007/S12264-016-0090-1/TABLES/1
32. Anwal L (2021) A comprehensive review on Alzheimer's disease. World J Pharm Pharm Sci 10:1170. https://doi.org/10.20959/wjpps20217-19427
33. Haake A, Nguyen K, Friedman L et al (2020) An update on the utility and safety of cholinesterase inhibitors for the treatment of Alzheimer's disease. Expert Opin Drug Saf 19:147–157. https://doi.org/10.1080/14740338.2020.1721456
34. Kuroda A, Setoguchi M, Uchino Y et al (2018) Effect of rivastigmine on plasma butyrylcholine esterase activity and plasma ghrelin levels in patients with dementia in Alzheimer's disease. Geriatr Gerontol Int 18:886–891. https://doi.org/10.1111/ggi.13275
35. Marotta G, Basagni F, Rosini M, Minarini A (2020) Memantine derivatives as multitarget agents in Alzheimer's disease. Molecules 25. https://doi.org/10.3390/molecules25174005
36. Matsunaga S, Kishi T, Nomura I et al (2018) The efficacy and safety of memantine for the treatment of Alzheimer's disease. Expert Opin Drug Saf 17:1053–1061. https://doi.org/10.1080/14740338.2018.1524870
37. Tanvir Kabir M, Sahab Uddin M, Al Mamun A et al (2020) Combination drug therapy for the management of alzheimer's disease. Int J Mol Sci 21. https://doi.org/10.3390/ijms21093272
38. Zhang X, Yu R, Wang H, Zheng R (2020) Effects of rivastigmine hydrogen tartrate and donepezil hydrochloride on the cognitive function and mental behavior of patients with Alzheimer's disease. Exp Ther Med 20:1789–1795. https://doi.org/10.3892/etm.2020.8872
39. Koola MM (2020) Galantamine-Memantine combination in the treatment of Alzheimer's disease and beyond. Psychiatr Res 293:113409. https://doi.org/10.1016/j.psychres.2020.113409
40. Cummings JL, Tong G, Ballard C (2019) Treatment Combinations for Alzheimer's Disease: Current and Future Pharmacotherapy Options. J Alzheimer's Dis 67:779–794. https://doi.org/10.3233/JAD-180766
41. Sang Z, Wang K, Shi J et al (2020) Apigenin-rivastigmine hybrids as multi-target-directed liagnds for the treatment of Alzheimer's disease. Eur J Med Chem 187:111958. https://doi.org/10.1016/j.ejmech.2019.111958
42. Cummings J, Lee G, Ritter A, Zhong K (2018) Alzheimer's disease drug development pipeline: 2018. Alzheimer's Dement Transl Res Clin Interv 4:195–214. https://doi.org/10.1016/j.trci.2018.03.009
43. Coric V, Salloway S, Van Dyck CH et al (2015) Targeting prodromal Alzheimer disease with avagacestat: a randomized clinical trial. JAMA Neurol 72:1324–1333. https://doi.org/10.1001/jamaneurol.2015.0607
44. Sultzer DL, Davis SM, Tariot PN et al (2008) Clinical symptom responses to atypical antipsychotic medications in Alzheimer's disease: Phase 1 outcomes from the CATIE-AD effectiveness trial. Am J Psychiatry 165:844–854. https://doi.org/10.1176/appi.ajp.2008.07111779
45. Farina N, Llewellyn D, Isaac MGEKN, Tabet N (2017) Vitamin E for Alzheimer's dementia and mild cognitive impairment. Cochrane Database Syst. Rev:2017
46. Tan RS, Pu SJ (2003) A pilot study on the effects of testosterone in hypogonadal aging male patients with Alzheimer's disease. Aging Male 6:13–17. https://doi.org/10.1080/tam.6.1.13.17
47. Schwarzbach C, Vauth C, Roll S et al (2010) Effects of Ginkgo biloba in dementia: systematic review and meta-analysis. BMC Geriatr 10:14
48. Massoud F, Léger GC (2011) Pharmacological treatment of Alzheimer disease. Can. J. Psychiatry 56:579–588

49. Dhana K, Evans DA, Rajan KB et al (2019) O1-06-01: Impact of healthy lifestyle factors on the risk of Alzheimer's dementia: findings from two prospective cohort studies. Alzheimer's Dement 15:P207–P207. https://doi.org/10.1016/J.JALZ.2019.06.4547
50. Brown BM, Peiffer JJ, Martins RN (2013) Multiple effects of physical activity on molecular and cognitive signs of brain aging: can exercise slow neurodegeneration and delay Alzheimer's disease? Mol Psychiatr 188(18):864–874. https://doi.org/10.1038/mp.2012.162
51. Jeong JH, Na HR, Choi SH et al (2016) Group- and home-based cognitive intervention for patients with mild cognitive impairment: a randomized controlled trial. Psychother Psychosom 85:198–207. https://doi.org/10.1159/000442261
52. Lyu J, Zhang J, Mu H et al (2018) The effects of music therapy on cognition, psychiatric symptoms, and activities of daily living in patients with Alzheimer's disease. J Alzheimers Dis 64:1347–1358. https://doi.org/10.3233/JAD-180183
53. Peck KJ, Girard TA, Russo FA, Fiocco AJ (2016) Music and memory in Alzheimer's disease and the potential underlying mechanisms. J Alzheimers Dis 51:949–959. https://doi.org/10.3233/JAD-150998
54. Rabey JM, Dobronevsky E (2016) Repetitive transcranial magnetic stimulation (rTMS) combined with cognitive training is a safe and effective modality for the treatment of Alzheimer's disease: clinical experience. J Neural Transm 123:1449–1455. https://doi.org/10.1007/S00702-016-1606-6
55. Yulug B, Hanoglu L, Khanmammadov E et al (2018) Beyond the therapeutic effect of rTMS in Alzheimer's disease: a possible neuroprotective role of hippocampal BDNF?: a minireview. Mini-Reviews Med Chem 18:1479–1485. https://doi.org/10.2174/1389557517666170927162537
56. Li X, Guo F, Zhang Q et al (2014) Electroacupuncture decreases cognitive impairment and promotes neurogenesis in the APP/PS1 transgenic mice. BMC Complement Altern Med 14:1–10. https://doi.org/10.1186/1472-6882-14-37/FIGURES/6
57. Yancheva S, Ihl R, Nikolova G, et al (2009) Ginkgo biloba extract EGb 761®, donepezil or both combined in the treatment of Alzheimer's disease with neuropsychiatric features: a randomised, double-blind, exploratory trial.13:183–190. https://doi.org/10.1080/13607860902749057
58. Miraj S, Rafieian-Kopaei KS (2017) Melissa officinalis L: a review study with an antioxidant prospective. J Evid Based Compl Altern Med 22:385–394. https://doi.org/10.1177/2156587216663433
59. Ghahghaei A, Bathaie SZ, Kheirkhah H, Bahraminejad E (2013) The protective effect of crocin on the amyloid fibril formation of aβ42 peptide in vitro. Cell Mol Biol Lett 183(18):328–339. https://doi.org/10.2478/S11658-013-0092-1
60. Sahoo AK, Dandapat J, Dash UC, Kanhar S (2018) Features and outcomes of drugs for combination therapy as multi-targets strategy to combat Alzheimer's disease. J Ethnopharmacol 215:42–73. https://doi.org/10.1016/j.jep.2017.12.015
61. Guo Y, Dong X, Zhang R et al (2020) Salvia miltiorrhiza improves Alzheimer's disease: A protocol for systematic review and meta-analysis. Medicine (Baltimore) 99:e21924. https://doi.org/10.1097/MD.0000000000021924
62. Sinha K, Sun C, Kamari R, Bettermann K (2020) Current status and future prospects of pathophysiology-based neuroprotective drugs for the treatment of vascular dementia. Drug Discov Today 25:793–799. https://doi.org/10.1016/j.drudis.2020.01.003
63. Wang M chao, Lin L hong (2020) Therapeutic efficacy and mechanism of heat-sensitive moxibustion for vascular dementia. J Acupunct Tuina Sci 18:47–52. https://doi.org/10.1007/s11726-020-1156-1
64. Kryscio RJ, Abner EL, Caban-Holt A et al (2017) Association of antioxidant supplement use and dementia in the prevention of Alzheimer's disease by Vitamin E and selenium trial (PREADViSE). JAMA Neurol 74:567–573. https://doi.org/10.1001/jamaneurol.2016.5778
65. Zhang T, Gu J, Wu L et al (2017) Neuroprotective and axonal outgrowth-promoting effects of tetramethylpyrazine nitrone in chronic cerebral hypoperfusion rats and primary hippocampal

neurons exposed to hypoxia. Neuropharmacology 118:137–147. https://doi.org/10.1016/j.neuropharm.2017.03.022
66. Jiang W, Gong L, Liu F, Mu J (2020) Stem cells and vascular dementia: from basic science to the clinic. Cell Tissue Bank 21:349–360. https://doi.org/10.1007/s10561-020-09829-0
67. Forbes D, Forbes SC, Blake CM et al (2015) Exercise programs for people with dementia. Cochrane database Syst Rev 2015. https://doi.org/10.1002/14651858.CD006489.PUB4
68. Bahar-Fuchs A, Clare L, Woods B (2013) Cognitive training and cognitive rehabilitation for mild to moderate Alzheimer's disease and vascular dementia. Cochrane database Syst Rev 2013. https://doi.org/10.1002/14651858.CD003260.PUB2
69. Orgeta V, Qazi A, Spector AE, Orrell M (2014) Psychological treatments for depression and anxiety in dementia and mild cognitive impairment. Cochrane database Syst Rev 2014. https://doi.org/10.1002/14651858.CD009125.PUB2
70. Man SC, Chan KW, Lu JH, et al (2012) Systematic review on the efficacy and safety of herbal medicines for vascular dementia. Evidence-based complement. Altern Med:2012
71. Chang D, Liu J, Bilinski K, et al (2016) Herbal medicine for the treatment of vascular dementia: an overview of scientific evidence. Evidence-based complement. Altern Med:2016
72. Xu QQ, Shan CS, Wang Y et al (2018) Chinese herbal medicine for vascular dementia: a systematic review and meta-analysis of high-quality randomized controlled trials. J Alzheimer's Dis 62:429–456. https://doi.org/10.3233/JAD-170856
73. Frank S (2014) Treatment of Huntington's disease. Neurotherapeutics 11:153–160
74. Richard A, Frank S (2019) Deutetrabenazine in the treatment of Huntington's disease. 9:31–37. https://doi.org/10.2217/NMT-2018-0040
75. Bashir H, Jankovic J (2018) Deutetrabenazine for the treatment of Huntington's chorea. 18:625–631. https://doi.org/10.1080/14737175.2018.1500178
76. Schultz JL, Killoran A, Nopoulos PC et al (2018) Evaluating depression and suicidality in tetrabenazine users with Huntington disease. Neurology 91:e202–e207. https://doi.org/10.1212/WNL.0000000000005817
77. Qureshi FH, Qureshi SH, Zia T, Khawaja F (2022) Huntington's disease (HD): a brief review. Eur J Public Heal Stud 5
78. Snowden JS (2017) The neuropsychology of Huntington's disease. Arch Clin Neuropsychol 32:876–887
79. Dey A, De JN (2015) Neuroprotective therapeutics from botanicals and phytochemicals against Huntington's disease and related neurodegenerative disorders. J Herb Med 5:1–19
80. Choudhary S, Kumar P, Malik J (2013) Plants and phytochemicals for Huntington's disease. Pharmacogn Rev 7:81. https://doi.org/10.4103/0973-7847.120505
81. Khan H, Ullah H, Tundis R, et al (2020) Dietary flavonoids in the management of Huntington's disease: mechanism and clinical perspective. eFood 1:38–52. https://doi.org/10.2991/EFOOD.K.200203.001
82. Khan A, Jahan S, Imtiyaz Z, et al (2020) Neuroprotection: targeting multiple pathways by naturally occurring phytochemicals. Biomed 8:284. https://doi.org/10.3390/BIOMEDICINES8080284
83. Connolly BS, Lang AE (2014) Pharmacological treatment of Parkinson disease: a review. JAMA 311:1670–1683. https://doi.org/10.1001/JAMA.2014.3654
84. Clarke CE (2002) Medical management of Parkinson's disease. Neurol Pract 72:590
85. Radhakrishnan DM, Goyal V (2018) Parkinson's disease: a review. Neurol India 66:26. https://doi.org/10.4103/0028-3886.226451
86. Balestrino R, Schapira AHV (2020) Parkinson disease. Eur J Neurol 27:27–42. https://doi.org/10.1111/ENE.14108
87. Bomasang-Layno E, Fadlon I, Murray AN, Himelhoch S (2015) Antidepressive treatments for Parkinson's disease: a systematic review and meta-analysis. Parkinsonism Relat Disord 21:833–842. https://doi.org/10.1016/J.PARKRELDIS.2015.04.018
88. Rios Romenets S, Creti L, Fichten C et al (2013) Doxepin and cognitive behavioural therapy for insomnia in patients with Parkinson's disease – a randomized study. Park Relat Disord 19:670–675. https://doi.org/10.1016/j.parkreldis.2013.03.003

89. Jankovic J, Poewe W (2012) Therapies in Parkinson's disease. Curr Opin Neurol 25:433–447. https://doi.org/10.1097/WCO.0B013E3283542FC2
90. Armstrong MJ, Okun MS (2020) Diagnosis and treatment of Parkinson disease: a review. JAMA 323:548–560
91. Auxiliadora de Paula Vasconcelos L (2020) Parkinson's disease rehabilitation: effectiveness approaches and new perspectives. In: Physical therapy effectiveness. IntechOpen
92. Rajendran PR, Thompson RE, Reich SG (2001) The use of alternative therapies by patients with Parkinson's disease. Neurology 57:790–794. https://doi.org/10.1212/WNL.57.5.790
93. Stallibrass C, Sissons P, Chalmers C (2002) Randomized controlled trial of the Alexander Technique for idiopathic Parkinson's disease. Clin Rehabil 16:695–708. https://doi.org/10.1191/0269215502cr544oa
94. Kapur SS, Stebbins GT, Goetz CG (2012) Vibration therapy for Parkinson's disease: charcot's studies revisited. J Parkinsons Dis 2:23–27. https://doi.org/10.3233/JPD-2012-12079
95. de Dreu MJ, van der Wilk ASD, Poppe E et al (2012) Rehabilitation, exercise therapy and music in patients with Parkinson's disease: a meta-analysis of the effects of music-based movement therapy on walking ability, balance and quality of life. Park Relat Disord 18. https://doi.org/10.1016/s1353-8020(11)70036-0
96. Song JX, Sze SCW, Ng TB et al (2012) Anti-Parkinsonian drug discovery from herbal medicines: WHAT have we got from neurotoxic models? J Ethnopharmacol 139:698–711
97. Li XZ, Zhang SN, Liu SM, Lu F (2013) Recent advances in herbal medicines treating Parkinson's disease. Fitoterapia 84:273–285
98. Manyam BV, Dhanasekaran M, Hare TA (2004) Neuroprotective effects of the antiparkinson drug Mucuna pruriens. Phyther Res 18:706–712. https://doi.org/10.1002/ptr.1514
99. Samoylenko V, Rahman MM, Tekwani BL et al (2010) Banisteriopsis caapi, a unique combination of MAO inhibitory and antioxidative constituents for the activities relevant to neurodegenerative disorders and Parkinson's disease. J Ethnopharmacol 127:357–367. https://doi.org/10.1016/j.jep.2009.10.030
100. Amro MS, Teoh SL, Norzana AG, Srijit D (2018) The potential role of herbal products in the treatment of Parkinson's disease. Clin Ter 169:e23–e33
101. Garnock-Jones KP (2016) Rotigotine transdermal patch: a review in restless legs syndrome. Drugs 76:1031–1040. https://doi.org/10.1007/s40265-016-0601-4
102. Faulkner MA (2018) Use of α2δ ligands for restless legs syndrome/willis ekbom disease. CNS Drugs 32:149–159
103. Sagheb MM, Dormanesh B, Fallahzadeh MK, et al (2012) Efficacy of vitamins C, E, and their combination for treatment of restless legs syndrome in hemodialysis patients: a randomized, double-blind, placebo-controlled trial. Sleep Med 13:542–545. https://doi.org/10.1016/j.sleep.2011.11.010
104. Roshi R, Tandon V, Mahajan A et al (2019) Comparative efficacy and safety of clonazepam versus nortriptyline in restless leg syndrome among forty plus women: a prospective, open-label randomized study. J Midlife Health 10:197–203. https://doi.org/10.4103/jmh.JMH_26_18
105. Kurlan R, Rabin M (2013) Augmentation in restless legs syndrome: poor response to sudden withdrawal of dopaminergic therapy. J Park Restless Legs Syndr 3:49. https://doi.org/10.2147/jprls.s47648
106. Brand S, Beck J, Hatzinger M, Holsboer-Trachsler E (2013) Patients suffering from restless legs syndrome have low internal locus of control and poor psychological functioning compared to healthy controls. Neuropsychobiology 68:51–58. https://doi.org/10.1159/000350957
107. Aurora RN, Kristo DA, Bista SR, et al (2012) Update to the AASM clinical practice guideline: "The treatment of restless legs syndrome and periodic limb movement disorder in adults – an update for 2012: Practice parameters with an evidence-based systematic review and meta-analyses." Sleep 35:1037
108. Bega D, Malkani R (2016) Alternative treatment of restless legs syndrome: An overview of the evidence for mind-body interventions, lifestyle interventions, and neutraceuticals. Sleep Med 17:99–105

109. Giannaki CD, Hadjigeorgiou GM, Karatzaferi C, et al (2013) A single-blind randomized controlled trial to evaluate the effect of 6 months of progressive aerobic exercise training in patients with uraemic restless legs syndrome. Nephrol Dial Transplant 28:2834–2840. https://doi.org/10.1093/ndt/gft288
110. Lin YC, Feng Y, Zhan SQ et al (2015) Repetitive transcranial magnetic stimulation for the treatment of restless legs syndrome. Chin Med J (Engl) 128:1728–1731. https://doi.org/10.4103/0366-6999.159344
111. Yan X, Wang WD, Walters AS, et al (2012) Traditional Chinese medicine herbal preparations in restless legs syndrome (RLS) treatment: a review and probable first description of RLS in 1529. Sleep Med Rev 16:509–518
112. Buck ML, Pharm D (2011) Trihexyphenidyl for the Management of Dystonia in Children. In: Pediatr. Pharmacother.
113. Fehlings D, Brown L, Harvey A et al (2018) Pharmacological and neurosurgical interventions for managing dystonia in cerebral palsy: a systematic review. Dev Med Child Neurol 60:356–366. 10.1111/DMCN.13652/ABSTRACT
114. Albanese A (2014) Deep brain stimulation for cervical dystonia. Lancet Neurol 13:856–857. https://doi.org/10.1016/S1474-4422(14)70178-4
115. Rodrigues FB, Duarte GS, Prescott D, et al (2019) Deep brain stimulation for dystonia. Cochrane Database Syst Rev. https://doi.org/10.1002/14651858.CD012405.PUB2/MEDIA/CDSR/CD012405/IMAGE_N/NCD012405-CMP-001.01.SVG
116. Bergenheim AT, Nordh E, Larsson E, Hariz MI (2015) Selective peripheral denervation for cervical dystonia: long-term follow-up. J Neurol Neurosurg Psychiatr 86:1307–1313. https://doi.org/10.1136/JNNP-2014-307959
117. Thenganatt MA, Jankovic J (2014) Treatment of Dystonia. Neurotherapeutics 11:139–152. https://doi.org/10.1007/s13311-013-0231-4
118. Zeuner KE, Bara-Jimenez W, Noguchi PS et al (2002) Sensory training for patients with focal hand dystonia. Ann Neurol 51:593–598. https://doi.org/10.1002/ana.10174
119. Lee M, Kim S, Woo J et al (2016) Case study of a patient with dystonia due to dystonic cerebral palsy treated with traditional Korean medicine. J Intern Korean Med 37:257–264
120. Okamoto H, Chino A, Hirasaki Y, et al (2012) Orengedoku-to augmentation in cases showing partial response to yokukan-san treatment: A case report and literature review of the evidence for use of these Kampo herbal formulae. Neuropsychiatr Dis Treat 9:151–155. https://doi.org/10.2147/NDT.S38318
121. Stüve O, Bennett JL, Hemmer B, et al (2008) Pharmacological treatment of early multiple sclerosis. Drugs 68:73–83
122. Montalban X, Gold R, Thompson AJ et al (2018) ECTRIMS/EAN Guideline on the pharmacological treatment of people with multiple sclerosis. Mult Scler 24:96–120. https://doi.org/10.1177/1352458517751049
123. Minagar A (2013) Current and Future Therapies for Multiple Sclerosis. Scientifica (Cairo) 2013:1–11. https://doi.org/10.1155/2013/249101
124. Mokhber N, Azarpazhooh A, Orouji E et al (2015) Therapeutic effect of Avonex, Rebif and Betaferon on quality of life in multiple sclerosis. Psychiatry Clin Neurosci 69:649–657. https://doi.org/10.1111/PCN.12308/SUPPINFO
125. Aharoni R (2013) The mechanism of action of glatiramer acetate in multiple sclerosis and beyond. Autoimmun Rev 12:543–553. https://doi.org/10.1016/J.AUTREV.2012.09.005
126. Sirbu CA, Furdu-Lungut E, Plesa CF et al (2016) Pharmacological treatment of relapsing remitting multiple sclerosis-where are we? Farmacia 64:651–655
127. Rafiee Zadeh A, Askari M, Azadani NN, et al (2019) Mechanism and adverse effects of multiple sclerosis drugs: a review article. Part 1. Int J Physiol Pathophysiol Pharmacol 11:95–104
128. Martinelli Boneschi F, Vacchi L, Rovaris M et al (2013) Mitoxantrone for multiple sclerosis. Cochrane Database Syst Rev 2013. https://doi.org/10.1002/14651858.CD002127.PUB3/MEDIA/CDSR/CD002127/IMAGE_N/NCD002127-CMP-004-02.PNG

129. Cocco E, Marrosu MG (2014) The current role of mitoxantrone in the treatment of multiple sclerosis. 14:607–616. https://doi.org/10.1586/14737175.2014.915742
130. Kieseier BC, Stüve O (2011) A critical appraisal of treatment decisions in multiple sclerosis—old versus new. Nat Rev Neurol 75 7:255–262. https://doi.org/10.1038/nrneurol.2011.41
131. Brandstadter R, Sand IK (2017) The use of natalizumab for multiple sclerosis. Neuropsychiatr Dis Treat 13:1691. https://doi.org/10.2147/NDT.S114636
132. Eggleton P, Smerdon GR, Holley JE, Gutowski NJ (2017) Manipulation of oxygen and endoplasmic reticulum stress factors as possible interventions for treatment of multiple sclerosis: evidence for and against. Adv Exp Med Biol 958:11–27. 10.1007/978-3-319-47861-6_2
133. Moore L, Eggleton P, Smerdon G, et al (2020) Engagement of people with multiple sclerosis to enhance research into the physiological effect of hyperbaric oxygen therapy. Mult Scler Relat Disord 43:102084. https://doi.org/10.1016/J.MSARD.2020.102084
134. Bensa C, Bodiguel E, Brassat D, et al (2012) Recommendations for the detection and therapeutic management of cognitive impairment in multiple sclerosis. Rev Neurol (Paris) 168:785–794. https://doi.org/10.1016/J.NEUROL.2012.02.009
135. Amatya B, Khan F, La Mantia L et al (2013) Non pharmacological interventions for spasticity in multiple sclerosis. Cochrane Database Syst Rev 2013
136. Giesser BS (2015) Exercise in the management of persons with multiple sclerosis. Ther Adv Neurol Disord 8:123. https://doi.org/10.1177/1756285615576663
137. Halabchi F, Alizadeh Z, Sahraian MA, Abolhasani M (2017) Exercise prescription for patients with multiple sclerosis; potential benefits and practical recommendations. BMC Neurol 17:1–11. https://doi.org/10.1186/S12883-017-0960-9/TABLES/2
138. Amatya B, Young J, Khan F (2018) Non-pharmacological interventions for chronic pain in multiple sclerosis. Cochrane Database Syst Rev:2018
139. Mojaverrostami S, Bojnordi MN, Ghasemi-Kasman M et al (2018) A review of herbal therapy in multiple sclerosis. Adv Pharm Bull 8:575. https://doi.org/10.15171/APB.2018.066
140. Venkatesan R, Ji E (2015) Kim SY (2015) Phytochemicals that regulate neurodegenerative disease by targeting neurotrophins: a comprehensive review. Biomed Res Int. https://doi.org/10.1155/2015/814068
141. Haniadka R, Saldanha E, Sunita V, et al (2013) A review of the gastroprotective effects of ginger (Zingiber officinale Roscoe). Food Funct 4:845–855. https://doi.org/10.1039/C3FO30337C
142. Jafarzadeh A, Mohammadi-Kordkhayli M, Ahangar-Parvin R et al (2014) Ginger extracts influence the expression of IL-27 and IL-33 in the central nervous system in experimental autoimmune encephalomyelitis and ameliorates the clinical symptoms of disease. J Neuroimmunol 276:80–88. https://doi.org/10.1016/J.JNEUROIM.2014.08.614
143. Mohajeri M, Sadeghizadeh M, Najafi F, Javan M (2015) Polymerized nano-curcumin attenuates neurological symptoms in EAE model of multiple sclerosis through down regulation of inflammatory and oxidative processes and enhancing neuroprotection and myelin repair. Neuropharmacology 99:156–167. https://doi.org/10.1016/J.NEUROPHARM.2015.07.013
144. Ghanaatian N, Lashgari NA, Abdolghaffari AH, et al (2019) Curcumin as a therapeutic candidate for multiple sclerosis: Molecular mechanisms and targets. J Cell Physiol 234:12237–12248. https://doi.org/10.1002/JCP.27965
145. Naziroglu M, Kutluhan S, Övey IS et al (2014) Modulation of oxidative stress, apoptosis, and calcium entry in leukocytes of patients with multiple sclerosis by Hypericum perforatum. 17:214–221. https://doi.org/10.1179/1476830513Y.0000000083
146. Cho IH (2012) Effects of Panax ginseng in Neurodegenerative Diseases. J Ginseng Res 36:342. https://doi.org/10.5142/JGR.2012.36.4.342
147. Etemadifar M, Sayahi F, Abtahi SH, et al (2013) Ginseng in the treatment of fatigue in multiple sclerosis: a randomized, placebo-controlled, double-blind pilot study. 123:480–486. https://doi.org/10.3109/00207454.2013.764499
148. Barp A, Gerardi F, Lizio A et al (2020) Emerging Drugs for the Treatment of Amyotrophic Lateral Sclerosis: A Focus on Recent Phase 2 Trials. Expert Opin Emerg Drugs 25:145–164

149. Park JM, Kim SY, Park D, Park JS (2020) Effect of edaravone therapy in Korean amyotrophic lateral sclerosis (ALS) patients. Neurol Sci 41:119–123. https://doi.org/10.1007/s10072-019-04055-3
150. Smith R, Pioro E, Myers K et al (2017) Enhanced bulbar function in amyotrophic lateral sclerosis: the nuedexta treatment trial. Neurotherapeutics 14:762–772. https://doi.org/10.1007/s13311-016-0508-5
151. Mora JS, Genge A, Chio A, et al (2020) Masitinib as an add-on therapy to riluzole in patients with amyotrophic lateral sclerosis: a randomized clinical trial. Amyotroph Lateral Scler Front Degener 21:5–14. https://doi.org/10.1080/21678421.2019.1632346
152. Andersen PM, Abrahams S, Borasio GD, et al (2012) EFNS guidelines on the Clinical Management of Amyotrophic Lateral Sclerosis (MALS) – revised report of an EFNS task force. Eur J Neurol 19:360–375. https://doi.org/10.1111/j.1468-1331.2011.03501.x
153. Rosenfeld J, King RM, Jackson CE et al (2008) Creatine monohydrate in ALS: Effects on strength, fatigue, respiratory status and ALSFRS. Amyotroph Lateral Scler 9:266–272
154. Sanjak M, Bravver E, Bockenek WL et al (2010) Supported treadmill ambulation for amyotrophic lateral sclerosis: a pilot study. Arch Phys Med Rehabil 91:1920–1929. https://doi.org/10.1016/j.apmr.2010.08.009
155. Drory VE, Goltsman E, Goldman Reznik J, et al (2001) The value of muscle exercise in patients with amyotrophic lateral sclerosis. J Neurol Sci 191:133–137. https://doi.org/10.1016/S0022-510X(01)00610-4
156. Fang J, Zhou M, Yang M, et al (2013) Repetitive transcranial magnetic stimulation for the treatment of amyotrophic lateral sclerosis or motor neuron disease. Cochrane Database Syst. Rev:2013
157. Zanette G, Forgione A, Manganotti P, et al (2008) The effect of repetitive transcranial magnetic stimulation on motor performance, fatigue and quality of life in amyotrophic lateral sclerosis. J Neurol Sci 270:18–22. https://doi.org/10.1016/j.jns.2008.01.011
158. Masrori P, Van Damme P (2020) Amyotrophic lateral sclerosis: a clinical review. Eur J Neurol 27:1918–1929. https://doi.org/10.1111/ENE.14393
159. Zhang X, Hong YL, Xu DS et al (2014) A review of experimental research on herbal compounds in amyotrophic lateral sclerosis. Phyther Res 28:9–21
160. Lascano AM, Lalive PH (2021) Update in immunosuppressive therapy of myasthenia gravis. Autoimmun Rev 20:102712
161. Alcantara M, Sarpong E, Barnett C, et al (2021) Chronic immunoglobulin maintenance therapy in myasthenia gravis. Eur J Neurol 28:639–646. https://doi.org/10.1111/ene.14547
162. Farmakidis C, Pasnoor M, Dimachkie MM, Barohn RJ (2018) Treatment of Myasthenia Gravis. Neurol Clin 36:311–337
163. Kaminski HJ, Alnosair E, Algahtani RM (2018) Clinical trials for myasthenia gravis: a historical perspective. Ann N Y Acad Sci 1413:5–10. https://doi.org/10.1111/nyas.13545
164. Morren J, Li Y (2020) Maintenance immunosuppression in myasthenia gravis, an update. J. Neurol. Sci. 410:116648
165. Brauner S, Eriksson-Dufva A, Hietala MA, et al (2020) Comparison between rituximab treatment for new-onset generalized myasthenia gravis and refractory generalized myasthenia gravis. JAMA Neurol 77:974–981. https://doi.org/10.1001/jamaneurol.2020.0851
166. Mantegazza R, Wolfe GI, Muppidi S et al (2021) Post-intervention Status in Patients With Refractory Myasthenia Gravis Treated With Eculizumab During REGAIN and Its Open-Label Extension. Neurology 96:e610–e618. https://doi.org/10.1212/WNL.0000000000011207
167. Waters MJ, Field D, Ravindran J (2019) Refractory myasthenia gravis successfully treated with ofatumumab. Muscle Nerve 60:E45–E47. https://doi.org/10.1002/mus.26707
168. Mantegazza R, Bonanno S, Camera G, Antozzi C (2011) Current and emerging therapies for the treatment of myasthenia gravis. Neuropsychiatr Dis Treat 7:151–160
169. Jayam Trouth A, Dabi A, Solieman N et al (2012) Myasthenia gravis: a review. Autoimmune Dis 1

170. Jia ZS, Ting WR, Yu YZ et al (2022) Chinese herbal medicine for myasthenia gravis: a systematic review and meta-analysis of randomized clinical trials. Integr Med Res 11:969
171. Thome-Souza S, Klehm J, Jackson M, et al (2016) Clobazam higher-evening differential dosing as an add-on therapy in refractory epilepsy. Seizure 40:1–6. https://doi.org/10.1016/J.SEIZURE.2016.05.014
172. Pernea M, Sutcliffe AG (2016) Clobazam and its use in epilepsy. Pediatr Reports 8:6516. https://doi.org/10.4081/PR.2016.6516
173. Dokkedal V, Scorza FA, Galduroz JCF et al (2016) Epilepsy comorbidities: Is clonazepam a friend or a foe? Epilepsy Behav 62:309–310. https://doi.org/10.1016/j.yebeh.2016.06.031
174. Riss J, Cloyd J, Gates J, Collins S (2008) Benzodiazepines in epilepsy: pharmacology and pharmacokinetics. Acta Neurol Scand 118:69–86. https://doi.org/10.1111/j.1600-0404.2008.01004.x
175. Beydoun A, DuPont S, Zhou D et al (2020) Current role of carbamazepine and oxcarbazepine in the management of epilepsy. Seizure 83:251–263. https://doi.org/10.1016/J.SEIZURE.2020.10.018
176. Grabenstatter HL, Edward Dudek F (2019) Effect of carbamazepine on spontaneous recurrent seizures recorded from the dentate gyrus in rats with kainate-induced epilepsy Present addresses. https://doi.org/10.1111/epi.14680
177. Zhang X, Zhong R, Chen Q et al (2020) Effect of carbamazepine on the bone health of people with epilepsy: a systematic review and meta-analysis. J Int Med Res 48:1–11. https://doi.org/10.1177/0300060520902608/ASSET/IMAGES/LARGE/10.1177_0300060520902608-FIG2.JPEG
178. Galiana GL, Gauthier AC, Mattson RH (2017) Eslicarbazepine acetate: a new improvement on a classic drug family for the treatment of partial-onset seizures. Drugs R D 17:329–339. https://doi.org/10.1007/S40268-017-0197-5/FIGURES/8
179. Jóźwiak S, Veggiotti P, Moreira J et al (2018) Effects of adjunctive eslicarbazepine acetate on neurocognitive functioning in children with refractory focal-onset seizures. Epilepsy Behav 81:1–11. https://doi.org/10.1016/J.YEBEH.2018.01.029
180. Panebianco M, Al-Bachari S, Hutton JL et al (2021) Gabapentin add-on treatment for drug-resistant focal epilepsy. Cochrane Database Syst. Rev. https://doi.org/10.1002/14651858.CD001415.pub4
181. Ijff DM, van Veenendaal TM, Majoie HJM, et al (2015) Cognitive effects of lacosamide as adjunctive therapy in refractory epilepsy. Acta Neurol Scand 131:347–354. https://doi.org/10.1111/ANE.12372
182. Alfaro A, Asensio M, García-Escrivá A, et al (2019) LAM study: Effects of lacosamide on behaviour and quality of life in patients with epilepsy. Neurol (English Ed 34:1–6. https://doi.org/10.1016/J.NRLENG.2018.10.006
183. Hasegawa N, Fukuda M (2020) The effect of lacosamide on psychiatric comorbidities in patients with epilepsy. Epilepsy Behav Reports 14:100402. https://doi.org/10.1016/J.EBR.2020.100402
184. Alabi A, Todd A, Husband A, Reilly J (2016) Safety profile of lamotrigine in overdose. 6:369–381. https://doi.org/10.1177/2045125316656707
185. De Liso P, Moavero R, Coppola G et al (2017) Current role of perampanel in pediatric epilepsy. Ital J Pediatr 43:1–6. https://doi.org/10.1186/S13052-017-0368-6/TABLES/2
186. Trinka E, Steinhoff BJ, Nikanorova M, Brodie MJ (2016) Perampanel for focal epilepsy: insights from early clinical experience. Acta Neurol Scand 133:160–172. https://doi.org/10.1111/ANE.12529
187. Potschka H, Trinka E (2019) Perampanel: Does it have broad-spectrum potential? Epilepsia 60:22–36. https://doi.org/10.1111/EPI.14456
188. Pati S, Alexopoulos AV (2010) Pharmacoresistant epilepsy: from pathogenesis to current and emerging therapies. Cleve Clin J Med 77:457–467
189. Javedan SP, Fisher RS, Eder HG et al (2002) Cooling abolishes neuronal network synchronization in rat hippocampal slices. Epilepsia 43:574–580. https://doi.org/10.1046/j.1528-1157.2002.40101.x

190. Mitchell JW, Seri S, Cavanna AE (2012) Pharmacotherapeutic and non-pharmacological options for refractory and difficult-to-treat seizures. J Cent Nerv Syst Dis 4:JCNSD.S8315. https://doi.org/10.4137/jcnsd.s8315
191. Alqahtani F, Imran I, Pervaiz H et al (2020) Non-pharmacological Interventions for Intractable Epilepsy. Saudi Pharm J 28:951–962
192. Banerjee S, Anand U, Ghosh S et al (2021) Bacosides from Bacopa monnieri extract: an overview of the effects on neurological disorders. Phyther Res 35:5668–5679. https://doi.org/10.1002/PTR.7203
193. Tirumalapura Shalini V, Jambappa Neelakanta S (2021) Sitaram Sriranjini J. Neuroprotection with Bacopa monnieri-A review of experimental evidence. 48:2653–2668. https://doi.org/10.1007/s11033-021-06236-w
194. Shah MS, Bansode SS, Gangarde SB et al (2020) Herbal medicines used in epilepsy: an overview and effect of Covid-19 on epileptic patients. 9:2680–2700. https://doi.org/10.20959/wjpr20207-18030

Pant et al. 2022 (2022) A Review on Endangered Medicinal Plant Nardostachys jatamansi : An Important A Review on Endangered Medicinal Plant Nardostachys jatamansi : An Important Himalayan Herb. Sci Temper 13:81–88. https://doi.org/10.58414/SCIENTIFICTEMPER.2022.13.1.12

195. Irvin L, Jackson C, Hill AL, et al (2019) Skullcaps (Scutellaria spp.): ethnobotany and current research. Med Plants From Farm to Pharm 141–168. https://doi.org/10.1007/978-3-030-31269-5_7/COVER
196. Auditeau E, Chassagne F, Bourdy G et al (2019) Herbal medicine for epilepsy seizures in Asia, Africa and Latin America: a systematic review. J Ethnopharmacol 234:119–153. https://doi.org/10.1016/J.JEP.2018.12.049
197. Biu A, Luka J, Wasagwa J et al (2018) Phytochemistry and toxicity studies on aqueous extracts of leaf and stem of mistletoe (Viscum album) in albino rats. Int J Homeopath Nat Med 4:1–5. https://doi.org/10.11648/j.ijhnm.20180401.11
198. Szurpnicka A, Zjawiony JK, Szterk A (2019) Therapeutic potential of mistletoe in CNS-related neurological disorders and the chemical composition of Viscum species. J Ethnopharmacol 231:241–252. https://doi.org/10.1016/J.JEP.2018.11.025
199. Fernandes TP, Shaqiri A, Brand A et al (2019) Schizophrenia patients using atypical medication perform better in visual tasks than patients using typical medication. Psychiatry Res 275:31–38. https://doi.org/10.1016/J.PSYCHRES.2019.03.008
200. Patel KR, Cherian J, Gohil K, Atkinson D (2014) Schizophrenia: overview and treatment options. P T 39:638–645
201. Citrome L, McEvoy JP, Todtenkopf MS et al (2019) A commentary on the efficacy of olanzapine for the treatment of schizophrenia: the past, present, and future. Neuropsychiatr Dis Treat 15:2559. https://doi.org/10.2147/NDT.S209284
202. Preda A, Shapiro BB (2020) A safety evaluation of aripiprazole in the treatment of schizophrenia. 19:1529–1538. https://doi.org/10.1080/14740338.2020.1832990
203. Carrithers B, El-Mallakh RS (2020) Transdermal Asenapine in Schizophrenia: A Systematic Review. Patient Prefer Adherence 14:1541. https://doi.org/10.2147/PPA.S235104
204. Ganguly P, Soliman A, Moustafa AA (2018) Holistic Management of Schizophrenia Symptoms Using Pharmacological and Non-pharmacological Treatment. Front Public Heal 6:166. https://doi.org/10.3389/FPUBH.2018.00166/XML/NLM
205. Ali SA, Mathur N, Malhotra AK, Braga RJ (2019) Electroconvulsive therapy and schizophrenia: a systematic review. Complex Psychiatr 5:75–83. https://doi.org/10.1159/000497376
206. Sinclair DJM, Zhao S, Qi F et al (2019) Electroconvulsive therapy for treatment-resistant schizophrenia. Cochrane Database Syst Rev 2019. https://doi.org/10.1002/14651858.CD011847.PUB2/MEDIA/CDSR/CD011847/IMAGE_N/NCD011847-CMP-007-01.PNG
207. Onaolapo OJ, Onaolapo AY (2021) Nutrition, nutritional deficiencies, and schizophrenia: An association worthy of constant reassessment. World J Clin Cases 9:8295. https://doi.org/10.12998/WJCC.V9.I28.8295

208. Tsigalou C, Konstantinidis T, Paraschaki A et al (2020) Mediterranean diet as a tool to combat inflammation and chronic diseases. An overview. Biomedicines 8. https://doi.org/10.3390/BIOMEDICINES8070201
209. Deng H, Adams CE (2017) Traditional Chinese medicine for schizophrenia: a survey of randomized trials. Asia-Pacific Psychiatr:9
210. Rathbone J, Zhang L, Zhang M et al (2005) Chinese herbal medicine for schizophrenia. Cochrane Database Syst Rev 2005. https://doi.org/10.1002/14651858.cd003444.pub2
211. Zirak N, Shafiee M, Soltani G et al (2019) Hypericum perforatum in the treatment of psychiatric and neurodegenerative disorders: current evidence and potential mechanisms of action. J Cell Physiol 234:8496–8508
212. Michel HE, Tadros MG, Abdel-Naim AB, Khalifa AE (2013) Prepulse inhibition (PPI) disrupting effects of Glycyrrhiza glabra extract in mice: a possible role of monoamines. Neurosci Lett 544:110–114. https://doi.org/10.1016/j.neulet.2013.03.055

Chapter 7
Antioxidants Mitigate Oxidative Stress: A General Overview

Javeria Umber, Muhammad Qasim, Saba Ashraf, Usman Ali Ashfaq, Mahmood-ur-Rahman, Arshia Iram, Rashid Bhatti, Muhammad Tariq, and Muhammad Shareef Masoud

7.1 Introduction

Oxidative stress (OS) is a term used to describe the condition that happens due to the imbalance between antioxidants and oxidants. The issue mainly arises due to the inappropriate functions of the antioxidant system or because of the increasing amount of reactive oxygen species (ROS) [1]. ROS are generally formed in less amount in mitochondria during respiration and act as requisite to perpetuate signal transduction, and migration, ensuring cell viability, proliferation, receptor activation, gene expression, and differentiation pathogen recognition, [2]. Neurotransmission, blood vessel relaxation, host defense, and wound healing, need ROS/RNS for cell signaling and various biological functions [3]. ROS are enzymatically generated to kill microbes, in macrophages, and it was demonstrated that the cells failed to distinguish between foreign bodies and host cells, and their high generation results in the degradation of host cells and respective organs [4, 5]. Disproportionate nitrogenated species/reactive oxygenated generation, which undermines the defense mechanism of the organism, led to the loss of the cellular buffering system to perpetuate the redox equilibrium characterize by oxidative stress and cause

J. Umber · M. Qasim · S. Ashraf · U. A. Ashfaq · Mahmood-ur-Rahman · A. Iram
M. S. Masoud (✉)
Department of Bioinformatics and Biotechnology, Government College University, Faisalabad, Pakistan
e-mail: masoudshareef@gcuf.edu.pk

R. Bhatti
Centre of Excellence in Molecular Biology, University of the Punjab, Lahore, Pakistan

M. Tariq
Department of Biotechnology, Mirpur University of Science and Technology, Mirpur, Azad Jammu and Kashmir, Pakistan

A. Imran, G. Hussain (eds.), *The Role of Natural Antioxidants in Brain Disorders*, Food Bioactive Ingredients, https://doi.org/10.1007/978-3-031-41188-5_7

pathogenesis[6]. Oxidative stress is closely related to cancer expansion, cardiovascular diseases, melanogenesis, kidney disease, diabetes, rheumatoid arthritis, and eye disease.

Hydrogen peroxide (H_2O_2), hydroxyl (OH), peroxynitrite (ONOO), and superoxide (O_2) are the most prevalent reacting oxygen species and reacting nitrogen species. [7]. The presence of endogenous as well as exogenous sources of ROS and RNS have been seen in the system. Mitochondria is the primary way of ROS production, during both pathological and physiological situations many environmental contaminants such as diesel exhaust particles, cigarette smoke, and asbestos, are harmful oxidants in the environment and contain electrophiles, carcinogens, free radicals, including quinones, as well as polycyclic aromatic hydrocarbons, that cause oxidative stress.

In this review, the focus is to debate the basic pathophysiological mutual relation between the emergence of neurodegenerative diseases and oxidative stress. Nearly 30 million people are affected by neurodegenerative diseases, which can cause death and disability. The brain eats up a lot amount of oxygen for regular operation and could be called a reservoir of ROS. ROS remarkably speeds up the degeneration of neuronal cells by controlling the activity of the biomolecule that helps to reduce the degeneration [8]. Necrosis or Apoptosis and deterioration of nerve cells are the key feature of neurodegenerative disorders, which have a detrimental impact on the nervous system. Although oxygen is vital for life it also negatively affects the biomolecules in the free radical form. These free radicals attack several biomolecules like lipids, DNA, proteins, and RNA and proceed to nucleic acid oxidation as well as peroxidation of lipids in the cell [8]. PUFA(polyunsaturated fatty acid) is found in high amounts in brain cell membranes so these fatty acids are more vulnerable to lipid peroxidation [9, 10]. The ROS that contributes to neurodegeneration includes the superoxide anion (O2), the highly reactive hydroxyl radical (HO•), and hydrogen peroxide (H2O2). Nitric oxide (NO) is among one the active nitrogen species (RNS) that also harm neurons and previous studies are rich evidence of the fact that there is a significant relation between neurodegenerative diseases and OS [11–14].

Therefore, establishing an antioxidant barrier is needed to control the amount of ROS/RNS to an equilibrium that is not alarming to the biological system. Antioxidants, a substance that inhibits oxidation, play a crucial role as scavengers that can regulate the ROS/RNS level at equilibrium.

Almost many species of organisms are well-protected as opposed to damage via free radicals either by compounds and enzymes, for example, α-tocopherol, glutathione, and ascorbic acid that helps to protect the human body from oxidation. Several reports have been reported related to antioxidant activity in vegetables, herbal compounds, and fruits [15]. Enzymes, for example, superoxide dismutase (SOD) and catalase (CAT), act as important factors in the process of the antioxidant defense system, these are liable for decreasing the level of ROS [16, 17]. CAT is responsible for the conversion of H_2O_2 into nontoxic substances, oxygen, and water while SOD work as a catalyst in the transformation of superoxide anion into $O_{2,}$ and H_2O_2required for proper functioning [18, 19].

Vitamin E and flavonoids are the most relative natural antioxidants and protect against oxidative stress. Flavonoids are polyphenolic groups found in plants that contain benzo-γ-pyrone which have pharmacological activities [20, 21]. Their action is influenced by the hydroxyl group which is a scavenger to free radicals [20, 22, 23]. Flavonoids determine ROS synthesis suppression, scavenging ROS, and up-regulate antioxidant defenses [24, 25].

Vitamin E contains a group of lipophilic molecules amalgamated by plants and is also present in seeds, edible oil, and artificially enriched in food [26–28]. Vitamin E regulates oxidative stress – upregulated, ox-LDL- induced foam cell formation, NF-κB pathway, matrix metalloprotease expression, and decrease c-jun phosphorylation [29–31].

7.1.1 Early Leads, Concept, and Some Aspects

The idea of oxidative stress and oxidative stress responses was formulated in 1985 in the book entitled "Oxidative Stress" [32]. In the textbook, a chapter is devoted to oxidative stress "Free Radicals in Biology and Medicine",[24]. Most important early contributions came from different fields of applied research such as food chemistry, and rubber chemistry. *Oxidative stress is defined as "an imbalance between oxidants and antioxidants in favor of the oxidants, leading to a disruption of redox signaling and control and/or molecular damage"*[33].

According to intensity, stress can be divided into oxidative stress (good stress) and oxidative distress(bad stress). Oxidative distress is characterized by an excessive buildup of reactive oxygen species (ROS)]. Whereas eustress is characterized by levels of ROS required for healthy redox signaling, which is crucial for normal cellular/tissue/organism functioning [34]. There are several endogenous and exogenous sources of oxidative stress. External factors such as radiation, pollution, medication, and cigarette smoke can disturb the ROS equilibrium while Intracellular sources include oxidases (NOX), mitochondria, NADPH (nicotinamide adenine dinucleotide phosphate) endoplasmic reticulum, lysosomes, peroxisomes, cytochrome P450, and others.

Oxidative stress and the onset of diseases like cancer, kidney disease, neurodegeneration, arthritis, cardiovascular diseases, diabetes, eye disease, and psoriasis are closely related. About 3% of the world's population suffers from psoriasis (Ps). It is important to understand the exact molecular mechanism underlying this disease. According to the most recent studies on the pathophysiology of Ps, oxidative stress is one of the major risk factors for this skin disease [35, 36].

7.1.1.1 Practical Aspects

Different cell types and organs exhibit oxidative stress in varying degrees and ways.

Table 7.1 oxidative stress biomarkers classification in terms of different classes [43]

Types	Biomarker
Type#0	For measuring ROS directly (e.g. H_2O_2)
Type#1	Hydroxynonenal, oxidized low-density lipoprotein, 8-oxo-deoxyguanosine, protein carbonyls, isoprostanes
Type#2	Allatonin, hypochlorous acid, uric acid
Type#3	Paraoxonase1,vitamin E and C, billirubin, catalase, dual oxidase
Type#4	Mutation and genetic factors

Clinical Aspects
As presented in relatable reviews, there are massive studies regarding oxidative stress expression in a clinical setting. One of the most reliable and promising newly diverted areas for research relates to the link between psychological stress as well as oxidative stress [37, 38]. In the clinical aspect, the future reporting is on the following topics, reactive oxygen species in mechanisms of aging and lifespan regulation, cardiovascular system, inflammation, immune system, wound repair, skeletal muscle; central nervous system, sensitivity against insulin, diabetes pathogenesis, cancer; and prospects for redox medicine [39, 40].

Biomarkers
Numerous oxidative stress biomarkers have been discovered [41, 42]. In Table 7.1, there are some biomarkers involved in various diseases.

7.1.2 *Mechanism of Oxidative Stress*

Even though oxygen is considered vital to life and required for the transcription of genes, signaling, along with additional cellular operations, as well as ultimately damages biomolecules by producing reactive oxygen species (ROS). Oxygen has negative effects due to its monovalent metabolic reduction status that results in the production of ROS. Nitric oxide (NO), is the second-most significant species that cause oxidative stress [8]. The primary source of the generation of ROS is oxidative stress which will affect the cells negatively. Cyclooxygenases (COX) and lipoxygenases that are involved in cellular respiration can produce O_2 since mitochondria are thought to be the main source of ROS in both normal and pathological situations (LOX). During the metabolism of arachidonic acid, endothelial and inflammatory cells there is significant production of oxygen free radicals [44]. Even though these mentioned organelles are capable of scavenging ROS on their own [45].

The processing and production of the respiratory chain, phagocytosis, prostaglandin, and P450 cytochrome pathway are among the reactions that work enzymatically and can produce ROS [46–50]. NADPH oxidase, xanthine oxidase, and peroxidases produced superoxide radicals (O_2•). Once created, it participates in many processes that result in the production of several chemicals, including hypochlorous acid (HOCl), peroxynitrite (ONOO), OH• hydroxyl oxide, and hydrogen peroxide. A series of oxidase enzymes, specifically xanthine and amino acid oxidase produces H_2O_2. The reaction of O_2• with H_2O_2, which generates the highly reactive radical among all free radical species that cause oxidative stress is hydroxyl radicals, highly responsible for ROS effects [46, 51, 52]. Nitric oxide synthase (NOS) transforms arginine into citrulline amino acid, generating the oxide of nitric acid, (NO•) free radical which has physiological functions [48, 52, 53].

Even nonenzymatic reactions could generate free radicals, such as in the reaction of oxygen with hydrocarbon organic materials or in a condition in which cells are exposed to ionizing radiation. During mitochondrial respiration, non-enzymatic free radical generation can also take place [49, 53]. Both internal and external processes can produce free radicals. Endogenous free radical generation is caused by aging, infection, cancer, mental stress, ischemia, immune cell activation, and inflammation and extracellular free radical generation can happen due to exposure to toxins present in the environment, heavy metals (Pb, As, Hg, Fe, and Cd), specific medications (cyclosporine, gentamycin, tacrolimus, bleomycin), chemical solvents, cigarette smoke, alcohol, radiation, and cooking includes (smoked meat, used oil, and fat) [53–55]. The entry of extracellular substances into the body led to breakdown or metabolism due to different mechanisms, resulting in the production of different substances as well as a byproduct, and free radicals are produced.

Limited numbers of ROS are necessary for maintaining cell homeostasis. However, excessive production of ROS/RNS that surpasses the antioxidant barrier's capacity disrupts the pro-/antioxidant balance, which ultimately leads to the emergence of the OS state. Distinct macromolecules (DNA, RNA, lipids, and proteins) and cellular activities are targeted by different ROS species. Oxidative stress results in the carbonylation of large masses of molecules for example proteins, lipids, and DNA at the cellular level [56]. The incapability of buffering system of body cells to keep the oxidation-reduction balance, results in a variety of biomolecule changes that are a hallmark of disease conditions.

7.1.3 Oxidative Stress as a Primary and Secondary Contributor

Numerous disorders, including cancer, atherosclerosis, Alzheimer's disease, and chronic obstructive pulmonary disease (COPD), have been linked to oxidative stress. While oxidants can harm cells in a variety of ways, the contribution of oxidative stress to disease etiology and treatment differs greatly. Additionally, in other circumstances, the increasing antioxidant defense may be ineffective because it may make the illness worse.

Oxidative Stress causes disease primarily through two different pathways, first one is macromolecule oxidation through direct methods, such as enzymes, lipids membrane nucleic acids, and proteins involved in structural development, by the generation of ROS during oxidative stress. The reactive oxygen species especially •OH, ONOO, and HOCl, are among those free radicals that are involved in abnormal cell function and death of cells [57]. Abnormal redox signaling is the other way of oxidative stress production states that the oxidants, especially H_2O_2, function as a second messenger that is produced as a result of physiological stimuli. The generation of oxidative stress in the non-physiological generation of H_2O_2 could result in the malfunctioning of redox signaling [57]. In diabetes, both types of oxidative stress pathways can happen which involve the accumulation of glycation products as well as activation of aberrant stress signaling mechanisms resulting in diabetic complications [58].

According to the etiology of the oxidative stress contributor, many disorders have been linked to it and are divided into two categories: first, oxidative stress is the major cause of pathology, and second, as the secondary contributor to the start of infection.

7.1.3.1 Oxidative Stress as the Main Pathological Factor

Oxidative stress is considered a major contributor to toxicity and disease but there are crucial caveats: if tissue damage has started, antioxidant therapy is frequently difficult to stop it from development because other pathological processes become dominant.

Atherosclerosis

Plaque accumulates in the innermost/outer layer of arteries due to atherosclerosis and after a long-term buildup the arteries narrow, causing infarction and stroke. Previous studies are evidence that oxidative stress acts as a key factor in the development of atherosclerosis. Numerous types of research have demonstrated the upregulation of oxidized lipids and atherosclerotic lesions in other oxidative stress markers [57]. One study suggested that 20% of cholesteryl linoleate in recently separated human disease was oxidized in comparison with healthy arteries where it is present in nearly diminished quantity [59]. Additionally, it was discovered that the plasma of patients with atherosclerosis had upregulated levels of HNE-modified low-density lipoprotein (LDL) that were 50% higher than those of healthy volunteers [60].

Furthermore, oxidized linoleic acid was solely found in human lesions, and isoprostanes, peroxidation products of arachidonic acid, are fivefold increase by comparing the human atherosclerotic lesions with human umbilical veins [61]. Oxidative stress performs the transformation of LDL cholesterol into the pathogenic form of oxidized LDL (OxLDL). This process is important in triggering and enhancing the inflammation responses and attracting the accumulation and activation of white blood cells to the site of the lesion [62].

Radiation-Induced Lung Injury

The treatment of malignancies of the lungs and esophagus through radiotherapy frequently generates the onset of pneumonitis and fibrosis as side negative effects of the process that is quite dangerous [63]. As a result of the hemolytic cleavage of H2O that occurs when radiation passes through the cell, •OH is produced, which oxidizes macromolecules, and causes inflammation which leads to drastic effects and results in the death of body cells along with the invasion of inflammatory cells into the lung (pneumonitis). The long-term buildup of collagen and lung fibrosis is brought on by abnormal redox signaling for the ongoing generation of cytokines [64]. Additionally, radiation generate lung injury in rats has been linked to greater levels of lipid peroxidation and DNA oxidation which can last for months [65].

Paraquat Toxicity

Oxidative stress is liable to harmful properties of the paraquat, a most applied chemical herbicide, and is instantly used up by alveolar type II cells after ingestion, which causes pneumonitis and gradual fibrosis of the lungs with a dreadful prognosis. The liver and kidney are among the other organs damaged by paraquat. Parkinson's disease is also linked to paraquat exposure over an extended period [66]. The ongoing redox cycling that results in the production of O2• is what starts the paraquat toxicity [67].

7.1.3.2 Oxidative Stress as a Secondary Contributor to the Progression of Disease

Oxidative stress act as secondary to the progression of pathology in most of the diseases by other factors such as it includes the oxidative stress brought on by increased O_2• /H_2O_2 production through NADPH oxidases (NOXs) during the inflammatory with the help of xanthine oxidase pursue early tissue injury during ischemia-repercussion [57]. By changing the proteins, displacing autophagy, fostering inflammatory response, triggering cell death, and malfunctioning mitochondrial function, along with other methods, oxidative stress could disturb several signaling mechanisms and affect numerous biological mechanisms. As stated in the situations below, these consequences usually have pathological development and aggravate disease symptoms.

Hypertension

Hypertension is caused by several risk factors, including nutrition, lifestyle genetics, comorbidities, and smoking. In more than 90% of cases, the cause of essential hypertension is unknown. However, oxidative stress is a prevalent aspect of this illness at the molecular level. According to experimental research, NOXs are the principal source of oxidants in hypertension [68]. Hypertension patients have significantly higher levels of oxidative indicators in their plasma, including H2O2, malondialdehyde, and 8-isoprostanes [69].

Alzheimer Disease

Alzheimer's disease is expressed through the continuous aggregation of neurofibrillary tangles and exogenous amyloid-β plaques inside neurons [57] and factors like age, trauma, genetics, sex, and air pollution are some of the risk factors for Alzheimer's disease, although actual reason is still unknown [57]. Numerous studies have shown that Alzheimer's patients experience more oxidative stress in their brains. These studies show elevated levels of F2-isoprostane-, acrolein, and HNE in the amygdala, hippocampus, inferior parietal lobule, as well as in the cerebrospinal fluid, frontal and temporal poles, and cortex [70]. Three lobes of the brain of Alzheimer's disease patients showed the presence of higher levels of mitochondrial DNA as well as nuclear oxidation. Additionally, protein carbonyls in the cerebral cortex and protein oxidation in the hippocampus are both involved in up-regulated in the brains of patients with Alzheimer's [71, 72].

Cancer

Oxidants play role in several stages of carcinogenesis, such as the conversion of healthy cells into tumor cells, aggressive cell growth, expansion, invasion, angiogenesis, and metastasis. On the other hand, oxidative stress leads to cell death and ferroptosis, which decreases the chances for conversion and stops the generation of tumor cells [73]. Consequently, oxidative stress is connected to practically every stage of cancer, as it generates more radicals than healthy cells which exposed them to more oxidative stress from the environment.

The extra oxidants involved in cancer cells are generally derived from mitochondria [74]. The healthy cells around the tumor body like inflammatory immune and endothelial cells may also provide oxidants to the locus. Different cancer forms have been linked to an increase in oxidative indicators. For instance, it has been demonstrated that lung cancer patients exhale more H2O2 than control subjects [75].

7.1.4 Oxidative Stress and Neurological Disorder

20% of the entire baseline oxygen (O2) is used by the human brain to support ATP-intensive neuronal activity [76]. The function of the central nervous system (CNS) depends on oxidation/reduction signaling because of the requirement of high oxygen amount and energy in the form of adenosine triphosphate (ATP) which is the most commonly utilized energy in the body [76]. It also contains significant amounts of iron and polyunsaturated fatty acids, which weaken its antioxidant defenses and make it more vulnerable to oxidative stress due to disruptions in the oxidation-reduction equilibrium [77]. Numerous neurodegenerative conditions, including Parkinson's disease (PD), Huntington's disease (HD), and Alzheimer's disease (AD), have been connected to increased reactive species generation and a lack of an antioxidant defense mechanism in the body [78]. Brain oxidative stress involves the peroxidation of lipids, and produces reactive aldehydes, such as 4-hydroxynonenal (4-HNE), which performed a crucial function in the emergence of different diseases, for example, AD and Parkinson's disease [79]. While 4-HNE is also physiologically synthesized by glial cells, undoubtedly neurotoxic and disrupts the blood-brain barrier in case of drastic damage [79]. The majority of lipid peroxidation in the brain is linked to ferrotopsis, which causes neurodegenerative diseases [80].

Numerous experimental and clinical research on AD has shown that oxidative stress plays a significant role in neuronal decline and dementia development [81].

Neurodegeneration is known to be caused, at least in part, by the toxic peptide -amyloid, which is frequently found in the brains of AD patients [82].

7.2 Antioxidants; Positive or Negative Actors

Antioxidants are compounds that are known t inhibit oxidation. A chemical process called oxidation has the potential to generate free radicals, which can set off a series of events that harm an organism's cells [83]. Antioxidants had rising attention due to their preventive effects in pharmaceutical and food products against oxidative stress as well as against disease processes caused by oxidative stress in human bodies. They are known to reduce, delay, or even remove oxidative damage [84]. Scientists have recently grown interested in understanding the mechanisms of action of various antioxidants due to the significance of protecting opposed to the effects of reactive nitrogen species and reactive oxygen species. Pro-oxidants and antioxidants produced from oxygen are now better understood to have important roles in both healthy metabolism and several clinical disease states. Antioxidants occasionally display pro-oxidant activity, depending on the particular circumstances. Their dose and redox conditions within the cell are of particular significance [85].

7.2.1 Early Concepts, and Classification of Antioxidants

7.2.1.1 Early Concepts

Originally, an agent that prevents the consumption of oxygen was referred to as an antioxidant. Numerous research accompanied in the 19th and 20th centuries demonstrated the usefulness of antioxidants in key industrial processes, such as the cyclization of fuels in the smear of internal combustion engines, the vulcanization of rubber, and the avoidance of metal corrosion [86]. Gershman presented the "free radical theory of oxygen toxicity" in 1954, which states that oxygen's ability to produce free radicals is what causes it to be poisonous. Loschen first suggested in 1971 that cellular metabolic respiration is where reactive oxygen species are produced. By activating the enzyme guanylate cyclase, Mittal and Murad discovered in 1977 that the hydroxyl radical, OH, causes the second messenger, cyclic GMP, to form. Reactive oxygen species (ROS), according to research by Halliwell and Gutteridge published later in 1989, include both free radicals and non-radical derivatives of oxygen. The role of reactive oxygen species in numerous pathophysiological situations has since been extensively studied [85, 87, 88]. Early studies on the function of antioxidants in biology focused on how to use them to prevent rancidity-causing oxidation of unsaturated lipids. By simply placing the fat in a container with oxygen and tracking the rate of oxygen consumption, the anti-oxidant effect can be easily calculated. However, the discovery of the antioxidant properties of the vitamins C, E, and A transformed the area and showed how crucial antioxidants are to the biochemistry of living things [89].

7.2.1.2 Classification of Antioxidants

Antioxidants are produced by the body in many ways to combat oxidative stress, either internally from natural sources (endogenous antioxidants) or externally from meals (exogenous antioxidants). Antioxidants have three main functions: counterbalancing additional ROS or free radicals, defending cells from their harmful effects, and helping to prevent disease. Generally, antioxidants are categorized into two main classes that are enzymatic and non-enzymatic antioxidants. Among them, many non-enzymatic antioxidants are being used nowadays as prescription drugs and nutritional supplements [90].

Primary Antioxidants vs Secondary Antioxidants

Primary antioxidants are those that work by hunting radical species and turning them into stable radicals or non-radical species. They have also referred to chain-breaking antioxidants (known to retard initiation or prevent propagation). Secondary antioxidants either absorb UV rays or quench singlet oxygen. They also convert peroxides into non-radical species, bind pro-oxidative ions of metal, and restrict oxidizing enzymes like lipoxygenase. When used in conjunction with the main antioxidants, they might display synergistic benefits. They bind metal ions, renew primary antioxidants by donating hydrogen, stabilize primary antioxidants by producing an acidic environment, quench molecular oxygen, or regenerate primary antioxidants [91] (Fig. 7.1).

Enzymatic Antioxidants

In cells, various antioxidant enzymes are present and work to balance out the harmful effects of cellular oxidative stress. Superoxide dismutase, catalase, and the glutathione system are the three main enzymatic antioxidants [92]. Table 7.2 summarizes enzymatic antioxidants and their applications:

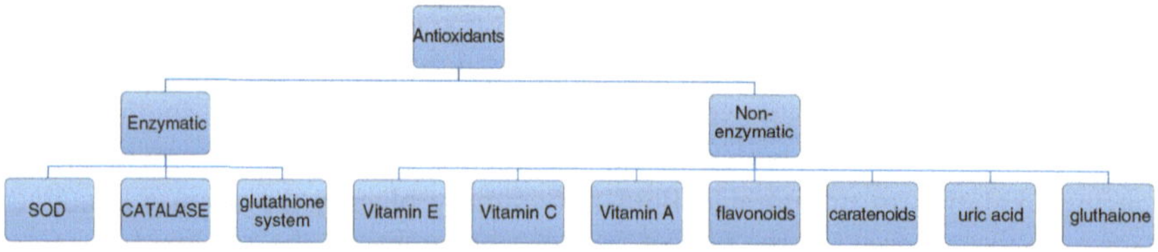

Fig. 7.1 Classification of antioxidants

Table 7.2 Enzymatic antioxidants

Antioxidant	Abbreviation	Applications	References
Super oxide dismutase	SOD	Triggers cellular repair accelerates wound healing, and helps to reduce oxidative damage	[93]
Catalase	CAT	Used to prevent food from oxidation inside wrappers, used in textile industry to remove peroxide from fabric, in aesthetic industries to make facial masks	[93]
Gluta-thione per-oxidase	GPx	Important antioxidant in body, play important role in immune regulation by keeping red blood cell intact and protects white blood cells, protective role for brain	[93]

Table 7.3 Non-enzymatic antioxidants

Antioxidant	Applications	References
Glutathione (γ-glutamyl-cysteinyl-glycine)	Therapeutic applications in aging and neurodegenerative disorders and Parkinson's disease	[94]
Vitamin-E	Have a role to protect lipid peroxidation, increases autoantibody levels against oxidized low-density lipoprotein, and helps to prevent cardiac diseases.	
Vitamin C	It plays a crucial role in the treatment of oxidative stress and several other physiological problem like degenerative diseases and autoimmune diseases	[95]
Carotenoids	Act as chemopreventive agents, also include in ROS scavenging	[96]
Flavonoids	Involved in quenching of ROS formation, also inhibits several enzymes, involved in various chain reactions	
Alpha Lipoic acid	Regenerate Vitamin C, E, and GSH, have a role in metal chelation	[97, 98]
Melatonin	A potent antioxidant protects the cell membrane from lipid peroxidation	

Non-enzymatic Antioxidants

Non-enzymatic antioxidants are crucial during the fight against oxidative stress. The majority of them, such as vitamin E, flavonoids, carotenoids, and vitamin C. However, some of the physiological antioxidants, such as glutathione, uric acid, and others, are produced by the cell itself. The following table provides a concise overview of non-enzymatic antioxidants and their applications (Table 7.3).

7.2.2 Mechanism

For antioxidants, two main modes of action have been reported. The basic principle is that an antioxidant gives an electron to a free radical in the system in the first phase, breaking the chain mechanism. The second method is quenching chain starting catalyst eliminating reactive nitrogen species/reactive oxygen species initiators. Co-antioxidants, chelation of metal ions, donation of electrons, and expression of genes as well as regulation of genes are some of the different ways that antioxidants can affect biological systems [99, 100].

As lipid peroxidation occurs in food or cell membranes, which includes the subsequent processes of chain initiation, chain progression, and termination of chain process, antioxidants can function at various phases of oxidative progression of radicals. It was stated that when oxygen is added to carbon-centered radicals, which takes place at or near the diffusion-controlled pace, initiates the peroxidation's propagation stage. Mostly oxidation involves a radical mechanism, the propagation happens at a typically slow rate and is symbolized by the addition of a hydrogen atom to the chain that is carrying the peroxyl radical. Free radicals from peroxyl compounds can combine with carbon-carbon double bonds. Particularly vulnerable to peroxyl addition could be conjugated dienes. Polyunsaturated lipids take part in peroxidation processes, although intramolecular radical substitution on peroxide and radical cyclization reactions might happen and produce cyclic peroxides. Antioxidants can act by hunting chain-starting radicals like alkoxyl RO, peroxyl ROO, hydroxyl OH, quenching singlet oxygen, breaking the chain of radical sequence, trapping aggressive reactive oxygen species like the hydrogen peroxide, superoxide anion radical, or any combination of these, and removing pro-oxidative metal ions [101, 102].

Preventive antioxidants ate those antioxidants that help to prevent lipid peroxidation by eliminating oxygen, quenching oxygen to reduce its concentration as well as removing pro-oxidative metal ions. Since they may be reduced catalytically by enzymes (such as glutathione peroxidase, superoxide dismutase, and CAT), those that can do so are also preventative and are not destroyed in the reaction. On the other hand, while performing their protective functions, antioxidants that act to break chain reactions, oxygen quenchers, and those that are involved in the chelation of metal use up their energy. In most circumstances, multiple modes of action can be followed by the same antioxidant [101]. It has been determined that primary antioxidants are antioxidants that break the chain and can scavenge radical species. Examples of secondary antioxidants include peroxide decomposers, singlet oxygen quenchers that generate non-radical species, oxidative enzyme inhibitors (like lipoxygenase), UV radiations absorber, and metal chelators [103, 104].

When combined with primary antioxidants, secondary antioxidants may demonstrate synergistic effects through one of the following mechanisms:

- Establishing an acidic environment to stabilize main antioxidants.
- Hydrogen donation for the regeneration of primary antioxidants
- Chelating transition metal cations that are pro-oxidative.
- putting out molecules of oxygen

The synthesis or renewal of non-enzymatic antioxidants has also been reported to be facilitated by antioxidant enzymes [91].

7.2.3 Defense Mechanisms of Antioxidants

The ability of an electron to neutralize any ROS or free radical is the foundation of the antioxidant activity principle. Additionally, the quantity and type of the pattern of hydroxylation on the aromatic rings are associated with antioxidant action [84]. Included in the defense systems against oxidative damage brought on by free radicals are: Examples of these processes: 1) the elimination of free radicals and ROS by the catalytic activity of enzymes like CAT, SOD, and per-oxidase, antioxidants; 2) the binding of proteins to pro-oxidant metal ions like copper and iron by transferrin, metallothionein, haptoglobins, and ceruloplasmin 3) the elimination of free radicals(ROS) by electron donors including glutathione, vitamin C, vitamin E), uric acid and bilirubin; protection against damage at the macromolecular level by proteins like heat shock or stress proteins [105].

The defense systems' antioxidants function on a variety of levels, including the fourth line of defense, or adaptation, as well as preventive, radical scavenging, repair, and de novo [89].

1st level of defense: the 1st line of defense includes those inhibitory antioxidants which stop the production of reactive oxygen species. Even though the specific method and location of radical generation in vivo are still not fully understood, one of the major sources must be the breakdown of hydro-peroxides and hydrogen-peroxide induced by metals.

2nd level of defense: In 2nd line of defense includes those antioxidants that are involved to prevent or halt the propagation of chain reactions or even used to stop these reactions by scavenging the active radicals. Numerous antioxidants are endogenous by nature and involved in scavenging the active radicals; among them some are lipophilic and others are hydrophilic. Examples of lipophilic antioxidants include ubiquinol and vitamin E, while others like bilirubin, thiols, albumin, uric acid, and ascorbic acid are hydrophilic antioxidants. Considered to be the most effective lipophilic antioxidant for scavenging free radicals is vitamin E.

3rd level of defense: The restoration and de novo antioxidants are the 3rd level of defense of antioxidants. The proteases, peptidases, and proteinases, which are present in the mitochondria as well as cytoplasm of cells of mammals, identify, break down, and eliminate proteins that are oxidatively as well as prevent the buildup of oxidized proteins [89].

7.2.4 Clinical Significance of Antioxidants

Chronic inflammation or chronic exposure to environmental stimuli would result in the production of ROS (and byproducts), which would operate as a constant source of DNA changes leading to mutations and raising the risk of cancer [106]. Smoking cigarettes is a significant contributor to oxidative stress that can be avoided, and it increases the risk of lung cancer and atherosclerosis. Numerous neonatal health issues are associated with maternal smoking during pregnancy. Premature retinopathy is exacerbated by oxygen therapy given to infants who are born prematurely or with very low body weight. Little ones are subjected to greater oxygen consumption and corresponding rises in ROS at birth. Children that are born prematurely are more susceptible to oxygen poisoning. When supplementary oxygen is eliminated, the retinal microvasculature is first blocked before accelerating [107]. A major risk factor for age-related macular degeneration is ROS production. Dry AMD is the most prevalent kind and is treated with diligent monitoring and antioxidant therapy. Several bacteria populate the gastrointestinal system, especially the large intestine, and can produce a lot of ROS, which may be linked to inflammatory bowel disorders and colorectal cancer. Neurological disorders are also impacted by ROS. Cu-Zn/SOD1 point mutations have been closely bound to familial ALS (amyotrophic lateral sclerosis) [108–110]. An LWHS(Iowa Women's Health Study) discovered that aged women using supplements with antioxidant activity have a higher rate of mortality, even though antioxidants are considered to accelerate the breakdown of ROS. As a result, it is unclear what clinical function dietary antioxidants or antioxidant supplements play in preventing disorders linked to OS [110].

7.3 Antioxidants and Oxidative Stress; How Antioxidants Influence Oxidative Stress

Emerging study data suggests that antioxidants can intercept the production and prevention of free radicals that result in a decrease in oxidative stress, boost immune system performance, and lengthen healthy lifespan by controlling autoxidation. These substances aid in quenching O2, quenching the auto-oxidation chain reaction, scavenging the species that cause peroxidation, and inhibiting the synthesis of peroxides. The capability of antioxidant to intervene in the free radical chain reactions are the most potent. During oxidation, these antioxidants can contribute H• to the free radicals because of the phenolic or aromatic rings that they contain [111]. In the aging of human brains, increasing oxidative damage has been noted. According to epidemiological research, consuming antioxidants may help to enhance cognitive abilities. Contrary to good in vivo results, several observational studies in people have revealed no appreciable reduction in cognitive deterioration related to dietary antioxidant intake. The disparate findings may be partially explained by taking insufficient amounts of antioxidant supplements, their sources and form, their period of usage and regularity, the difficulties in estimating the necessary optimum amount of food intake of antioxidants, and the person's lifestyle [112]. In a natural infection, on the action of the immune system, in autoimmunity, hypersensitivity, etc., the equilibrium among the quantity of reactive nitrogen and reactive oxygen species and antioxidants is crucial. RNS/ROS have a direct role in the innate immune system's ability to destroy infectious agents, and it has been discovered that administering too many antioxidants interferes with the immune system's ability to respond. However, it has been discovered that neurodegenerative illnesses are linked to elevated RNS/ROS and, consequently, the OS. The autoimmune diseases exhibit elevated OS and noticeably low levels of endogenous antioxidants. Therefore, dietary antioxidant supplementation can be employed therapeutically to lessen the pathophysiological conditions induced by oxidative stress [113]. Oxidative stress irreversibly damages cellular macromolecules and triggers the progression of some physiological problems such as diabetes, the start of carcinogenesis, atherosclerosis, ischemic heart disorders, and liver disorders. Antioxidants prevent the synthesis of ROS and the removal of free radical [114].

7.4 Conclusion

The presence of ROS and antioxidants must typically coexist in a delicate balance for the cell to operate normally. ROS are produced throughout the cell, particularly in the mitochondria. By not only neutralizing free radicals but also protecting them from starting the chain reactions that cause a variety of health problems and premature aging, antioxidants fight oxidative stress. A natural antioxidant contributes to scavenging reactive oxygen species and upholding the typical environment within the cells when things are normal. Initiation of OS (oxidative stress) results in the initiation of reactive oxygen species which negatively affect bio-molecules by causing DNA damage, mutations, protein misfolding, and aggregation. The delicate organ of the body is the brain because of its high oxygen consumption and enrichment in PUFA. Neurodegenerative disorders are caused by ROS, which accumulates in the brain and harms neurons. The excess level of metals in the aging brain and mutations in mitochondrial DNA cause oxidative stress even though the metals are crucial for the events that are mediated by an enzyme in cell signaling as well as in cellular metabolism. A series of events lead to the eventual impairment of neuronal proteins, which causes neuro-inflammation and some other neurological conditions characterized by loss of cognitive ability (PD, HD, ALS, AD). The disturbance between the antioxidant and reactive oxygen species inside the cells is mostly caused by the mitochondrial malfunction. Antioxidants are also designed to combat free radical scavenging and neural inflammation. Saffron is an intriguing spice that has been shown to have antioxidant properties, protecting against CNS illnesses. It is a good candidate to treat a variety of disorders because of its low cytotoxicity, commercial viability, and capacity to pass the blood-brain barrier. The only option for regional restoration is stem-cell-centered therapy, which aims to arrest the loss of neurons in the brain. Therefore, it is essential to keep the equilibrium between the production of ROS and anti-oxidants ability to scavenge ROS. Antioxidant therapy's efficacy for neurodegenerative illnesses in humans is still debatable, despite preclinical studies' encouraging findings. Globally, current research is revealing quite new directions and giving hope for effective treatment options to cope with such neuro-degenerative illnesses. Additional research would also explore the benefits of oxidative stress phenomena on biology and therapeutics [8, 115–117]. Antioxidants with neuroprotective properties are thought to be a viable strategy for reducing the rate and degree of neuronal cell death in certain illnesses. However, there is still a dearth of clinical data supporting antioxidant chemicals' ability to act as preventative medications in neurodegenerative illness (Moosmann and Behl 2002).

More and more evidence points to the importance of antioxidant systems in the defensive mechanisms against oxidative stress. For our health and the prevention of many illnesses, it will soon be necessary to increase the potency of antioxidants, develop molecules with intrinsic antioxidant activity, or discover substances that will directly or indirectly increase the number of endogenous antioxidant systems.

References

1. Chiurchiù V, Orlacchio A, Maccarrone M (2016) Is modulation of oxidative stress an answer? The state of the art of redox therapeutic actions in neurodegenerative diseases. Oxidative Med Cell Longev 2016
2. Hussain T, et al (2016) Oxidative stress and inflammation: What polyphenols can do for us? Oxid Med Cell Longev 2016:7432797
3. Henrik S, István B, David M, Jan V, Zoltán K, Ferenc F, Tibor K, László S, Ádám D, Odilia M (2022) The innate immune system and fever under redox control: a narrative review. Curr Med Chem 29:4324–4362
4. Sorci G, Faivre B (2009) Inflammation and oxidative stress in vertebrate host–parasite systems. Philosophical Trans R Soc B Biol Sci 364(1513):71–83
5. Yadav DK, Rai R, Kumar N, Singh S, Misra S, Sharma P, Shaw P, Pérez-Sánchez H, Mancera RL, Choi EH (2016) New arylated benzo [h] quinolines induce anti-cancer activity by oxidative stress-mediated DNA damage. Sci Rep 6(1):1–13
6. Pisoschi AM, Pop A, Iordache F, Stanca L, Predoi G, Serban AI (2021) Oxidative stress mitigation by antioxidants-an overview on their chemistry and influences on health status. Eur J Med Chem 209:112891
7. Prasad A, Sedlářová M, Balukova A, Rác M, Pospíšil P (2020) Reactive oxygen species as a response to wounding: in vivo imaging in Arabidopsis thaliana. Front Plant Sci 10:1660
8. Singh A, Kukreti R, Saso L, Kukreti S (2019) Oxidative stress: a key modulator in neurodegenerative diseases. Molecules 24(8):1583
9. Berg D, Youdim MB, Riederer P (2004) Redox imbalance. Cell and tissue research 318 (1):201–213
10. McQuillen PS, Ferriero DM (2004) Selective vulnerability in the developing central nervous system. Pediatr Neurol 30(4):227–235
11. Von Arnim CA, Gola U, Biesalski HK (2010) More than the sum of its parts? Nutrition in Alzheimer's disease. Nutrition 26(7-8):694–700
12. Yu YC, Kuo CL, Cheng WL, Liu CS, Hsieh M (2009) Decreased antioxidant enzyme activity and increased mitochondrial DNA damage in cellular models of Machado-Joseph disease. J Neurosci Res 87(8):1884–1891
13. Mandel S, Grünblatt E, Riederer P, Gerlach M, Levites Y, Youdim MB (2003) Neuroprotective strategies in Parkinson's disease. CNS Drugs 17 (10):729–762
14. El-Bachá RS, De-Lima-Filho J, Guedes RC (1998) Dietary antioxidant deficiency facilitates cortical spreading depression induced by photoactivated riboflavin. Nutr Neurosci 1(3):205–212
15. Akbari B, Baghaei-Yazdi N, Bahmaie M, Mahdavi Abhari F (2022) The role of plant-derived natural antioxidants in reduction of oxidative stress. Biofactors 48(3):611–633

16. Shindo Y, Witt E, Han D, Epstein W, Packer L (1994) Enzymic and non-enzymic antioxidants in epidermis and dermis of human skin. J Investig Dermatol 102(1):122–124
17. Kadam DP, Suryakar AN, Ankush RD, Kadam CY, Deshpande KH (2010) Role of oxidative stress in various stages of psoriasis. Indian J Clin Biochem 25(4):388–392
18. Baek J-O, Byamba D, Wu WH, Kim T-G, Lee M-G (2012) Assessment of an imiquimod-induced psoriatic mouse model in relation to oxidative stress. Arch Dermatol Res 304(9):699–706
19. Abdel-Mawla MY, Nofal E, Khalifa N, Abdel-Shakoor R, Nasr M (2013) Role of oxidative stress in psoriasis: an evaluation study. J Am Sci 9:151–155
20. Pandey AK (2007) Anti-staphylococcal activity of a pan-tropical aggressive and obnoxious weed Parthenium histerophorus: an in vitro study. Natl Acad Sci Lett 30(11/12):383–386
21. Mahomoodally MF, Gurib-Fakim A, Subratty AH (2005) Antimicrobial activities and phytochemical profiles of endemic medicinal plants of Mauritius. Pharm Biol 43(3):237–242
22. Rice-evans CA, Miller NJ, Bolwell PG, Bramley PM, Pridham JB (1995) The relative antioxidant activities of plant-derived polyphenolic flavonoids. Free Radic Res 22(4):375–383
23. Cook NC, Samman S (1996) Flavonoids—chemistry, metabolism, cardioprotective effects, and dietary sources. J Nutr Biochem 7(2):66–76
24. Halliwell B, Gutteridge JM (2015) Free radicals in biology and medicine. Oxford University Press, Oxford
25. Mishra A, Kumar S, Pandey AK (2013) Scientific validation of the medicinal efficacy of Tinospora cordifolia. Sci World J:2013
26. Mène-Saffrané L, DellaPenna D (2010) Biosynthesis, regulation and functions of tocochromanols in plants. Plant Physiol Biochem 48(5):301–309
27. Sheppard A, Pennington J, Weihrauch J (1993) Vitamin E in health and disease. Marcel-Dekker, New York
28. Sundl I, Murkovic M, Bandoniene D, Winklhofer-Roob BM (2007) Vitamin E content of foods: comparison of results obtained from food composition tables and HPLC analysis. Clin Nutr 26(1):145–153
29. Catalgol B, Ziaja I, Breusing N, Jung T, Hoehn A, Alpertunga B, Schroeder P, Chondrogianni N, Gonos ES, Petropoulos I (2009) The proteasome is an integral part of solar ultraviolet a radiation-induced gene expression. J Biol Chem 284 (44):30076–30086
30. Podszun MC, Grebenstein N, Spruss A, Schlueter T, Kremoser C, Bergheim I, Frank J (2014) Dietary α-tocopherol and atorvastatin reduce high-fat-induced lipid accumulation and down-regulate CD36 protein in the liver of guinea pigs. J Nutr Biochem 25(5):573–579
31. Barella L, Muller PY, Schlachter M, Hunziker W, Stöcklin E, Spitzer V, Meier N, de Pascual-Teresa S, Minihane A-M, Rimbach G (2004) Identification of hepatic molecular mechanisms of action of alpha-tocopherol using global gene expression profile analysis in rats. Biochimica et Biophysica Acta (BBA) Mol Basis Dis 1689(1):66–74
32. Sies H (1985) Oxidative stress: introductory remarks. In: Sies H (ed) Oxidative Stress. Academic, London,
33. Sies H, Berndt C, Jones DP (2017) Oxidative stress. Annu Rev Biochem 86:715–748
34. Niki E (2016) Oxidative stress and antioxidants: Distress or eustress? Arch Biochem Biophys 595:19–24. https://doi.org/10.1016/j.abb.2015.11.017
35. Lin X, Huang T (2016) Oxidative stress in psoriasis and potential therapeutic use of antioxidants. Free Radic Res 50(6):585–595
36. Shilov V, Sergienko V (2000) Oxidative stress in keratinocytes as an etiopathogenetic factor of psoriasis. Bull Exp Biol Med 129(4):309–313
37. Aschbacher K, O'Donovan A, Wolkowitz OM, Dhabhar FS, Su Y, Epel E (2013) Good stress, bad stress and oxidative stress: insights from anticipatory cortisol reactivity. Psychoneuroendocrinology 38(9):1698–1708
38. Siegrist J, Sies H (2017) Disturbed redox homeostasis in oxidative distress: a molecular link from chronic psychosocial work stress to coronary heart disease? Circ Res 121(2):103–105

39. Dennis KK, Go Y-M, Jones DP (2019) Redox systems biology of nutrition and oxidative stress. J Nutr 149(4):553–565
40. Aschbacher K, Mason AE (2020) Eustress, distress, and oxidative stress: promising pathways for mind-body medicine. In: Oxidative stress. Elsevier, pp 583–617
41. Griffiths HR, Møller L, Bartosz G, Bast A, Bertoni-Freddari C, Collins A, Cooke M, Coolen S, Haenen G, Hoberg A-M (2002) Biomarkers. Mol Asp Med 23(1-3):101–208
42. Frijhoff J, Winyard PG, Zarkovic N, Davies SS, Stocker R, Cheng D, Knight AR, Taylor EL, Oettrich J, Ruskovska T (2015) Clinical relevance of biomarkers of oxidative stress. Antioxid Redox Signal 23(14):1144–1170
43. Ghezzi P (2020) Environmental risk factors and their footprints in vivo – a proposal for the classification of oxidative stress biomarkers. Redox Biol 34:101442
44. Al-Gubory KH, Garrel C, Faure P, Sugino N (2012) Roles of antioxidant enzymes in corpus luteum rescue from reactive oxygen species-induced oxidative stress. Reprod BioMed Online 25(6):551–560
45. Hansen JM, Go Y-M, Jones DP (2006) Nuclear and mitochondrial compartmentation of oxidative stress and redox signaling. Annu Rev Pharmacol Toxicol 46:215–234
46. Kumar S, Pandey AK (2013) Chemistry and biological activities of flavonoids: an overview. Sci World J 2013
47. Valko M, Izakovic M, Mazur M, Rhodes CJ, Telser J (2004) Role of oxygen radicals in DNA damage and cancer incidence. Mol Cell Biochem 266(1):37–56
48. Willcox JK, Ash SL, Catignani GL (2004) Antioxidants and prevention of chronic disease. Crit Rev Food Sci Nutr 44(4):275–295
49. Genestra M (2007) Oxyl radicals, redox-sensitive signalling cascades and antioxidants. Cell Signal 19(9):1807–1819
50. Bahorun T, Soobrattee M, Luximon-Ramma V, Aruoma O (2006) Free radicals and antioxidants in cardiovascular health and disease. Intern J Med Update 1(2):25–41
51. Kumar S, Pandey AK (2015) Free radicals: health implications and their mitigation by herbals. Br J Med Med Res 7(6):438–457
52. Pacher P, Beckman JS, Liaudet L (2007) Nitric oxide and peroxynitrite in health and disease. Physiol Rev 87(1):315–424
53. Valko M, Leibfritz D, Moncol J, Cronin MT, Mazur M, Telser J (2007) Free radicals and antioxidants in normal physiological functions and human disease. Int J Biochem Cell Biol 39(1):44–84
54. Frei B (1994) Reactive oxygen species and antioxidant vitamins: mechanisms of action. Am J Med 97(3):S5–S13
55. Young I, Woodside J (2001) Antioxidants in health and disease. J Clin Pathol 54(3):176–186
56. Pizzino G, Irrera N, Cucinotta M, Pallio G, Mannino F, Arcoraci V, Squadrito F, Altavilla D, Bitto A (2017) Oxidative stress: harms and benefits for human health. Oxid Med Cell Longev:8416763. https://doi.org/10.1155/2017/8416763
57. Forman HJ, Zhang H (2021) Targeting oxidative stress in disease: promise and limitations of antioxidant therapy. Nat Rev Drug Discov 20(9):689–709. https://doi.org/10.1038/s41573-021-00233-1
58. Evans JL, Goldfine ID, Maddux BA, Grodsky GM (2002) Oxidative stress and stress-activated signaling pathways: a unifying hypothesis of type 2 diabetes. Endocr Rev 23(5):599–622
59. Suarna C, Dean RT, May J, Stocker R (1995) Human atherosclerotic plaque contains both oxidized lipids and relatively large amounts of α-tocopherol and ascorbate. Arterioscler Thromb Vasc Biol 15(10):1616–1624
60. Salomon RG, Kaur K, Podrez E, Hoff HF, Krushinsky AV, Sayre LM (2000) HNE-derived 2-pentylpyrroles are generated during oxidation of LDL, are more prevalent in blood plasma from patients with renal disease or atherosclerosis, and are present in atherosclerotic plaques. Chem Res Toxicol 13(7):557–564

61. Gniwotta C, Morrow JD, Roberts LJ, Kühn H (1997) Prostaglandin F2-like compounds, F2-isoprostanes, are present in increased amounts in human atherosclerotic lesions. Arterioscler Thromb Vasc Biol 17(11):3236–3241
62. Yang X, Li Y, Li Y, Ren X, Zhang X, Hu D, Gao Y, Xing Y, Shang H (2017) Oxidative stress-mediated atherosclerosis: mechanisms and therapies. Front Physiol 8:600
63. Giuranno L, Ient J, De Ruysscher D, Vooijs MA (2019) Radiation-induced lung injury (RILI). Front Oncol 9:877
64. Barnes PJ (2020) Oxidative stress-based therapeutics in COPD. Redox Biol 33:101544
65. Fleckenstein K, Zgonjanin L, Chen L, Rabbani Z, Jackson IL, Thrasher B, Kirkpatrick J, Foster WM, Vujaskovic Z (2007) Temporal onset of hypoxia and oxidative stress after pulmonary irradiation. Int J Radiat Oncol Biol Phys 68(1):196–204
66. Rappold PM, Cui M, Chesser AS, Tibbett J, Grima JC, Duan L, Sen N, Javitch JA, Tieu K (2011) Paraquat neurotoxicity is mediated by the dopamine transporter and organic cation transporter-3. Proc Natl Acad Sci 108(51):20766–20771
67. Bus JS, Gibson JE (1984) Paraquat: model for oxidant-initiated toxicity. Environ Health Perspect 55:37–46
68. Montezano AC, Touyz RM (2014) Reactive oxygen species, vascular Noxs, and hypertension: focus on translational and clinical research. Antioxid Redox Signal 20(1):164–182
69. Lacy F, Kailasam MT, O'connor DT, Schmid-Schönbein GW, Parmer RJ (2000) Plasma hydrogen peroxide production in human essential hypertension: role of heredity, gender, and ethnicity. Hypertension 36(5):878–884
70. Lovell M, Ehmann W, Mattson M, Markesbery W (1997) Elevated 4-hydroxynonenal in ventricular fluid in Alzheimer's disease. Neurobiol Aging 18(5):457–461
71. Ansari MA, Scheff SW (2010) Oxidative stress in the progression of Alzheimer disease in the frontal cortex. J Neuropathol Exp Neurol 69(2):155–167
72. Wang J, Xiong S, Xie C, Markesbery W, Lovell M (2005) Increased oxidative damage in nuclear and mitochondrial DNA in Alzheimer's disease. J Neurochem 93(4):953–962
73. Hayes JD, Dinkova-Kostova AT, Tew KD (2020) Oxidative stress in cancer. Cancer Cell 38(2):167–197
74. Raimondi V, Ciccarese F, Ciminale V (2020) Oncogenic pathways and the electron transport chain: a dangeROS liaison. Br J Cancer 122(2):168–181
75. Chan HP, Tran V, Lewis C, Thomas PS (2009) Elevated levels of oxidative stress markers in exhaled breath condensate. J Thorac Oncol 4(2):172–178
76. Cobley JN, Fiorello ML, Bailey DM (2018) 13 reasons why the brain is susceptible to oxidative stress. Redox biology 15:490–503
77. Halliwell B (2001) Role of free radicals in the neurodegenerative diseases. Drugs Aging 18(9):685–716
78. Li J, Wuliji O, Li W, Jiang Z-G, Ghanbari HA (2013) Oxidative stress and neurodegenerative disorders. Int J Mol Sci 14(12):24438–24475
79. Jaganjac M, Milkovic L, Zarkovic N, Zarkovic K (2022) Oxidative stress and regeneration. Free Radic Biol Med 181:154–165. https://doi.org/10.1016/j.freeradbiomed.2022.02.004
80. Jaganjac M, Milkovic L, Gegotek A, Cindric M, Zarkovic K, Skrzydlewska E, Zarkovic N (2020) The relevance of pathophysiological alterations in redox signaling of 4-hydroxynonenal for pharmacological therapies of major stress-associated diseases. Free Radic Biol Med 157:128–153. https://doi.org/10.1016/j.freeradbiomed.2019.11.023
81. Christen Y (2000) Oxidative stress and Alzheimer disease. Am J Clin Nutr 71(2):621S–629S
82. Allan Butterfield D (2002) Amyloid β-peptide (1-42)-induced oxidative stress and neurotoxicity: implications for neurodegeneration in Alzheimer's disease brain. A review. Free Rad Res 36(12):1307–1313
83. Salehi B, Martorell M, Arbiser JL, Sureda A, Martins N, Maurya PK, Sharifi-Rad M, Kumar P, Sharifi-Rad J (2018) Antioxidants: positive or negative actors? Biomolecules 8(4):124
84. Gulcin İ (2020) Antioxidants and antioxidant methods: An updated overview. Arch Toxicol 94(3):651–715

85. Kurutas EB (2015) The importance of antioxidants which play the role in cellular response against oxidative/nitrosative stress: current state. Nutr J 15(1):1–22
86. Mattill H (1947) Antioxidants. Annu Rev Biochem 16(1):177–192
87. Kusch M (2005) Psychological knowledge: A social history and philosophy. Routledge, London
88. NOHL H, HEGNER D (1978) Do mitochondria produce oxygen radicals in vivo? Eur J Biochem 82(2):563–567
89. Lobo V, Patil A, Phatak A, Chandra N (2010) Free radicals, antioxidants and functional foods: impact on human health. Pharmacogn Rev 4(8):118
90. Panda SK (2012) Assay guided comparison for enzymatic and non-enzymatic antioxidant activities with special reference to medicinal plants. Antioxid Enzyme 14:382–400
91. Pisoschi AM, Pop A (2015) The role of antioxidants in the chemistry of oxidative stress: a review. Eur J Med Chem 97:55–74
92. Flieger J, Flieger W, Baj J, Maciejewski R (2021) Antioxidants: Classification, natural sources, activity/capacity measurements, and usefulness for the synthesis of nanoparticles. Materials 14(15):4135
93. Aziz MA, Majeed GH, Diab KS, Al-Tamimi RJ (2016) The association of oxidant–antioxidant status in patients with chronic renal failure. Ren Fail 38(1):20–26
94. Hunaiti AA (2020) The therapeutic potential of glutathione supplement: a review of clinical trials. Eur J Med Health Sci 2(2)
95. Skowyra M, Gallego MG, Segovia F, Almajano MP (2014) Antioxidant properties of Artemisia annua extracts in model food emulsions. Antioxidants 3(1):116–128
96. Martí R, Roselló S, Cebolla-Cornejo J (2016) Tomato as a source of carotenoids and polyphenols targeted to cancer prevention. Cancer 8(6):58
97. Aslani BA, Ghobadi S (2016) Studies on oxidants and antioxidants with a brief glance at their relevance to the immune system. Life Sci 146:163–173
98. Zhang J, McCullough PA (2016) Lipoic acid in the prevention of acute kidney injury. Nephron 134(3):133–140
99. Krinsky NI (1992) Mechanism of action of biological antioxidants. Proc Soc Exp Biol Med 200(2):248–254
100. Rice E, Diplock A (1993) Current status of antioxidant therapy. Free Radic Biol Med 15:77–96
101. Martysiak-Żurowska D, Wenta W (2012) A comparison of ABTS and DPPH methods for assessing the total antioxidant capacity of human milk. Acta Sci Pol Technol Aliment 11(1):83–89
102. Halliwell B (2001) Free radicals and other reactive species in disease. e LS
103. Tessutti L, Macedo D, Kubota L, Alves A (2013) Measuring the antioxidant capacity of blood plasma using potentiometry. Anal Biochem 441(2):109–114
104. Pisoschi AM, Pop A, Cimpeanu C, Predoi G (2016) Antioxidant capacity determination in plants and plant-derived products: a review. Oxidat Med Cell Longe:2016
105. Limón-Pacheco J, Gonsebatt ME (2009) The role of antioxidants and antioxidant-related enzymes in protective responses to environmentally induced oxidative stress. Mutati Res/ Genetic Toxicol Environ Mutagen 674(1-2):137–147
106. Reuter S, Gupta S, Chaturvedi M, Aggarwal BB (2010) Oxidative stress, inflammation, and cancer: how are they linked. Free Radic Biol Med 49:1603–1616
107. Stone WL, Shah D, Hollinger SM (2016) Retinopathy of prematurity: an oxidative stress neonatal disease. Front Biosci Landmark 21(1):165–177
108. Al-Zamil WM, Yassin SA (2017) Recent developments in age-related macular degeneration: a review. Clin Interv Aging 12:1313
109. Perez S, Talens-Visconti R, Rius-Perez S, Finamor I, Sastre J (2017) Redox signaling in the gastrointestinal tract. Free Radic Biol Med 104:75–103
110. Tarafdar A, Pula G (2018) The role of NADPH oxidases and oxidative stress in neurodegenerative disorders. Int J Mol Sci 19(12):3824

111. Tan BL, Norhaizan M, Liew W Sulaiman Rahman H (2018) Antioxidant and oxidative stress: a mutual interplay in age-related diseases. Front Pharmacol 9:1162
112. Chen X, Guo C, Kong J (2012) Oxidative stress in neurodegenerative diseases. Neural Regen Res 7(5):376–385. https://doi.org/10.3969/j.issn.1673-5374.2012.05.009
113. Mittal M, Siddiqui MR, Tran K, Reddy SP, Malik AB (2014) Reactive oxygen species in inflammation and tissue injury. Antioxid Redox Signal 20(7):1126–1167
114. Adwas AA, Elsayed A, Azab A, Quwaydir F (2019) Oxidative stress and antioxidant mechanisms in human body. J Appl Biotechnol Bioeng 6(1):43–47
115. Purushothuman S (2015) Oxidative stress in neurodegenerative conditions and the protective potential of a natural antioxidant, dietary saffron. Oxid Antioxid Med Sci 4:1–7
116. Liu XL, Sato S, Dai W, Yamanaka N (2001) The protective effect of hepatocyte growth-promoting factor (pHGF) against hydrogen peroxide-induced acute lung injury in rats. Med Electron Microsc 34(2):92–102. https://doi.org/10.1007/s007950170003
117. Wang L, Wang X, Wang Q (2017) Mulberry seeds perform high hypoglycaemic effect partially by inhibition of α-glucosidase activity. Biomed Res (0970-938X) 28(8)

Chapter 8
Role of Endogenous and Dietary Antioxidants in Brain Disorders

Rabia Akram, Humaira Muzaffar, Haseeb Anwar, Shoaib Ahmad Malik, Faiqa Sajid, Tehreem Iman, Ayesha Ahsan, Mobina Manzoor, Ikram Ullah Khan, and Ghulam Hussain

8.1 Introduction

Oxidative stress (OS) occurs as a result of the disturbance in the production of pro-oxidant and antioxidant species. It can be brought by a decline of antioxidant species and an increase in oxidative metabolism that can occur due to many other factors, such as drinking alcohol, being exposed to the cold, taking medications, being injured, ingesting toxins, being exposed to radiation, engaging in strenuous exercise, and eating poorly. In addition to harming lipids, reactive oxygen species

R. Akram
Neurochemicalbiology and Genetics Laboratory (NGL), Department of Physiology, Faculty of Life Sciences, Government College University, Faisalabad, Pakistan

Department of Neuromuscular Diseases, UCL Institute of Neurology, University College London, London, United Kingdom

H. Muzaffar · H. Anwar · F. Sajid · T. Iman · A. Ahsan (✉)
Neurochemicalbiology and Genetics Laboratory (NGL), Department of Physiology, Faculty of Life Sciences, Government College University, Faisalabad, Pakistan

G. Hussain
Department of Physiology, Government College University Faisalabad, Faisalabad, Pakistan
e-mail: ghulamhussain@gcuf.edu.pk

S. A. Malik
Department of Biochemistry, Sargodha Medical College, University of Sargodha, Sargodha, Pakistan

M. Manzoor
Institute of Pharmacy, Lahore College for Women University, Lahore, Pakistan

I. U. Khan
Department of Pharmaceutics, Faculty of Pharmaceutical Sciences, Government College University, Faisalabad, Pakistan

A. Imran, G. Hussain (eds.), *The Role of Natural Antioxidants in Brain Disorders*, Food Bioactive Ingredients, https://doi.org/10.1007/978-3-031-41188-5_8

(ROS) can also result in cell death. They are produced during regular cellular processes like arachidonic acid metabolism, phagocytic digestion, mitochondrial respiration, ovulation, and fertilization. When there are pathological situations, however, their production multiplies many times. Hydrogen peroxide, Superoxide ions, peroxynitrite, and nitric oxide play a part in tissue destruction in pathological conditions.

8.1.1 Oxidative Stress and the Nervous System

The Central Nervous System (CNS) is particularly prone to oxidative injury for some reasons:

1. The extremely active oxidative metabolism of brain tissue results in high amounts of intra-cellular superoxides.
2. It has a limited capacity for anaerobic respiration, which results in elevated superoxide levels in anoxic environments.
3. High iron concentration, decrement of antioxidant species, and membrane elaborations in cellular characteristics make the oligodendrocyte population more susceptible to oxidative damage.
4. Myelin is a preferred target for ROS because of its high protein/lipid ratio. Due to their high mitochondrial density and increased rate of oxygen utilization, CNS tissues are particularly susceptible to OS. As a result of their regular oxidative metabolism, mitochondria invariably produce free radicals, that can harm the DNA of mitochondria. It leads to the production of defective proteins that reduces the generation of mitochondrial elements of the electron-transport chain, which can then increase free radical generation and further mitochondrial damage.

Additionally, CNS is abundant in iron and unsaturated fatty acids. Neural tissue is especially susceptible to oxidative injury due to high aerobic metabolic activity and lipid content. Iron is a crucial component for the development of the brain, yet brain cell damage can release iron ions that cause OS by catalyzing the production of ROS. The production of free radicals can cause serious damage to particular catecholamine-rich brain regions. Endogenous, as well as dietary antioxidants (Fig. 8.1), can shield the nervous system from harm brought on by OS, which is a contributing factor in the development of brain disorders [1–4].

8.1.2 Antioxidants and the Nervous System

Antioxidants are those species that remove free radicals, scavenge ROS or their precursors, and prevent ROS synthesis. Neurons are especially susceptible to damage brought on by OS because of their reduced antioxidant defense system, high

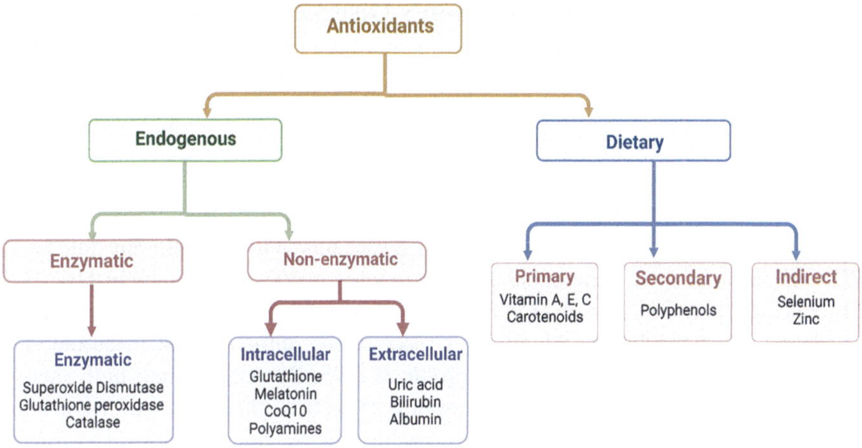

Fig. 8.1 Classification of endogenous and dietary antioxidants

need for oxygen utilization, and high contents of polyunsaturated fatty acids (PUFAs) in their cell membranes [5]. By quenching/scavenging free radical intermediates and halting the spread of oxidative chain events, antioxidants can alleviate OS. These antioxidants are mostly made up of exogenous (natural and synthetic) antioxidant sources that keep the biological system's redox balance in check as well as a variety of endogenous antioxidant enzymes and their coenzymes or substrates [6, 7].

All cells, including neurons, contain potent antioxidant enzymes that can assist detoxify ROS. Catalases, superoxide dismutases (SOD), and glutathione peroxidases (GPx) are the three main types of antioxidant enzymes. These antioxidant enzymes prevent cellular damage brought on by ROS. The antioxidant defense mechanisms in the brain itself, however, seem to be somewhat underwhelming. The majority of brain areas, except the substantia nigra and the hypothalamus, have relatively low amounts of catalase. Additionally, any catalase that is present is housed in micro peroxisomes, where it is unable to reduce the H_2O_2 generated in other subcellular spaces. Therefore, it appears that the brain's endogenous antioxidant defense system is easily overpowered if ROS generation rises too quickly. External antioxidant supplementation or herbal treatments may help maintain strict homeostatic regulation of ROS and prevent OS. A growing body of research indicates that consuming antioxidants including vitamin E, ascorbate, carotenoids, and plant phenols may lower the risk of some neurodegenerative illnesses. These antioxidants can be consumed naturally or as supplements [8, 9]. The neuroprotective effects of dietary and endogenous antioxidants in brain disorders have been discussed in detail in the next sections of this chapter:

8.2 Antioxidants and Parkinson's Disease

Parkinson's Disease (PD) is characterized by the death of dopaminergic neurons in the substantia nigra pars compacta (SNpc). Rigidity, tremor, bradykinesia, bradyphrenia, gait impairment, and postural instability are the main effects. Dopaminergic neurons in the SN provide signals to the adult hippocampus dentate gyrus (DG). Therefore, the decline of dopaminergic neurons could have a direct impact on adult hippocampus neurogenesis. Lewy bodies (LB) are linked to the pathophysiology of PD, and the main LB component that aggregates in PD is α-synuclein [10, 11]. Mitochondrial malfunction and OS are key factors in the development of the illness. Oxidative phosphorylation, which takes place at the mitochondrial level and is a by-product of aerobic respiration and produces ROS. Because of their high energy requirements and huge oxygen consumption as well as their post-mitotic origin, neurons are thought to be particularly vulnerable to ROS-induced injury. That's why neural tissues are susceptible to long-term and degenerative illnesses, such as PD [12].

8.2.1 Endogenous Antioxidants

8.2.1.1 Glutathione

Reduced Glutathione (GSH) levels and a lower GSH/GSSG ratio in the blood, lymphoblastoid cells and brain tissues have been seen in PD patients. The cerebellum and temporal cortex of people with PD have significantly lower GSH total contents and GSH/GSSG ratios. Patients with PD exhibit higher levels of circulating Hcy and lower levels of cellular GSH as a result of elevated OS and impaired methionine synthase activity. The brains of PD patients also show more chronic inflammatory responses, mitochondrial superoxide, as well as oxidative damage to proteins and DNA, due to the increased OS and decreased GSH/GSSG activity [13, 14]. Mice given glut amyl cysteine ethyl ester, dipeptide precursor of GSH, in a lipid-soluble form can cross the blood-brain barrier (BBB) and showed some resistance to the 1-methyl-4-phenyl-1,2,3,6-tetrahydropyridine (MPTP) induced loss of dopaminergic neurons [15].

8.2.1.2 Coenzyme Q10

Coenzyme Q10 (CoQ10) is ubiquinone that is found in almost all cells and is a vital part of the oxidative phosphorylation process in the mitochondria. In contrast to healthy persons, postmortem brain tissues from people with PD showed significantly lower levels of total plasma CoQ10 as well as significantly higher amounts of the oxidized form of ubiquinone. More particular, PD patients' platelets had lower levels of mitochondrial CoQ10 than controls of similar ages and sexes. In vitro PD models have shown the potency of CoQ10 as protection against cell toxicity. When adult mice's striatal slices were grown and treated with MPP+, the co-incubation with CoQ10 resulted in a dramatically reduced loss of dopaminergic cells. In vitro PD models have shown the potency of CoQ10 as protection against cell toxicity [16]. A decline in CoQ10 status contributes to the pathogenesis of the disease by aggravating MRC function and impairing cellular antioxidants [17].

8.2.1.3 Uric Acid

Uric acid (UA) shields dopaminergic neurons against H_2O_2 or MPP+ induced apoptosis in cultured SN neurons from mice. In 6 hydroxydopamine (6-OHDA) lesioned rats, elevated cerebral uric acid improves parkinsonian phenotypes. Higher serum UA concentration is strongly linked with a slower rate of PD progression. In contrast to those without cognitive impairment, PD patients with cognitive defects also have lower serum uric acid levels [18].

8.2.1.4 Alpha Lipoic Acid

Alpha Lipoic Acid (ALA) has potential therapeutic utility since it has anti-inflammatory, anti-oxidative, and free radical formation-inhibiting properties. It can lessen dyskinesia by increasing GSH activity and decreasing malondialdehyde (MDA), a byproduct of lipid peroxidation. ALA treatment significantly improves motor dysfunctions, causes a decrease in α-synuclein accumulation, and a reduction in the activation of pro-inflammatory molecules [19].

8.2.2 Dietary Antioxidants

8.2.2.1 Vitamin C

Vitamin C, also known as ascorbate or ascorbic acid can be obtained from fruits and vegetables (Fig. 8.2). The majority of mammals can produce vitamin C internally, but because humans lack the enzyme L-gulonolactone oxidase, they must consume this vital component through diet or supplements. By giving electrons to counteract the harmful effects of free radicals, vitamin C functions as an antioxidant. Vitamin C increased antioxidant enzyme activity and reduced the PD-related phenotype by reducing the antioxidant enzyme ubiquitin c-terminal hydrolase (UCH), which increases the age-related deterioration of dopaminergic neurons and lowers dopamine levels in the brain [20].

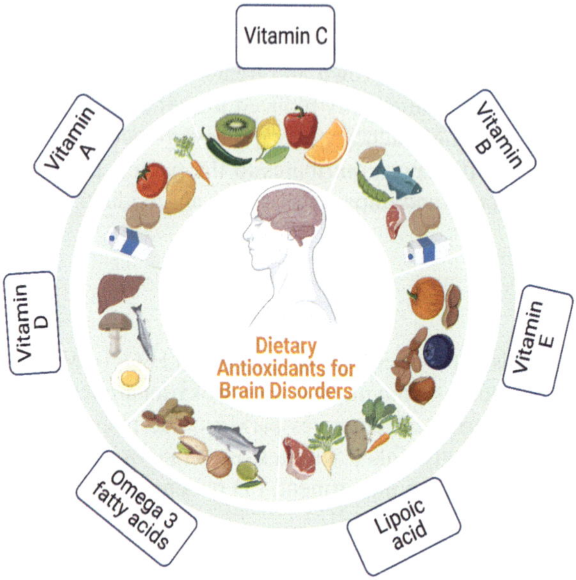

Fig. 8.2 Sources of dietary antioxidants for brain disorders

8.2.2.2 Vitamin E

Microtubule-associated protein tau (MAPT) gene polymorphisms are associated with an increased risk of PD, and MAPT methylation is linked negatively to MAPT expression. Vitamin E reduces the incidence of PD by reducing MAPT expression through raising MAPT gene methylation. Vitamin E administration is found to prevent PD and ameliorate its prognosis [21].

8.2.2.3 Phenols

Curcumin is found to enhance locomotion, lessen severe neurodegeneration, and lower OS markers in both 6-OHDA-induced PD in rats and *Drosophila milanogaster* models. Curcumin protects SN neurons by increasing dopamine levels in the nigrostriatal tract and lowering Fe^{3+} levels via chelation in the 6-OHDA rat model of PD. This is because the phenolic rings and diketone groups on the curcumin moiety function as an electron trap, reducing the production of superoxide, H_2O_2, and hydroxyl ions. It reduces ROS production and NF-κB overexpression and increases SOD expression to prevent 6-OHDA-actuated cell damage [22].

Resveratrol (RV) reduces astroglial activation in mice exposed to MPTP's nigrostriatal pathway. It shows synergistic effects when administered along with the dopamine precursor L-DOPA, hence, it reduces its harmful effects in the treatment of PD [23].

Tyrosol, found in extra-virgin oil, has been shown to slow down the aggregation of α-synuclein in PD. It decreases ROS levels while enhancing the expression of antioxidant enzymes and particular chaperones.

Chrysin, a naturally occurring flavonoid, is found in honey, bee propolis, and various plants. It reverses neurochemical impairments, behavioral abnormalities, and OS in 6-OHDA and MPTP-induced PD model.

Acteoside, a flavonoid, is known to lessen or even stop brain damage against the 6-OHDA zebrafish model of PD. Acteoside pretreatment may also increase the expression of antioxidants by triggering the nuclear factor erythroid 2-related factor 2 (Nrf2) signaling pathway.

Pinostrobin was also employed in the MPTP zebra fish model of PD with positive outcomes because it greatly increases Nrf2 expression and upregulates the expression of heme oxygenase-1 (HO-1).

Genistein activates estrogen receptors as well as NF-E2L2 channels and reduces OS damage and cell death in human SH-SY5Y cells, which reveals a mutant type of α-synuclein.

Salidroside was given to 6-OHDA-induced PD rats, and the results showed that it protected the brain from OS. This effect was most likely caused by the control of the Wnt/β-catenin signaling pathway [24].

Oleuropein (OL) reduces the harmful effects of α-synuclein-induced stress on dopaminergic neurons to treat and prevent PD. The toxicity is reduced by OL and other structurally related chemicals such as verbascoside, dihydro Oleuropein, 3-hydroxytyrosol, oleanolic acid, oleuropein glycosides, and rutin. This is accomplished by converting α-synuclein oligomers into small monomers that have less harmful effects. Moreover, OLA binds with α-synuclein's N-terminal region, preventing it from reacting with lipid membranes and preventing the creation of toxic aggressiveness. Additionally, OL protects against microglial inflammation-mediated dopaminergic neurons by reducing the pro-inflammatory action of activated microglia cells by blocking mitochondrial fission. A substantial reduction in OS, apoptosis, and cell damage was observed in adrenal pheochromocytoma (PC12) cells and 6-OHDA induced PD when different formulations of OL were supplemented into the diet. Additionally, OL lowers the levels of DNA denaturation, mitochondrial ROS generation, and superoxide anion [25].

Carvacrol promotes significant neuroprotection in the 6-OHDA model of PD via its general blocking impact upon TRPM7 cation channels, that are involved in causing neurodegeneration [26].

8.2.2.4 Asiatic Acid

Asiatic acid (AA) inhibits OS, preserves the MMP, and controls the expression of Bcl-2, Bax, and caspases to prevent rotenone-induced apoptosis in SH-SY5Y cells as well as prevented the MPP+ induced apoptosis of dopaminergic neurons.

Additionally, AA provides neuroprotection against MPP+-induced loss of neuronal cells by activating the ERK and PI3K/Akt/mTOR/GSK-3β pathways [27].

8.3 Antioxidants and Huntington's Disease

Huntington's disease (HD) occurs due to expanded polyglutamine (poly Q) in the huntingtin (Htt) protein that leads to the death of striatal neurons and eventually damages cortical regions [28]. mHtt increases OS by attaching to PGC1α's promoter region and lowers the transcriptional level of PGC1α. It also inhibits the production of antioxidant enzymes and mitochondrial uncoupling proteins by directly inactivating PGC1α. By interfering with Drp1's functionality, mHtt upsets the equilibrium between the fission-fusion process in the mitochondria. The mitochondrial permeability transition pore (mPTP) opens as a result of mHtt's induction of calcium ion leakage through the calcium channel ryanodine receptors, which also causes mitochondrial OS. Due to the downregulation of ROS defense genes like SOD1, SOD2, and GPx, oxidative damage and neuronal death are enhanced in HD [29].

Nrf2 maintains the expression of numerous antioxidant enzymes, phase I and phase II enzymes, and a number of mediators that reduce inflammation. To protect neurons and glial cells from OS, neuroinflammation, and other pathogenic insults, Nrf2 serves as a crucial defensive mechanism in HD [30–33].

8.3.1 Endogenous Antioxidants

8.3.1.1 Dichloroacetate

Dichloroacetate (DCA) is found to boost pyruvate dehydrogenase complex (PDHC) activity and reduces lactate levels in the brain. It dramatically extended survival in the R6/2 and N171-82Q transgenic mice models of HD, enhanced motor performance, postponed weight loss, reduced the onset of striatal neuron atrophy, and shielded against diabetes [33].

8.3.1.2 L-Carnitine

It plays a part in facilitating the transport of fatty acids into mitochondria and also shields cells from oxidative damage. It decreases both the loss of neurons and the number of intranuclear aggregates in neurons and exerts neuroprotective effects against HD [34].

8.3.1.3 Melatonin

Melatonin lowers lipid peroxidation levels as well as protein carbonyl content and boosts succinate dehydrogenase and SOD activity against 3-NP-induced OS in an animal model of HD. It increases neuronal survival and decreases DNA damage, reducing the rise in SOD activity, protein carbonyls, and lipid peroxidation within the striatum against the 3-NP model of HD [35].

8.3.1.4 Glutathione, Catalase, and Superoxide Dismutase

Mutant cells (GRed) displayed elevated glutathione levels in the intracellular space, as well as elevated glutathione reductase and GPx activities [36]. The striatum regions of the brains of HD patients have been detected with OS and significant activation of antioxidative stress enzymes [37, 38].

8.3.2 Dietary Antioxidants

8.3.2.1 Vitamin C

Vitamin C reverses neurodegeneration and reduces the behavioral phenotype of HD [35]. Preclinical research suggests that HD is linked to ascorbic acid insufficiency and inhibits cortical afferents' ability to release glutamate (Fig. 8.3). Moreover, sodium ascorbate has been found to restore the level of striatal extracellular ascorbate in R6/2 mice [39].

8.3.2.2 Phenols

Polyphenol such as green tea is linked to a reduction in early HD pathogenesis events, including Huntington's misfolding. The combination of fish oil and quercetin has also been said to provide defense against HD brought on by 3-nitropropionic acid [40]. Grape seed polyphenolic extract (GSPE) therapy greatly reduces polyQ aggregation in the PC12 cells and decreases motor impairments in the R6/2 mice as well as improves lifespan in both the fly and R6/2 mouse models of HD [41].

Flavonoids like hesperidin, quercetin, naringin, and EGCG in various concentrations are efficient components in both the prevention as well as treatment of HD. Flavonoids focus on a variety of pathways.

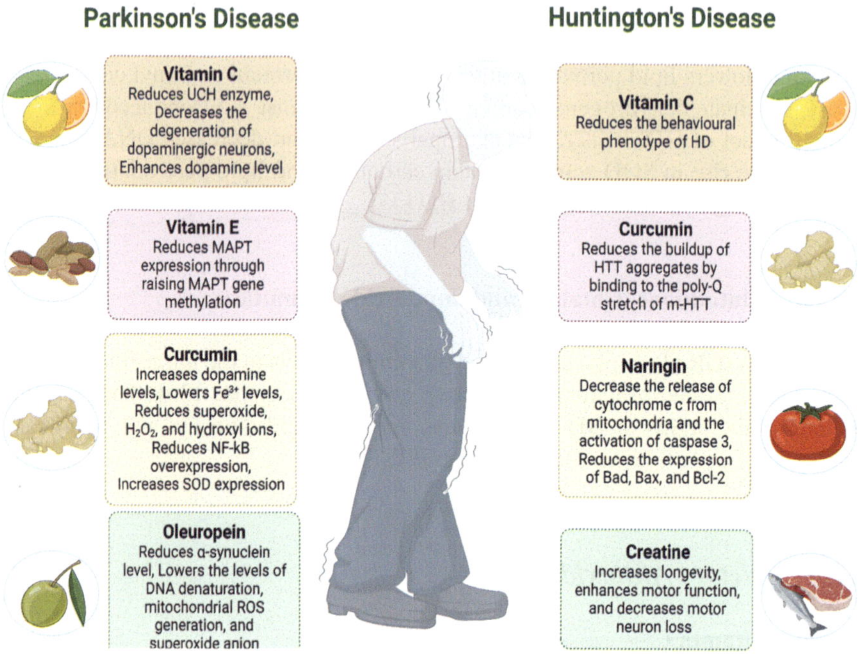

Fig. 8.3 Mechanism of Action of dietary antioxidants in Parkinson's and Huntington's disease (*UCH*: Ubiquitin C-terminal hydrolase; *MAPT*: Microtubule Associated Protein Tau; H_2O_2: Hydrogen peroxide; *NFκB*: Nuclear factor kappa B; *SOD*: Superoxide dismutase; *DNA*: Deoxyribonucleic acid; *ROS*: Reactive oxygen species; *HD*: Huntington's disease; *HTT*: Huntingtin protein; *Bad/Bax/Bcl-2*: apoptosis markers)

1. They can reduce the generation of ROS and boost the generation of glutathione, which reduces OS.
2. They can chelate metal ions and lessen metal ion toxicity which increases OS in neural tissues.

Both of these methods aid in reducing OS, which in turn causes the downregulation of inflammatory mediators and, as a result, a decrease in neuroinflammation and neuroprotection [42].

A flavonoid called **myricetin** works as the key player in an interaction with the CAG motif that stops the translation of mutant huntingtin protein and the sequestration of MBNL1. Additionally, myricetin was discovered to lessen the proteo-toxicity caused by the aggregation of polyglutamine, and its supplementation also helped to improve the HD mice model's neurobehavioral impairments [43]. **Naringin** is found to decrease the 3-NP-induced apoptosis by lowering the activation of caspase 3 and the release of cytochrome c from mitochondria. The use of Naringin also reduced the expression of Bad and Bax, two pro-apoptotic indicators. It also prevented the 3-NP-induced reduction in Bcl-2 mRNA expression [44, 45].

Curcumin at dosages of 0.01 μM in human bone marrow neuroblast cells, modulates HSP70 and HSP90 expression, reducing the buildup of HTT aggregates in HD [46]. Curcumin stimulated-HSP70 inhibits the formation of aggregates by binding to the poly-Q stretch of m-HTT, which is the mechanism behind its anti-amyloidogenic effect. Curiously, curcumin does not affect HSP90's activity on Akt, which lowers the apoptotic stimuli [47].

8.3.2.3 Creatine

It improves motor function, prolongs survival, attenuates brain and body weight loss, and lessens neuronal atrophy in N171-82Q and both R6/2 mouse models of HD, as well as the size of striatal lesions and behavioral changes brought on by neurotoxins (malonate and 3-NP). It also reduces the elevated 8-OHdG levels in the blood as well as the brain's ATP [48]. In addition, creatine treatment increases longevity, enhances motor function, and decreases motor neuron loss in N-171-82Q HD and R6/2 mice models [49].

8.4 Antioxidants and Alzheimer's Disease

Alzheimer's disease (AD) impairs memory abilities and the ability to do even the most common tasks [50, 51]. It occurs due to the buildup of β amyloid plaques and neurofibrillary tangles of hyperphosphorylated tau [52–54]. Several oxidative damage markers have been linked to AD, including nitration, advanced glycation end products, lipid peroxidation adduction products, free carbonyls, and carbonyl-modified neurofilament protein. In AD patients, the plasma levels of antioxidants like uric acid, bilirubin, albumin, lycopene, vitamin E, vitamin C, and A were lowered [55]. It was also discovered that various AD brain regions, particularly the frontal and temporal cortices, exhibit significantly reduced activity of antioxidant enzymes like glutathione peroxidase, catalase, heme oxygenase, and superoxide dismutase [56].

8.4.1 Endogenous Antioxidants

8.4.1.1 Glutathione

Glutathione functions by causing a reduction of the protein's sulfenic ion through covalent adduction and maintains the balance of protein sulfhydryl molecules in eukaryotic cells [57]. The reduction of free radicals in mitochondria is accomplished by N-acetyl-l-cysteine choline ester and glutathione choline ester, which are produced due to the formation of the ester link between GSH and choline. It was shown to efficiently reduce protein oxidation and prevent DNA breakage caused by the Aβ peptide in the mitochondria of AD cells or neurons damaged by mutant APP (amyloid precursor protein) [58]. Reduced GSH antioxidant activity leads to initiate the progression of AD [59].

8.4.1.2 Superoxide Dismutase

Superoxide dismutase (SOD), an enzyme, catalyzes the conversion of superoxide radicals into H_2O_2. Three separate forms of SOD are present in the body: manganese-SOD, copper/zinc-SOD (mainly in the mitochondria and cytoplasm), and extracellular SOD. Its expression and activity in the hippocampus region of the brain are decreased in AD [60]. This antioxidant is thought to play a crucial role in human aging and AD. It was seen that SOD knockout mice show an increase in tau phosphorylation, and deposition of Aβ plaques and aggravate behavioral impairments [61]. In 3X-Tg-AD, an aggressive AD mouse model, antioxidant treatment such as EUK-207, SOD mimetic, diminished the spread of tau phosphorylation and thereby reduced clinical symptoms [62].

8.4.1.3 Catalase

The peroxisomes contain the enzyme catalase, which converts approximately 6 million molecules of H_2O_2 into O_2 and H_2O per minute. Reduced catalase activity is seen in rats with AD caused by Streptozotocin (STZ) [63]. The Aβ in AD inhibits the enzyme catalase, causing H_2O_2 to build up in the hippocampal neurons [64]. It is possible for Aβ to directly or indirectly inhibit the catalase activity and encourages OS [65].

8.4.1.4 Methionine Sulfoxide Reductase

Methionine sulfoxide reductase-A (MsrA) is in charge of turning methionine sulfoxide into methionine and its expression is reduced in various areas of the brain in AD. Decreased MsrA activity alters the solubility characteristics of Aβ in AD and results in mitochondrial dysfunction; hence, increasing MsrA activity can help slow the progression of AD [66, 67].

8.4.1.5 Uric Acid

It is one of the most significant antioxidants in human bodily fluids and has antioxidant, anti-inflammatory, and neuroprotective properties. AD patients have much lower levels of UA than healthy individuals [68]. By quenching superoxide and singlet oxygen, UA can scavenge up to 60% of free radicals in the peripheral system [69].

8.4.2 Dietary Antioxidants

8.4.2.1 Vitamin E

A well-known antioxidant called vitamin E has been shown to provide neuroprotection in AD [70]. The best sources of vitamin E include sunflower seeds, almonds, wheat germ oil, and hazelnuts [71]. Vitamin E's capacity as an antioxidant is a result of the hydroxyl group's presence in the phenolic ring structure. According to a recent meta-analysis, it has the strongest protective effects against AD [72].

8.4.2.2 Vitamin B

Vitamin B plays a protective role against AD by altering the level of phosphorylated tau, and OS, modifying the brain energy metabolism and improving cognitive function [73]. Supplementing with B vitamins is said to lessen the risk of AD with elevated homocysteine (Hcy) levels [74]. Vitamins B6, B9, and B12 cause a reduction in Hcy levels, and aid in the management of this modifiable risk factor for AD [75].

8.4.2.3 Vitamin A

Vitamin A (retinol, retinal, and retinoic acid) suppresses the synthesis, extension, and destabilizing effects of β-amyloid fibrils. It prevents the oligomerization of Aβ, reduced Aβ accumulation and tau phosphorylation, slowed neuronal degeneration, and enhanced spatial learning and memory [76, 77].

8.4.2.4 Phenols

Natural anti-oxidants called polyphenols have anti-AD effects via a variety of biological processes, including interactions with transition metals, suppression of the inflammatory response, blockage of free radicals, and modification of enzymes' activity [78]. They lower Hcy in AD patients seen in a clinical trial [79–81].

β-carotene belongs to the group of hydrocarbons called carotenoids and acts as an antioxidant. It can help with memory improvement in AD patients [82]. The diagnosis of AD was correlated with plasma β-carotene levels. Lower plasma β-carotene levels are linked to an AD diagnosis [78].

The polyphenol benzopyrene compounds known as **coumarins** have anti-inflammatory and anti-cancer properties [83]. They work by boosting signaling through the antioxidant response element and Nrf2 pathways [84]. The Nrf2/ARE pathway protects the cell by reducing OS which leads to apoptosis. Daphentin, a type of coumarin, is capable of preventing cytochrome c leakage in mitochondria, and loss of membrane potential as well as t-BHP-initiated NLRP3 inflammatory pathways [83]. Coumarins inhibit acetylcholinesterase (AChE) and lead to the reduction in Aβ deposition and β-secretase inhibition [85, 86].

Flavonoids can be found in a variety of fruits, vegetables, wine, tea, cereals, and other plant-based beverages and foods, such as chocolate. They can promote neurogenesis, activate neural regeneration, enhance current neuronal function, and prevent malfunctioning of neurons. They can encourage the removal of Aβ-peptides and prevent tau from being phosphorylated by the mTOR/autophagy signaling. Flavonoids can also be promising symptomatic anti-medicines for AD because of their ability to inhibit cholinesterase [87].

The turmeric, derived from *Curcuma longa*, contains **curcumin**, which has been shown to have anti-amyloid effects in AD. Curcumin can diffuse the BBB and binds to Aβ due to its lipophilic nature. Its neuroprotective efficacy against Aβ toxicity was demonstrated by its ability to inhibit Aβ fibril formation from Aβ monomer, suppress Aβ aggregation, and dismantle the fibril form of Aβ (Fig. 8.4). Additionally, it disaggregated Aβ deposits, inhibited the formation of new Aβ deposits, and decreased the size of the residual Aβ aggregations [88, 89]. It has been shown to suppress β-secretase and AChE activity as well as Aβ aggregation and Aβ-induced

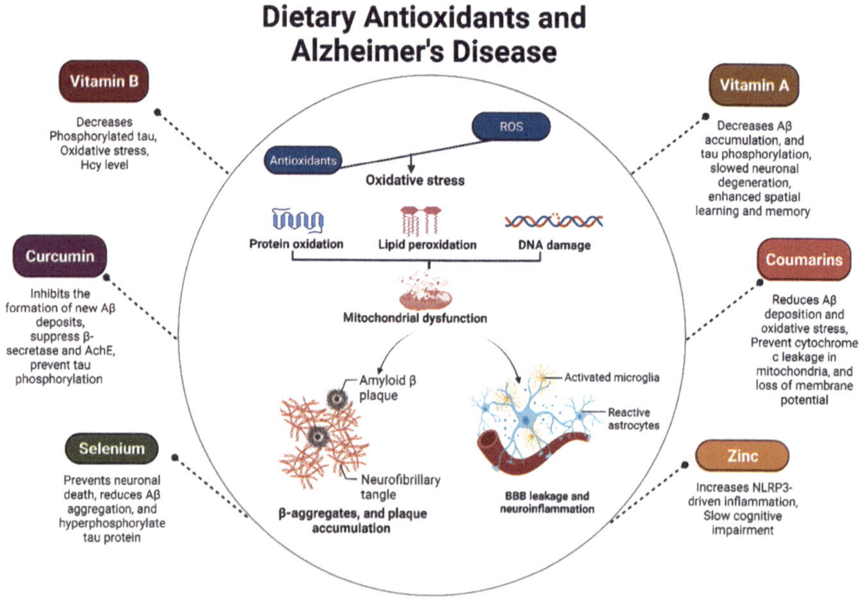

Fig. 8.4 Role of dietary antioxidants in Alzheimer's disease. (*Aβ:* Amyloid β; *AchE:* Acetylcholine esterase; *H_2O_2*: Hydrogen peroxide; *NLRP-3:* NLR family pyrin domain containing 3; *BBB:* Blood brain barrier; *DNA:* Deoxyribonucleic acid; *ROS:* Reactive oxygen species)

inflammation. Oral administration of curcumin reduces behavioral impairment in animal models of AD and prevents Aβ deposition, Aβ oligomerization, and tau phosphorylation [90].

8.4.2.5 Zinc

Zinc plays multiple roles in AD as both the enzymatic breakdown of the Aβ peptide and the non-amyloidogenic processing of the APP. Zinc binds to Aβ, encouraging its aggregation into neurotoxic species, and zinc homeostasis disruption in the brain causes deficiencies in synaptic function and memory. Consequently, zinc dyshomeostasis may be crucial in the development of AD, and zinc chelation is a potential treatment strategy [91, 92]. Its deficiency exacerbated cognitive impairment in an animal model of AD via increasing NLRP3-driven inflammation [93].

8.4.2.6 Selenium

Selenium (Se) is a biological trace element significant for the functioning of the brain. Consolidated evidence from meta-analyses shows that selenium status is much lower in AD brains than in controls [94]. Se's biological actions are mostly carried out by selenoproteins, which are essential for preserving healthy brain function. Selenoproteins particularly those linked with brain function have a role in the development of AD. The impact of the ER (endoplasmic reticulum)-resident protein SELENOK on Ca^{2+} equilibrium, the receptor-associated synaptic activities, and the role of GPX4 in ferroptosis are explored as putative roles of these selenoproteins in AD [95]. It prevents neuronal death, reduces amyloid β aggregation, and hyperphosphorylate tau protein in the fight against AD [96, 97].

8.5 Antioxidants and Epilepsy

Epilepsy is a chronic, dynamic neurological condition that causes continuing brain damage. The development of epilepsy is influenced by an oxidative injury that leads to neuronal death [98, 99]. Glutamate excitotoxicity, OS, and mitochondrial dysfunction are all contributing factors to epilepsy [100, 101]. Patients with epilepsy suffer from chronic neurological conditions such as spontaneous recurrent seizures and deficits in learning and memory [102, 103]. Epileptogenesis is a collection of events that transforms a normal brain into one that experiences recurrent seizure activity. Neurodegeneration, damage to the BBB, and dysfunction of the glutamatergic system, which is caused by neuroinflammation, are important variables that contribute to epileptogenesis. Hypoxia and OS are also thought to involve in the epigenetic alteration of DNA. In addition, hypoxia can cause the complement system and cytokines to activate, both of which support neuroinflammation. In a feedback loop, the neuroinflammation in turn triggers the cytokine and complements system production. The convergence of all these mechanisms causes epilepsy to develop [104].

8.5.1 Endogenous Antioxidants

8.5.1.1 Alpha Lipoic Acid

It is an endogenous thiol that occurs naturally in mammalian tissues and functions as a cofactor for α-ketodehydrogenase complexes [105]. ALA can produce endogenous antioxidants such as vitamins C and E and GSH in the body. It prevents DNA damage brought on by peroxynitrite and the production of hydroxyl radicals [106]. After pilocarpine-induced convulsions, antioxidant therapy dramatically decreased nitrite content and lipid peroxidation as well as elevated catalase and SOD activity in the hippocampus of rats [107]. Administration of ALA considerably reduces the frequency of spontaneous seizures [102].

8.5.1.2 Melatonin

One of the most enigmatic compounds made by the human body is melatonin, an indoleamine derivative of serotonin produced in the pineal gland, a region of the epithalamus. By increasing GABA-ergic neurotransmission and decreasing glutamatergic neurotransmission, melatonin also regulates the electrical activity of neurons. Melatonin may reduce seizures in people, and it works best for treating juvenile intractable epilepsy. Additionally, melatonin reduces electron leakage from mitochondria, reducing the production of free radicals. All of these procedures lessen DNA damage, lipid peroxidation, and protein peroxidation [108]. Melatonin and its analogs, which bind to melatonin receptors, are used to control seizures [100]. It lessens the activation of certain proteins in the hippocampal region, including the transient receptor potential (TRP), and glutamate receptors, and regulates excessive OS products, as well as mitochondrial and Ca^{2+} dysregulations in epilepsy [109].

8.5.1.3 Coenzyme Q10

CoQ10 pretreatment reduces spontaneous recurring seizures and prevents hippocampus neuronal death and abnormal mossy fiber sprouting (MFS) by reducing the burden of OS [110]. In epilepsy mouse models, CoQ10 is used as an adjuvant for Anti-epileptic drugs (AEDs) therapy, suggesting that it may lessen seizure severity and guard against seizure-induced oxidative damage that contributes to the cognitive impairment linked to long-term use of AEDs [111]. In rats experiencing pilocarpine-induced status epilepticus, CoQ10 supplementation reduced RNA oxidation, seizure onset, and neuronal death [112].

8.5.2 *Dietary Antioxidants*

8.5.2.1 Vitamin E

Vitamin E increases catalase activity in mouse epilepsy models using pilocarpine. A drop in the level of vitamin E has been seen in the cerebral cortex after pilocarpine-induced seizures. Vitamin E and glutathione treatment can decrease neuronal mortality and lipid peroxidation in kindling rat models of epilepsy [71, 113].

8.5.2.2 Vitamin C

Vitamin C can cross the BBB with ease and lessens hippocampal damage during seizures. It strengthens cell membranes and reduces lipid peroxidation and inhibits seizure activity that can lower mortality, depending on the type of seizure [71, 114]. It increases the hippocampal SOD and catalase activities, lengthens the time between the onset of the first seizure, suppresses behavioral seizure episodes, and reduces brain damage [115]. Vitamin E and C supplementation causes a considerable decrease in serum MAD levels and an increase in serum total antioxidant status (TAS), which was accompanied by a reduction in seizure frequency of more than 70% [116].

8.5.2.3 Flavonoids

Flavonoids are present in vegetables, fruits, nuts, and beverages made from plants, traditional eastern remedies, and herbal nutritional supplements. In epilepsy, flavonoids may provide neuroprotection [117]. The flavonoids eugenol, naringin, silibinin, naringenin, hesperetin, and morin have been found to lessen the symptoms of epilepsy. These effects appeared to be occurring via two main mechanisms:

1. The amelioration of hippocampal structural modifications, including through dentate gyrus (DG), granule cell dispersion (GCD), and
2. The inhibition of pro-inflammatory cytokine expression [118].

Naringenin, one of the most prevalent flavanones found in citrus fruits, is found to inhibit OS biomarkers in a rodent model of epilepsy by activating numerous signaling pathways [119, 120]. Naringenin therapy can minimize the intensity of seizures, decrease lipid peroxidation and ROS production, and restore antioxidant enzymes in the hippocampus of epileptic mice [121].

Polyphenol **resveratrol** lowers ROS production by stifling mitochondrial complex II activity and cytochrome c leakage. It indirectly inhibits OS, apoptosis, and inflammation by activating sirtuin 1, a class III histone deacetylase. It causes inhibition of neurodegeneration, mossy fiber sprouting, astro- and microgliosis, and spontaneous recurrent seizures [112]. By opening voltage-gated sodium channels and activating calcium-activated potassium channels, resveratrol can reduce the activity of cortical neurons and a reduction in the rate at which neurons fire an action potential [122].

Curcumin can lessen seizures, reduce several markers of OS, and stop hippocampus neuronal loss and MFS [123]. It lessens the severity of spontaneous recurring seizures and acts as an inhibitor of NFκB and a strong inducer of the HO-1 protein. These two elements are crucial in the body's OS and inflammation. Through its capacity to reduce the production of several inflammatory indicators, including COX-2, lipoxygenase, and inducible nitric oxide synthase (NOS), curcumin can help treat epilepsy [124]. It also inhibits the expression of α-synuclein and the Wnt/β-catenin, apoptosis, and autophagy pathways in brain regions [125] [126]. Combined usage of curcumin and carbamazepine demonstrates that the powerful antioxidant curcumin can be utilized as an adjuvant in antiepileptic medication [127].

8.6 Antioxidants and Amyotrophic Lateral Sclerosis

Amyotrophic Lateral Sclerosis (ALS) is first reported by Joffrey and Jean-Martin Charcot. Upper motor neurons (UMN) of the motor cortex, as well as lower motor neurons (LMN) of the spinal cord and brainstem, are selectively lost in ALS [128, 129]. Patients with ALS experience respiratory failure, dysphagia, and muscle atrophy [130–132]. Increased OS is linked to the etiology of ALS, and neuroinflammation is brought on at the pathogenic level by the activation of astrocytes, microglia, and peripheral immune cells [133].

8.6.1 Endogenous Antioxidants

8.6.1.1 Coenzyme Q10

It is known to prevent OS by scavenging free radicals in ALS. When given 50 days after birth, CoQ10 (200 mg/kg daily) dramatically enhanced the cerebral cortex's mitochondrial CoQ10 contents and lengthened the longevity of SOD1G93A transgenic mice [134]. It also boosts brain mitochondrial concentration and prolonged longevity in the SOD1 transgenic mice [135].

8.6.1.2 Melatonin

Melatonin lowers OS and inhibits apoptotic pathways in ALS [136]. High dosages of melatonin slowed the course of the disease and improved survival rates in ALS [137, 138]. Melatonin (30 mg/kg) is found to delay the onset of disease, neurological degeneration, and death rate [139]. However, melatonin-treated mice displayed increased motor neuron loss, 4-HNE (levels of the lipid peroxidation marker), and upregulation of SOD1 level, indicating that melatonin worsens the disease phenotype in the SOD1G93A model by enhancing toxic SOD1 [140].

8.6.1.3 Glutathione

The redox imbalance of GSH is linked as a significant modulator of enhanced ROS production and death in motor neurons and astrocytes. Its depletion and neuronal toxicity have been linked to mutations in the ALS-causing genes [141]. The GSH reduction in motor neuron cells and the spinal cord is associated with caspase 3 activation, apoptosis-inducing factor (AIF) translocation, and motor neuron deterioration during the progression of ALS-like disease [142]. The activation of ALS-causing genes triggers numerous pathways and regulators that result in a GSH redox imbalance [143]. In SOD1G93A transgenic mice, Nrf2 overexpression in astrocytes improved survival and postponed neuromuscular denervation [141, 144, 145].

8.6.1.4 Superoxide Dismutase

More than 180 mutations in the SOD1 gene's coding area and several others in its non-coding regions have been found in ALS patients [146]. These mutations cause a decrease, maintenance, or increase in dismutase activity when compared to SOD1 in its wild-type state [140]. Motor neurons that express mutant SOD1 are vulnerable to OS-induced cell death [147, 148]. The calcium-binding ER chaperone calreticulin is present at reduced levels in motor neurons. The activation of the Fas/NO pathways in motor neurons requires a reduction in the expression of this protein, which is both required and sufficient [149].

8.6.1.5 Catalase

In transgenic fALS mice, catalase was modified with putrescine to help it better traverse the BBB, delaying the development of disease signs [150]. This antioxidant did not delay the onset of the disease in SOD1G93A animals, but it does show that avoiding peroxide-mediated mitochondrial damage stops the disease [151].

8.6.2 Dietary Antioxidants

8.6.2.1 Vitamin E

Long-term vitamin E supplementation has been linked to lower ALS rates, and a study found a non-significant reduction in ALS risk in men who received α-tocopherol supplementation (50 mg/day) [132]. Vitamin E is mostly obtained from legumes, and regular use of this vitamin is linked to a lower risk of death in ALS patients [152, 153]. Preclinical research with SOD1G93A transgenic mice revealed that vitamin E supplementation (200 UI/kg) reduced the beginning of the disease and delayed its progression, but had no effect on survival time [154]. Patients taking α-tocopherol and riluzole for 3 months showed a decline in plasma TARS and an elevation in GSH levels [130].

8.6.2.2 Carotenes

Carotenes are natural pigments that give fruits and vegetables their orange, red, yellow, or green color. They also have antioxidant and ROS-neutralizing capabilities [155, 156]. Carotenoid intake delays the onset of ALS; nevertheless, case-control research involving 77 Koreans who had been diagnosed with the disease found a negative correlation between ALS and dietary intake of carotenes [132]. β-carotene can be used to treat apoptosis and neuroinflammation in ALS patients [157]. Patients with ALS who regularly take carotenoid supplements have higher survival times [153].

8.6.2.3 Phenols

A strawberry extract rich in anthocyanins, with the main ingredient being callistephin, delayed the onset of ALS, preserved grip strength, and extended longevity in SOD1G93A mice [158, 159]. **Fisetin** (9 mg/kg) improves motor capabilities, delayed the start of the disease, and increased longevity in SOD1G85R *Drosophila melanogaster*, and SOD1G93A mice. The ERK pathway is stimulated to control cell survival and appears to be the main signaling pathway behind the activity of fisetin [160]. **Quercetin** is found to lessen mitochondrial damage and reduces neuronal death and inhibit the assemblage and misfolding of SOD1 linked to ALS [161].

Resveratrol exhibits beneficial effects through increasing sirtuin 1 (SIRT1) expression in ALS [162]. It maintains lower and upper motor neuron function and enhances the mitochondrial activity of muscle fibers. It restores the down-regulated AMPK/SIRT1 signaling that was present in the bone marrow mesenchymal stem cells (BMSCs) of ALS patients [132]. It increases survival and postpones the start of ALS [160, 163].

Fruits, coffee, tea, and grains contain phenolic acids. They make intriguing possibilities for improved ALS therapy because of their variety of neuroprotective properties. **Protocatechuic acid** (100 mg/kg) increases survival, enhances motor function, and lowers gliosis in SOD1G93A mice [164]. **Gallic acid** and **wedelolactone** can enhance motor learning capacities and locomotor function in ALS. Both work by decreasing inflammatory cytokines, causing normalization of L-glutamate levels, and reducing caspase-3 activation [165]. **Rosmarinic acid**, the primary component of rosemary extract, improved motor function, prolonged longevity of SOD1G93A mice, and decreased weight loss [166]. **Caffeine acid phenethyl ester** (CAPE) stimulated the antioxidant response element while deactivating the OS-associated NFκB release and delayed the course of symptoms and lengthened survival which leads to a decrease in phospho-p38 levels and glial activation [167].

8.6.2.4 N-Acetyl-L-Cysteine

N-acetyl-L-cysteine (NAC) restores depleted GSH pools, and plasma levels of cysteine, and reduces the effects of free radical damage. It also reduces mitochondrial ROS production, restores the MTT level, and also boosts ATP levels in SH-SY5Y cell lines with the G93A SOD1 mutation. Furthermore, NAC (2 mg/Kg/day) treatment dramatically increased motor function and prolonged survival in SOD1G93A transgenic mice [140].

8.7 Antioxidants and Multiple Sclerosis

The CNS is affected by the chronic inflammatory autoimmune illness known as multiple sclerosis (MS). ROS are crucial in several processes that underlie the pathophysiology of MS. The CNS is equipped with a defense mechanism that includes enzymatic and non-enzymatic antioxidants to counteract the harmful effects of ROS. Antioxidants are used in MS and other autoimmune and inflammatory illnesses because OS is one of the most significant elements of the inflammatory process, which causes myelin breakdown and axonal damage [168, 169].

Clinically, inflammatory and OS mediators, including cytokines like IL-6, IL-1β, IL-17, INF-γ, and TNF-α have been linked to the progression of MS. Dietary antioxidants are found to control immune-inflammatory cell activation, which would reduce inflammation. They can also reduce OS, which would stop persistent demyelination and axonal damage [168, 170].

8.7.1 Endogenous Antioxidants

8.7.1.1 Glutathione

Active demyelinating MS lesions have considerably higher GPx gene expression. Elevated ROS is linked to the concentration of oxidized glutathione (GSSG) and the concurrent decline in α-tocopherol levels in the blood. The blood of MS patients with the progressive form had significantly higher levels of GSH in addition to GSSG as a compensatory mechanism that protects cells from further oxidative damage. During MS exacerbations, GSH oxidation is also enhanced in patients' cerebrospinal fluid (CSF). Oligodendrocytes are more susceptible to oxidative injury due to their inherently low GSH levels.

The loss of GSH, iron accumulation, mitochondrial dysfunction, and increased ROS production lead to elevated levels of protein carbonylation in MS. The substantial carbonylation of brain proteins can be produced by rapid GSH depletion. This effect is caused by the iron-catalyzed production of hydroxyl radicals from H_2O_2. As a result, the absence of GSH alone results in OS that is sufficient to generate protein carbonyls in addition to lipid peroxides. The findings imply that glutathione therapy is an effective treatment for neuroinflammatory illnesses like MS.

8.7.1.2 Superoxide Dismutase

Inflammatory circumstances that result in excessive TNF-α production are linked to the increased OS and antioxidant enzyme inhibition, most notably reduced SOD1 expression. SOD1 gene expression has been noticeably increased in actively demyelinating lesions in MS. The SOD1 activity in the erythrocytes of MS patients was also significantly reduced, which points to weaker enzymatic defense mechanisms against OS. In guinea pigs with Experimental autoimmune encephalomyelitis (EAE), SOD2 was found to have increased expression, but not SOD1.

8.7.1.3 Catalase

H_2O_2 affects oligodendroglia and can travel in the perivascular space and cause myelin and lipid peroxidation at distant locations in the interstitial optic nerve. Catalase in the CNS prevents the buildup of H_2O_2 and demyelination. Catalase treatment markedly decreased demyelination of the optic nerves and reduced neurological EAE symptoms. It has recently been revealed that combining the removal of superoxide by extracellular SOD (EC-SOD) and H_2O_2 significantly reduced experimental ocular neuritis in EAE [169].

8.7.1.4 Melatonin

Melatonin, which the pineal gland normally produces at night, is created outside the body from tryptophan. Meat, salmon, milk, eggs, nuts, seeds, soy products, and almonds are the main sources of melatonin. It promotes the production of SOD and glutathione peroxidase (GPx), particularly in SPMS (secondary progressive MS) patients [138, 171].

8.7.2 Dietary Antioxidants

8.7.2.1 Phenols

Curcumin can suppress proinflammatory cytokines and the infiltration of inflammatory cells into the CNS [172, 173] Additionally, macrophages and monocytes contribute to the production of COX-2, iNOS, macrophage inflammatory protein (MIP-1α), monocyte chemoattractant protein 1 (MCP-1) in the presence of curcumin while IL-12, IL-8, IL-6, IL-2, and IL-1 are inhibited by curcumin supplementation. It prevents the cytokines from mediating NF-κB pathway activation by inhibiting Akt (protein kinase B) and IκB (inhibitor of kappa B) through a variety of inflammatory stimuli. It also reduces the expression of NF-κB regulated gene products, such as IL-17, IL-1β, prostaglandin E2, MIP-1α, and TNF-α. It reduces the BBB disruption brought on by Th17 cells due to its role as an NFκB pathway inhibitor [170].

Resveratrol prevents neutrophils from producing the pro-inflammatory metabolites 5-LO and 15-LO, which are part of the arachidonic pathway. In EAE-induced mice, resveratrol was found to significantly reduce the levels of several cytokines and chemokines, including IFN-γ, TNF-α, IL-17, IL-12, IL-9, IL-2, as well as MCP-1, MIP-1α, and chemokine (C-C motif) ligand 5 (CCL5). It alters the synthesis of eicosanoids or blocks the COX-2 and iNOS pathways by inhibiting AP-1 or NF-κB. It was shown that in an EAE animal model, resveratrol decreased the inflammatory responses and clinical symptoms, which was mainly due to the reduction of pro-inflammatory mediators and triggering the apoptosis in activated T cells in the spinal cord [170].

Of the biologically active catechins found in green tea, **EGCG** is the most prevalent. Purified EGCG (95 percent) lessens the severity of EAE by lowering the severity scores, which were linked to lowered immune cell infiltrates, decreased demyelination in the spinal cord, and lowered levels of inflammatory cytokines that support Th1 and Th17 differentiation. Additionally, it improves hippocampus cell survival and development [174].

8.7.2.2 Vitamin D

In addition to supporting calcium homeostasis and bone health, vitamin D is important for immune regulation and the reduction of OS. A low measure of vitamin D leads to an increased possibility of MS development and relapse. Vitamin D possesses immunomodulatory and anti-inflammatory effects on the pathogenetic pathways of MS by preventing the development of CD4+ T cells, hence reducing the likelihood of developing MS and slowing the course of the illness.

8.7.2.3 Vitamin A

It is a fat-soluble substance that plays a number of roles in immunity, skin, and vision. Retinoids and carotenoids, which are components of vitamin A, can be found in milk, liver, cheese, green leaves, oil, fruits, and vegetables. A low quantity of vitamin A in plasma is associated with an increased probability of MS development. Patients with MS who are supplemented with high dosages of vitamin A (400 IU/day), show improvements in their fatigue, depression, and cognitive state [175].

8.7.2.4 Vitamin E

Vitamin E is found to have immunomodulatory effects on several immune cells. By lowering macrophages' production of the T cell inhibitory prostaglandin E2, it improves naive T cell activity. It is necessary for the proper operation and communication between T regulatory cells, dendritic cells, and CD8+T cells. It also appears to downregulate certain adhesion molecules (molecules that allow lymphocytes to travel past the BBB) and lowers the chance of getting MS [1].

8.7.2.5 Alpha Lipoic Acid

It scavenges ROS, chelating copper and iron, raising vitamin C and GSH levels, and healing OS damage. It has immunomodulatory properties as well. It increases the cAMP synthesis and decreases IFN-γ generation. It can also prevent the migration of macrophages, inflammatory T cells, and monocytes into the spinal cord and brain, possibly by lowering the expression of ICAM-1 and VICAM-1 by CNS endothelial cells, inhibiting the enzymes known as matrix metalloproteinases (MMPs), and lowering BBB permeability, thus can be used as a therapeutic strategy in several disorders, especially MS, AD, and diabetic neuropathy [1]. It inhibits monocyte infiltration into the CNS by lowering monocyte migratory potential and enhancing BBB integrity against OS attacks. A recent clinical trial that was double-blind, randomized, and controlled showed an improved TAC level after the daily intake of ALA (1200 mg/day) for 12 weeks in a group of MS patients [176].

8.7.2.6 Fatty Acids

One study found that MS incidence was low in those who consumed diets high in PUFAs. According to the findings of meta-analyses, PUFAs are ineffective at stopping the progression of the disease but may lessen the number of relapses. In human research, PUFAs are linked to a better quality of life, a minor improvement in relapse rat health, and a lower quantity of disability as measured by the Expanded Disability Status Scale (EDSS) [175, 177]. In contrast to the quality of life, EDSS score, or fatigue, a different study showed that PUFAs improved certain markers associated with inflammation and/or neurodegeneration in MS patients [178].

One PUFA with a low incidence of MS is α-linolenic acid. It can support the immune system by lowering inflammation-related indicators. MMP-9 levels in MS patients can also be reduced by eicosapentaenoic and docosahexaenoic acids (EPAs and DHAs). Fish oil supplements that are high in omega-3 fatty acids help MS patients by reducing their levels of MMP-9 and inhibiting its expression and also reducing inflammation and OS. By reducing proinflammatory cytokines and free radicals, omega-3 fatty acid supplementation enhances the quality of life of MS patients by lowering relapse rates [179, 180].

8.8 Antioxidants and Schizophrenia

Schizophrenia is a severe and crippling mental illness with an estimated 0.75 percent lifetime prevalence worldwide. The long-term effects of this condition are frequently negative, and even receiving treatment, people with schizophrenia have a three times higher risk of dying young than the general population. Positivity, negativity, and disorganization are signs of schizophrenia. Hallucinations and delusions are examples of positive symptoms and motivational decline, apathy, and social retrieval are negative signs. Numerous cellular structures have been observed to suffer oxidative damage as a result of elevated ROS levels and depleted antioxidant defenses. It has been demonstrated that patients with non-medicated, medicated, first-episode and chronic schizophrenia have reduced levels of TAC and glutathione in their plasma. Additionally, schizophrenia patients' peripheral tissues have been discovered to have higher quantities of ROS in addition to lower levels of SOD and GPx [181].

A higher quantity of 8-hydroxydeoxyguanosine, a marker of DNA damage and cell death, as well as protein carbonylation, have all been seen in schizophrenic patients. Increased OS can result in from impairments in catalase, SOD, glutathione, and GPx, as well as thiobarbituric acid reactive substances (TARS) and MDA, and decreased antioxidant levels in the red blood cells (RBC), cerebrospinal fluid, plasma, and serum. Additionally, ROS production can also rise due to mitochondrial dysfunction, dopamine auto-oxidation, and the pro-oxidant properties of several antipsychotic drugs [182].

8.8.1 Endogenous Antioxidants

8.8.1.1 Superoxide Dismutase

Schizophrenic patients have faced both enhanced and decreased SOD activity. It's likely that when the condition worsens, SOD levels increase as a protective measure against OS. Its activity is considerably reduced in RBC specimens from schizophrenia patients. An altered antioxidant defense system in addition to abnormalities in the peripheral activity of SOD has been shown in schizophrenic patients. The frontal and temporal cortex has been found to have an elevated level of Mn-SOD with no change in Zn or Cu-SOD. An increase in Cu, Zn, and Mn-SOD was seen in the substantia innominata regions and frontal brains of schizophrenia patients.

8.8.1.2 Glutathione Peroxidase

It is associated with the elimination of H_2O_2 and other peroxides using GSH. When compared to control subjects, first-episode schizophrenia patients with drug naiveness had significantly higher plasma GPx activity. Additionally, GPx activity was shown to be lower in RBC samples from schizophrenic patients but plasma samples from both neuroleptic-naive and long-term neuroleptic-free showed higher GPx activity.

8.8.1.3 Catalase

Catalase activity did not change in leukocytes, it was shown to be much higher in the erythrocytes of schizophrenia patients. Additionally, compared to control participants, drug-naive first-episode schizophrenia patients had significantly lower plasma CAT activity [183, 184].

8.8.2 Dietary Antioxidants

8.8.2.1 Vitamin E and Vitamin C

Vitamin E and C are seen to avoid oxidative damage and the aggravation of symptoms in schizophrenia. The majority of ROS are produced in the nucleus, mitochondria, cytosolic proteins, and nucleus, where vitamin E has a limited ability to counteract oxidative damage. Vitamin C protects neurons from OS, ensures normal neurotransmission control, reduces inflation, and modifies neuronal development and epigenetic function. Vitamin C can not only reduce membrane phospholipid peroxidation but also improve vitamin E regeneration. It's interesting to note that vitamin C levels in the brain are 10 times greater than in serum, and it can pass past the BBB and remain there by way of the glucose transmitter GLUT1. SOD and vitamin C measures are considerably lower in schizophrenia patients than in healthy controls.

Schizophrenia patients show higher serum MDA levels and lower plasma ascorbic acid levels. Vitamin C supplementation combined with atypical antipsychotics can reduce OS, and raise ascorbic acid levels. Vitamin C treatment alone or in conjunction with vitamin E significantly lowers total dyskinetic movement scores and improves Brief Psychiatric Rating Scale (BPRS).

8.8.2.2 Vitamin D

Vitamin D is frequently inadequate in schizophrenia patients. Vitamin D modulates neurotrophin synthesis, calcium homeostasis, neuro mediators synthesis, and reduces oxidative damage. It is seen that lower levels of vitamin D lead to cognitive dysfunction and more severe symptoms in schizophrenia [184].

8.9 Antioxidants and Stroke

A stroke is a neurological deficit caused by an acute, focused injury to the CNS caused by intracerebral hemorrhage (ICH), subarachnoid hemorrhage (SAH), and cerebral infarction. It is a leading cause of disability and death worldwide [185]. An important mechanism of cell damage during cerebral ischemia is OS, which is caused by excessive formation of ROS or defective metabolism [186]. According to research and clinical investigations, OS is a major factor in brain damage that occurs after a stroke [187]. Overproduction of ROS occurs in OS and causes damage to neurons and kills cells [188, 189].

8.9.1 Endogenous Antioxidants

8.9.1.1 Bilirubin

It is the most powerful endogenous antioxidant and increases in various OS conditions, including stroke. It is a byproduct of heme metabolism and can affect the occurrence and prognosis of ischemic stroke [190]. Although all bilirubins have some antioxidant qualities, only unbound, bioactive bilirubin can pass the BBB and be used to treat ischemic strokes [191]. Using tertiles of bilirubin, this study included 13,214 patients and evaluated the risk of stroke linked with an increase in total bilirubin level of 1.71 mol/L [192]. This study found that total bilirubin levels were positively correlated with stroke outcomes in participants with a history of stroke and negatively correlated with stroke prevalence [193].

8.9.1.2 Uric Acid

It scavenges hydroxyl radicals, superoxide anion, hydrogen peroxide, and peroxynitrite and acts as a powerful antioxidant in human plasma. Preclinical investigations in hyperglycemic mice showed improved stroke outcomes after UA therapy. In healthy volunteers, IV injection of UA enhanced serum free-radical scavenging ability both at rest and during intense physical activity and eliminated lipid peroxidation [194, 195].

8.9.2 Dietary Antioxidants

8.9.2.1 Vitamin E

People with high dietary vitamin E consumption have a 17 percent lower risk of stroke compared to those with low dietary vitamin E intake. Vitamin E exhibits its capacity to protect cell membranes by scavenging ROS and inhibiting lipid peroxidation [196, 197].

8.9.2.2 Vitamin D

Individuals who obtained vitamin D experienced a notable improvement in their stroke prognosis after three months [198]. Vitamin D insufficiency is linked to an increased risk of ischemic stroke, with hypertension, diabetes mellitus, hyperlipidemia, and ischemic heart disease as contributing factors [199]. Vitamin D insufficiency was independently linked to ischemic stroke in both major artery atherosclerosis and cardiac embolic stroke [200].

8.9.2.3 Vitamin C

In view of the randomized controlled trials findings that vitamin C had no adverse impact on preventing stroke [201]. Plasma vitamin C levels are inversely correlated with the incidence of stroke and can be used as a preventive component [202].

8.9.2.4 Omega-3 Fatty Acids

Since mammals cannot produce omega-3 fatty acids, they must obtain them from their diet. Three different kinds of omega-3 PUFAs are present: α-linolenic acid, DHA, and EPA. In both adult and newborn animal models, all fatty acids exhibit a neuroprotective effect against brain damage brought on by experimental stroke [203]. They are particularly crucial for the human brain, and a lack of them increases the chance of developing several illnesses [204]. The severe deficiency of omega-3 fatty acids in the diet can enhance the probability of stroke [205].

8.9.2.5 Phenol

They work by blocking xanthine oxidase, reducing the production of hypoxanthine, xanthine, oxygen radicals, raising the levels of MDA, reducing glutathione, and leading to cause a reduction in OS in stroke patients [206]. They prevent stroke by protecting the integrity of the endothelium and counteracting the harmful consequences brought on by ionic imbalance, excitotoxicity, and the production of ROS [207]. The risk of ischemic stroke is inversely correlated with flavanone intake [208]. Specific flavonoids and their physiologically active metabolites have positive effects on platelet function, inflammation, thrombosis, and protection against ischemia-reperfusion injury and arrhythmia [209].

8.10 Antioxidants and Brain Cancer

The growth and survival of primary CNS cancers such as medulloblastoma, glioblastoma multiforme, and ependymoma depend on the presence of cancer propagating cells (CPCs). These cells also referred to as BCPCs (brain cancer propagating cells), can regenerate and multiply. The evidence is mounting that neural stem cells (NSCs) and their progenitors may undergo metamorphosis to become BCPCs. [210] An intracranial neoplasm known as a brain tumor can develop in either the brain or the central spinal canal. Most adult brain tumors are secondary or metastatic tumors, meaning that they can develop from cancers that are primarily found in other organs but have moved to the brain [211, 212].

The development of brain tumors has been linked to OS, which is expressed by an imbalance between the generation of free radicals and antioxidant defenses [213]. In these circumstances, both endogenous sources (peroxisomes and mitochondria, but also neurotransmitter oxidation or inflammatory cell activation) and exogenous sources (environmental factors, medications, irradiation, and chemicals) may produce excessive amounts of free radicals [214]. ROS may play a role in a variety of stages of carcinogenesis, including initiation, progression, angiogenesis, and metastasis [215]. The increasing quantity of ROS that results in tumor formation involves not only oxidative aggression but also a diminished response to antioxidant defenses. Both endogenous enzyme and non-enzymatic antioxidant systems work to avoid or lessen the harm done by too many free radicals [213].

8.10.1 *Endogenous Antioxidants*

8.10.1.1 Superoxide Dismutase

The majority of brain tumor types have elevated Mn-SOD expression, which is associated with a bad prognosis. In the proliferative stage, it appears to be a tumor suppressor. It is increased when a tumor develops more quickly [216] [217]. ROS levels are necessary for tumorigenesis and metastasis. Low Mn-SOD levels make cells more susceptible to OS, which can cause them to develop into tumor cells [218].

8.10.1.2 Glutathione

The main endogenous neuroprotectant for the brain is GSH. GSH shields brain cells from oxidative damage caused by peroxynitrite and lipid peroxidation in neuron cells [217]. Some brain tumors respond more favorably to certain chemotherapeutic agents than others, and this sensitivity is influenced by GSH and the GSH enzyme-linked system. The GST-p isoform, GSH metabolic pathway enzyme, has received the most attention as a crucial indicator of the effectiveness of chemotherapy in treating brain tumors. Drug resistance in primary brain tumors is significantly influenced by the interaction between Mrp-facilitated efflux of the GSH-drug conjugate and GSH/GST-mediated drug detoxification. This interaction provides a promising target for therapeutic approaches aimed at selectively modulating drug sensitivity [219].

8.10.1.3 Catalase

Catalase has a protective and anti-apoptotic effect in cells by removing ROS [220]. Brain tissue from rats with N-Ethyl-N-nitrosourea-induced gliomas had decreased levels of CAT activity. On the other hand, many brain cancers have shown much-increased catalase activity [221].

8.10.1.4 Glutamate

Numerous neurological diseases have been linked to glial Glutamate transport deficiency. Its uptake into astrocytes was compared to that of their cancerous counterparts, it was shown that Glutamate uptake into gliomas was virtually absent [222]. It activates metabotropic glutamate (mGlu) receptors that control the proliferation of BSPCs (brain stem-progenitor cells). Specific mGlu receptor subtypes are fresh prospective targets for the therapy of several malignant cancers, such as brain tumors [223, 224].

8.10.2 *Dietary Antioxidants*

8.10.2.1 Vitamin-E

All individuals with Grade III malignant gliomas have higher survival rates when they consume more vitamin E. Antioxidants, such as vitamins C and E, have been proven to lower the risk of brain cancers in children whose mothers took these nutrients throughout pregnancy [225]. Except for individuals with acoustic neurinoma, most patients with brain tumors tended their plasma levels of vitamins A and E to decline.

8.10.2.2 Retinoid

Retinol's ability to scavenge free radicals has long been recognized as making it an excellent antioxidant. Low amounts of retinol and β-carotene were seen in cancer patients. Initially, the theory that certain malignancies were related to an underlying vitamin A shortage led to the use of retinoids as a treatment for those illnesses [226]. In gliomas, retinoid receptor expression may become imbalanced as a result of environmental stimuli that boost glial cells' endogenous production of retinoic acid (RA). The promising novel therapeutic approach for gliomas is the combination use of RAR-agonist and RAR-antagonist, maybe even at a late stage of the disease. This theory predicts that the RAR-antagonist would prevent RAR-induced gliomas, while the RAR-agonist would slow the growth of tumors and aid in the regeneration of healthy glia [227]. Additionally, fat-soluble vitamins like vitamin A and vitamin D played a part in prevention by controlling cell differentiation and reducing the growth of cancer cells [228, 229].

8.10.2.3 Vitamin C

By removing free radicals and promoting apoptosis, several vitamins with antioxidant qualities, like vitamin C and vitamin E, have been found to slow the growth of tumors [230]. A recent meta-analysis indicates that larger intakes of vitamin C, β-carotene, and folate significantly reduced the risk of developing brain tumors [231]. On the other hand, the rat experiment discovered that rats administered vitamin C had higher levels of indicators linked to the proliferation of brain tumors, such as platelet-derived growth factor receptor (PDGF-R) [232].

8.11 Conclusion

Oxidative stress is a chief player in the pathophysiology of several brain disorders, including AD, HD, PD, ALS, MS, epilepsy, schizophrenia, stroke, and brain cancer. It causes protein, lipid, as well as DNA damage due to the creation of highly reactive chemicals like hydroxyl and peroxynitrite radicals. In the brain, endogenous enzymatic and non-enzymatic antioxidants such as glutathione, glutathione peroxidase, superoxide dismutase, melatonin, uric acid, and bilirubin act as a strong defense system against these processes. In addition to endogenous enzymes, all dietary antioxidants effectively maintain neuronal morphology and cell viability by restoring mitochondrial activity and lowering ROS levels in the brain. Fruits and vegetables including grapes, oranges, cherries, blueberries, lemon, tomatoes, and dairy products such as eggs, milk, fish, meat, and nuts are rich sources of dietary antioxidants. The intake of fruits, vegetables, grains, and nuts in a balanced fashion act as the most efficient strategy for people to boost their antioxidant as well as anti-inflammatory capability, and lower their chance of acquiring brain disorders. Hence, dietary antioxidants appear to be useful component for both therapeutic and preventive approaches to a variety of brain disorders.

References

1. Zuliani C, Baroni L (2015) Antioxidants for the prevention and treatment of multiple sclerosis: an overview. Bioact Nutraceut Diet Suppl Neurol Brain Dis Prev Ther:341–353. https://doi.org/10.1016/B978-0-12-411462-3.00035-7
2. Naz S, Farooq T, Hameed A et al (2022) Thuja Occidentalis l. Expedites functional recovery after sciatic nerve crush injury in mice. Pakistan J Med Heal Sci 16:396–396. https://doi.org/10.53350/PJMHS22163396
3. Zafar S, Rasul A, Iqbal J, et al (2021) Calotropis procera (leaves) supplementation exerts curative effects on promoting functional recovery in a mouse model of peripheral nerve injury. Food Sci Nutr 00:fsn3.2455. https://doi.org/10.1002/FSN3.2455
4. Zafar S, Anwar H, Qasim M, et al (2020) Calotropis procera (root) escalates functions rehabilitation and attenuates oxidative stress in a mouse model of peripheral nerve injury Calotropis procera (root) escalates functions rehabilitation and attenuates oxidative stress in a mouse model of p. 2801–2807. https://doi.org/10.36721/PJPS.2020.33.6.SUP.2801-2807.1
5. Liu Z, Ren Z, Zhang J et al (2018) Role of ROS and nutritional antioxidants in human diseases. Front Physiol 9:477. https://doi.org/10.3389/FPHYS.2018.00477/BIBTEX
6. Ashok A, Andrabi SS, Mansoor S et al (2022) Antioxidant therapy in oxidative stress-induced neurodegenerative diseases: role of nanoparticle-based drug delivery systems in clinical translation. Antioxidants 11:408. https://doi.org/10.3390/ANTIOX11020408
7. Anwar H, Rasul A, Iqbal J et al (2021) Dietary biomolecules as promising regenerative agents for peripheral nerve injury: an emerging nutraceutical-based therapeutic approach. J Food Biochem 45:e13989. https://doi.org/10.1111/JFBC.13989
8. Guest JA, Grant RS (2012) Effects of dietary derived antioxidants on the central nervous system. Int J Nutr Pharmacol Neurol Dis 2:185. https://doi.org/10.4103/2231-0738.99470

9. Winiarska-Mieczan A, Baranowska-Wójcik E, Kwiecień M et al (2020) The role of dietary antioxidants in the pathogenesis of neurodegenerative diseases and their impact on cerebral oxidoreductive balance. Nutrients 12
10. Hussain G, Rasul A, Anwar H et al (2018) Role of plant derived alkaloids and their mechanism in neurodegenerative disorders. Int J Biol Sci 14:341. https://doi.org/10.7150/IJBS.23247
11. Hussain G, Zhang L, Rasul A, et al (2018) Role of plant-derived flavonoids and their Mechanism in attenuation of Alzheimer's and Parkinson's diseases: an update of recent data. Molecules 23:814. https://doi.org/10.3390/MOLECULES23040814
12. Filograna R, Beltramini M, Bubacco L, Bisaglia M (2016) Send orders for reprints to reprints@benthamscience.ae anti-oxidants in Parkinson's disease therapy: a critical point of view
13. Bjørklund G, Peana M, Maes M et al (2021) The glutathione system in Parkinson's disease and its progression. Neurosci Biobehav Rev 120:470–478. https://doi.org/10.1016/j.neubiorev.2020.10.004
14. Espinosa-Diez C, Miguel V, Mennerich D et al (2015) Antioxidant responses and cellular adjustments to oxidative stress. Redox Biol 6:183. https://doi.org/10.1016/J.REDOX.2015.07.008
15. Martin HL, Teismann P (2009) Glutathione—a review on its role and significance in Parkinson's disease. FASEB J 23:3263–3272. https://doi.org/10.1096/fj.08-125443
16. Duarte-Jurado AP, Gopar-Cuevas Y, Saucedo-Cardenas O et al (2021) Antioxidant therapeutics in Parkinson's disease: current challenges and opportunities. Antioxidants 10:453. https://doi.org/10.3390/ANTIOX10030453
17. Mantle D, Heaton RA, Hargreaves IP (2022) Coenzyme Q10, ageing and the nervous system: An overview. Antioxidants 11
18. Chang KH, Chen CM (2020) The role of oxidative stress in Parkinson's disease. Antioxidants 9:1–32. https://doi.org/10.3390/ANTIOX9070597
19. Ciulla M, Marinelli L, Cacciatore I, Di Stefano A (2019) biomolecules role of dietary supplements in the management of Parkinson's disease. https://doi.org/10.3390/biom9070271
20. Park HA, Ellis AC (2020) Dietary antioxidants and Parkinson's disease. Antioxidants 9:1–23. https://doi.org/10.3390/ANTIOX9070570
21. Icer MA, Arslan N, Gezmen-Karadag M (2021) Effects of vitamin e on neurodegenerative diseases: an update. Acta Neurobiol Exp (Wars) 81:21–33. https://doi.org/10.21307/ane-2021-003
22. Mandal M, Jaiswal P, Mishra A (2020) Role of curcumin and its nanoformulations in neurotherapeutics: a comprehensive review. J Biochem Mol Toxicol 34. https://doi.org/10.1002/JBT.22478
23. Barreto GE, Wang J, Md Ashraf G et al (2020) Resveratrol derivatives as potential treatments for Alzheimer's and Parkinson's disease. https://doi.org/10.3389/fnagi.2020.00103
24. Percário S, Da Silva BA, Varela ELP et al (2020) Oxidative stress in Parkinson's disease: potential benefits of antioxidant supplementation. Oxid Med Cell Longev:2020. https://doi.org/10.1155/2020/2360872
25. Sadiq Butt M, Tariq U, Rizwan M (2021) Neuroprotective effects of oleuropein: recent developments and contemporary research. J Food Biochem 45:13967. https://doi.org/10.1111/jfbc.13967
26. Dati LM, Ulrich H, Real CC et al (2017) Carvacrol promotes neuroprotection in the mouse hemiparkinsonian model. Neuroscience 356:176–181. https://doi.org/10.1016/j.neuroscience.2017.05.013
27. Chen D, Zhang XY, Sun J et al (2019) Asiatic acid protects dopaminergic neurons from neuroinflammation by suppressing mitochondrial ROS production. Biomol Ther 27:442–449. https://doi.org/10.4062/biomolther.2018.188
28. Ribeiro M, Silva AC, Rodrigues J et al (2013) Oxidizing effects of exogenous stressors in Huntington's disease knock-in striatal cells—protective effect of cystamine and creatine. Toxicol Sci 136:487–499. https://doi.org/10.1093/TOXSCI/KFT199

29. Jayaram S, Thaggikuppe Krishnamurthy P, Joshi M, Kumar V (2022) Nrf2 as a potential therapeutic target for treatment of Huntington's disease. From Pathophysiol to Treat Huntington's Dis [Working Title]. https://doi.org/10.5772/intechopen.103177
30. Deshmukh P, Unni S, Krishnappa G, Padmanabhan B (2017) The Keap1–Nrf2 pathway: promising therapeutic target to counteract ROS-mediated damage in cancers and neurodegenerative diseases. Biophys Rev 9:41. https://doi.org/10.1007/S12551-016-0244-4
31. Jang M, Cho IH (2016) Sulforaphane ameliorates 3-nitropropionic acid-induced striatal toxicity by activating the Keap1-Nrf2-ARE pathway and inhibiting the MAPKs and NF-κB pathways. Mol Neurobiol 53:2619–2635. https://doi.org/10.1007/S12035-015-9230-2
32. Jang D, Lee HJ, Lee K, et al (2019) White ginseng ameliorates depressive behavior and increases hippocampal 5-HT level in the stressed ovariectomized rats. Biomed Res Int:2019. https://doi.org/10.1155/2019/5705232
33. Chaturvedi RK, Hennessey T, Johri A et al (2012) Transducer of regulated CREB-binding proteins (TORCs) transcription and function is impaired in Huntington's disease. Hum Mol Genet 21:3474–3488. https://doi.org/10.1093/HMG/DDS178
34. Vamos E, Voros K, Vecsei L, Klivenyi P (2010) Neuroprotective effects of L-carnitine in a transgenic animal model of Huntington's disease. Biomed Pharmacother 64:282–286. https://doi.org/10.1016/j.biopha.2009.06.020
35. Johri A, Beal MF (2012) Antioxidants in Huntington's disease. Biochim Biophys Acta Mol Basis Dis 1822:664–674. https://doi.org/10.1016/j.bbadis.2011.11.014
36. Ribeiro M, Rosenstock TR, Cunha-Oliveira T et al (2012) Glutathione redox cycle dysregulation in Huntington's disease knock-in striatal cells. Free Radic Biol Med 53:1857–1867. https://doi.org/10.1016/J.FREERADBIOMED.2012.09.004
37. Eddaikra A, Eddaikra N (2021) Endogenous enzymatic antioxidant defense and pathologies. Antioxidants Benefits, Sources, Mech Action. https://doi.org/10.5772/intechopen.95504
38. Bono-Yagüe J, Gómez-Escribano AP, Millán JM, Vázquez-Manrique RP (2020) Reactive Species in Huntington disease: are they really the radicals you want to catch? Antioxidants 9:577. https://doi.org/10.3390/ANTIOX9070577
39. Moretti M, Fraga DB, Rodrigues ALS (2017) Preventive and therapeutic potential of ascorbic acid in neurodegenerative diseases. CNS Neurosci Ther 23:921–929. https://doi.org/10.1111/CNS.12767
40. Denny Joseph KM, Muralidhara (2013) Enhanced neuroprotective effect of fish oil in combination with quercetin against 3-nitropropionic acid induced oxidative stress in rat brain. Prog Neuropsychopharmacol Biol Psychiatr 40:83–92. https://doi.org/10.1016/J.PNPBP.2012.08.018
41. Wang J, Pfleger CM, Friedman L et al (2010) Potential application of grape derived polyphenols in huntington's disease. Transl Neurosci 1:95–100. https://doi.org/10.2478/v10134-010-0022-y
42. Solanki I, Parihar P, Mansuri ML, Parihar MS (2015) Flavonoid-based therapies in the early management of neurodegenerative diseases. Adv Nutr 6:64–72. https://doi.org/10.3945/an.114.007500
43. Khan E, Tawani A, Mishra SK et al (2018) Myricetin reduces toxic level of CAG repeats RNA in Huntington's disease (HD) and spino cerebellar ataxia (SCAs). ACS Chem Biol 13:180–188. https://doi.org/10.1021/ACSCHEMBIO.7B00699
44. Gopinath K, Prakash D, Sudhandiran G (2011) Neuroprotective effect of naringin, a dietary flavonoid against 3-nitropropionic acid-induced neuronal apoptosis. Neurochem Int 59:1066–1073. https://doi.org/10.1016/J.NEUINT.2011.08.022
45. Gopinath K, Sudhandiran G (2015) Protective effect of naringin on 3-nitropropionic acid-induced neurodegeneration through the modulation of matrix metalloproteinases and glial fibrillary acidic protein. 94:65–71. 101139/cjpp-2015-0035. https://doi.org/10.1139/CJPP-2015-0035

46. Xia C, Cai Y, Li S et al (2015) Curcumin Increases HSP70 Expression in Primary Rat Cortical Neuronal Apoptosis Induced by gp120 V3 Loop Peptide. Neurochem Res 40:1996–2005. https://doi.org/10.1007/S11064-015-1695-X
47. Maiti P, Manna J, Veleri S (2014) Frautschy S (2014) Molecular chaperone dysfunction in neurodegenerative diseases and effects of curcumin. Biomed Res Int. https://doi.org/10.1155/2014/495091
48. Hersch SM, Schifitto G, Oakes D et al (2017) The CREST-E study of creatine for Huntington disease: a randomized controlled trial. Neurology 89:594. https://doi.org/10.1212/WNL.0000000000004209
49. Essa MM, Moghadas M, Ba-Omar T et al (2019) Protective effects of antioxidants in Huntington's disease: an extensive review. Neurotox Res 35:739–774
50. Reddy PH, Williams J, Smith F et al (2017) MicroRNAs, aging, cellular senescence, and Alzheimer's disease. Prog Mol Biol Transl Sci 146:127–171. https://doi.org/10.1016/BS.PMBTS.2016.12.009
51. Pais M, Martinez L, Ribeiro O et al (2020) Early diagnosis and treatment of Alzheimer's disease: new definitions and challenges. Brazilian J Psychiatr 42:431–441. https://doi.org/10.1590/1516-4446-2019-0735
52. Weller J, Budson A (2018) Current understanding of Alzheimer's disease diagnosis and treatment. F1000Research 7. https://doi.org/10.12688/F1000RESEARCH.14506.1
53. Hussain G, Rasul A, Anwar H et al (2018) Role of plant derived alkaloids and their mechanism in neurodegenerative disorders. Int J Biol Sci 14:341–357. https://doi.org/10.7150/ijbs.23247
54. Hussain G, Huang J, Rasul A, et al (2019) Putative roles of plant-derived tannins in neurodegenerative and neuropsychiatry disorders: an updated review. Molecules:24. https://doi.org/10.3390/MOLECULES24122213
55. Kim TS, Pae CU, Yoon SJ et al (2006) Decreased plasma antioxidants in patients with Alzheimer's disease. Int J Geriatr Psychiatr 21:344–348. https://doi.org/10.1002/GPS.1469
56. Venkateshappa C, Harish G, Mahadevan A et al (2012) Elevated oxidative stress and decreased antioxidant function in the human hippocampus and frontal cortex with increasing age: Implications for neurodegeneration in Alzheimer's disease. Neurochem Res 37:1601–1614. https://doi.org/10.1007/S11064-012-0755-8/TABLES/2
57. Emanuele S, D'Anneo A, Calvaruso G et al (2018) The double-edged sword profile of redox signaling: oxidative events as molecular switches in the balance between cell physiology and cancer. Chem Res Toxicol 31:201–210
58. Reddy PH, Beal MF (2008) Amyloid beta, mitochondrial dysfunction and synaptic damage: implications for cognitive decline in aging and Alzheimer's disease. Trends Mol Med 14:45–53. https://doi.org/10.1016/J.MOLMED.2007.12.002
59. Baldeiras I, Santana I, Proença MT et al (2008) Peripheral oxidative damage in mild cognitive impairment and mild Alzheimer's disease. J Alzheimer's Dis 15:117–128. https://doi.org/10.3233/JAD-2008-15110
60. Walia V, Kaushik D, Mittal V et al (2022) Delineation of neuroprotective effects and possible benefits of antioxidantstherapy for the treatment of Alzheimer's diseases by targeting mitochondrial-derived reactive oxygen species: bench to bedside. Mol Neurobiol 59:657–680. https://doi.org/10.1007/S12035-021-02617-1/FIGURES/3
61. Balendra V, Singh SK (2021) Therapeutic potential of astaxanthin and superoxide dismutase in Alzheimer's disease. Open Biol 11:2021. https://doi.org/10.1098/RSOB.210013
62. Clausen A, Bi X, Baudry M (2012) Effects of the superoxide dismutase/catalase mimetic EUK-207 in a mouse model of Alzheimer's disease: protection against and interruption of progression of amyloid and tau pathology and cognitive decline. J Alzheimers Dis 30:183–208. https://doi.org/10.3233/JAD-2012-111298
63. Sofic E, Salkovic-Petrisic M, Tahirovic I et al (2015) Brain catalase in the streptozotocin-rat model of sporadic Alzheimer's disease treated with the iron chelator–monoamine oxidase

inhibitor, M30. J Neural Transm 122:559–564. https://doi.org/10.1007/S00702-014-1307-Y/TABLES/1
64. Habib LK, Lee MTC, Yang J (2010) Inhibitors of catalase-amyloid interactions protect cells from β-amyloid-induced oxidative stress and toxicity. J Biol Chem 285:38933–38943. https://doi.org/10.1074/JBC.M110.132860/ATTACHMENT/76EDBCCA-1D8A-416E-9534-51D825029B29/MMC1.PDF
65. Hane F, Tran G, Attwood SJ, Leonenko Z (2013) Cu^{2+} Affects Amyloid-β (1–42) aggregation by increasing peptide-peptide binding forces. PLoS One 8:e59005. https://doi.org/10.1371/JOURNAL.PONE.0059005
66. Allani S, Weissbach H, Lopez-Toledano M (2021) Role of methionine sulfoxide reductase in Alzheimer's disease. FASEB J 35. https://doi.org/10.1096/FASEBJ.2021.35.S1.03557
67. Moskovitz J, Du F, Bowman CF, Yan SS (2016) Methionine sulfoxide reductase A affects β-amyloid solubility and mitochondrial function in a mouse model of Alzheimer's disease. Am J Physiol Endocrinol Metab 310:E388–E393. https://doi.org/10.1152/AJPENDO.00453.2015
68. Wojtunik-Kulesza KA, Oniszczuk A, Oniszczuk T, Waksmundzka-Hajnos M (2016) The influence of common free radicals and antioxidants on development of Alzheimer's disease. Biomed Pharmacother 78:39–49. https://doi.org/10.1016/J.BIOPHA.2015.12.024
69. Chen X, Guo X, Huang R et al (2014) Serum uric acid levels in patients with Alzheimer's disease: A meta-analysis. PLoS One 9:e94084. https://doi.org/10.1371/journal.pone.0094084
70. Guan JZ, Guan WP, Maeda T, Makino N (2012) Effect of Vitamin E administration on the elevated oxygen stress and the telomeric and subtelomeric status in Alzheimer's disease. Gerontology 58:62–69. https://doi.org/10.1159/000327821
71. Puttachary S, Sharma S, Stark S, Thippeswamy T (2015, 2015) Seizure-induced oxidative stress in temporal lobe epilepsy. Biomed Res Int. https://doi.org/10.1155/2015/745613
72. Li FJ, Shen L (2012) Dietary intakes of Vitamin E, Vitamin C, and β-carotene and risk of Alzheimer's disease: a meta-analysis. J Alzheimer's Dis 31:253–258. https://doi.org/10.3233/JAD-2012-120349
73. Calderón-Ospina CA, Nava-Mesa MO (2020) B Vitamins in the nervous system: current knowledge of the biochemical modes of action and synergies of thiamine, pyridoxine, and cobalamin. CNS Neurosci Ther 26:5–13. https://doi.org/10.1111/CNS.13207
74. Collins AE, Saleh TM, Kalisch BE (2022) Naturally occurring antioxidant therapy in Alzheimer's disease. Antioxidants 11. https://doi.org/10.3390/ANTIOX11020213
75. Hashim A, Wanga L, Junej K et al (2011) Vitamin B6s inhibit oxidative stress caused by Alzheimer's disease-related CuII-β-amyloid complexes—cooperative action of phospho-moiety. Bioorg Med Chem Lett 21:6430–6432. https://doi.org/10.1016/J.BMCL.2011.08.123
76. Ono K, Yamada M (2012) Vitamin A and Alzheimer's disease. Geriatr Gerontol Int 12:180–188. https://doi.org/10.1111/J.1447-0594.2011.00786.X
77. Hu N, Yu JT, Tan L et al (2013) Nutrition and the risk of Alzheimer's disease. Biomed Res Int 2013:12. https://doi.org/10.1155/2013/524820
78. Boccardi V, Arosio B, Cari L et al (2020) Beta-carotene, telomerase activity and Alzheimer's disease in old age subjects. Eur J Nutr 59:119–126. https://doi.org/10.1007/S00394-019-01892-Y/TABLES/3
79. Galasko DR, Peskind E, Clark CM et al (2012) Antioxidants for Alzheimer disease: a randomized clinical trial with cerebrospinal fluid biomarker measures. Arch Neurol 69:836–841. https://doi.org/10.1001/ARCHNEUROL.2012.85
80. Colizzi C (2019) The protective effects of polyphenols on Alzheimer's disease: a systematic review. Alzheimer's Dement Transl Res Clin Interv 5:184. https://doi.org/10.1016/J.TRCI.2018.09.002
81. Malar DS, Devi KP (2014) Dietary polyphenols for treatment of Alzheimer's disease – future research and development. Curr Pharm Biotechnol 15:330–342. https://doi.org/10.2174/1389201015666140813122703
82. Hira S, Saleem U, Anwar F, et al (2019) β-carotene: a natural compound improves cognitive impairment and oxidative stress in a mouse model of streptozotocin-induced Alzheimer's disease. Biomolecules:9. https://doi.org/10.3390/BIOM9090441

83. Lv H, Liu Q, Zhou J et al (2017) Daphnetin-mediated Nrf2 antioxidant signaling pathways ameliorate tert-butyl hydroperoxide (t-BHP)-induced mitochondrial dysfunction and cell death. Free Radic Biol Med 106:38–52. https://doi.org/10.1016/J.FREERADBIOMED.2017.02.016
84. Johnson JA, Johnson DA, Kraft AD et al (2008) The Nrf2-ARE pathway: an indicator and modulator of oxidative stress in neurodegeneration. Ann N Y Acad Sci 1147:61. https://doi.org/10.1196/ANNALS.1427.036
85. Anand P, Singh B, Singh N (2012) A review on coumarins as acetylcholinesterase inhibitors for Alzheimer's disease. Bioorg Med Chem 20:1175–1180. https://doi.org/10.1016/J.BMC.2011.12.042
86. Hamulakova S, Kozurkova M, Kuca K (2017) Coumarin derivatives in pharmacotherapy of Alzheimer's disease. Curr Org Chem 21:602–612. https://doi.org/10.2174/1385272820666160601155411
87. Uddin MS, Kabir MT, Niaz K, et al (2020) Molecular insight into the therapeutic promise of flavonoids against Alzheimer's disease. Molecules 25:1267. https://doi.org/10.3390/MOLECULES25061267
88. Reddy PH, Manczak M, Yin X et al (2018) Protective effects of Indian spice curcumin against Amyloid-ß in Alzheimer's disease. J Alzheimer's Dis 61:843–866. https://doi.org/10.3233/JAD-170512
89. Akram R, Anwar H, Rasul A, et al (2021) Natural compounds as versatile potential therapeutic agents of lung cancer. Med Plants Lung Dis 229–256. https://doi.org/10.1007/978-981-33-6850-7_10
90. Begum AN, Jones MR, Lim GP et al (2008) Curcumin structure-function, bioavailability, and efficacy in models of neuroinflammation and Alzheimer's disease. J Pharmacol Exp Ther 326:196–208. https://doi.org/10.1124/jpet.108.137455
91. Ventriglia M, Brewer GJ, Simonelli I et al (2015) Zinc in Alzheimer's disease: a meta-analysis of serum, plasma, and cerebrospinal fluid studies. J Alzheimer's Dis 46:75–87. https://doi.org/10.3233/JAD-141296
92. Hooper NM, Watt NT, Whitehouse IJ (2011) The role of zinc in Alzheimer's disease. Int J Alzheimers Dis 2011:. https://doi.org/10.4061/2011/971021
93. Rivers-Auty J, Tapia VS, White CS et al (2021) Zinc status alters Alzheimer's disease progression through NLRP3-dependent inflammation. J Neurosci 41:3025–3038. https://doi.org/10.1523/JNEUROSCI.1980-20.2020
94. Varikasuvu SR, Prasad VS, Kothapalli J, Manne M (2019) Brain selenium in Alzheimer's disease (BRAIN SEAD Study): a systematic review and meta-analysis. Biol Trace Elem Res 189:361–369. https://doi.org/10.1007/S12011-018-1492-X
95. Zhang ZH, Song GL (2021) Roles of selenoproteins in brain function and the potential mechanism of selenium in Alzheimer's disease. Front Neurosci 15:215. https://doi.org/10.3389/FNINS.2021.646518/BIBTEX
96. Hertzog da Silva Leme AG, Cardoso BR (2020) Selenium and Alzheimer's disease. Genet Neurol Behav Diet Dement:739–748. https://doi.org/10.1016/B978-0-12-815868-5.00047-5
97. do Nascimento CQ, Barros-Neto JA, Vieira NFL et al (2021) Selenium concentrations in elderly people with Alzheimer's disease: a cross-sectional study with control group. Rev Bras Enferm 74:e20200984. https://doi.org/10.1590/0034-7167-2020-0984
98. Costello DJ, Delanty N (2014) Oxidative injury in epilepsy: potential for antioxidant therapy? 4:541–553. https://doi.org/10.1586/14737175.4.3.541
99. Shekh-Ahmad T, Lieb A, Kovac S et al (2019) Combination antioxidant therapy prevents epileptogenesis and modifies chronic epilepsy. Redox Biol 26:101278. https://doi.org/10.1016/J.REDOX.2019.101278
100. Vishnoi S, Raisuddin S, Parvez S (2016) Glutamate excitotoxicity and oxidative stress in epilepsy: modulatory role of melatonin. J Environ Pathol Toxicol Oncol 35:365–374. https://doi.org/10.1615/JENVIRONPATHOLTOXICOLONCOL.2016016399
101. Shehta N, Kamel AE, Sobhy E, Ismail MH (2022) Malondialdehyde and superoxide dismutase levels in patients with epilepsy: a case–control study. Egypt J Neurol Psychiatr Neurosurg 58:1–7. https://doi.org/10.1186/S41983-022-00479-5/TABLES/4

102. Baluchnejadmojarad T, Roghani M, Kamran M, Karimi N (2015) Downloaded from bcn.iums.ac.ir at 22:08 CET on Friday. 3
103. Martinc B, Grabnar I, Vovk T (2014) Antioxidants as a preventive treatment for epileptic process: a review of the current status. Curr Neuropharmacol 12:527–550. https://doi.org/10.2174/1570159X12666140923205715
104. Ashrafi MR, Shams S, Nouri M et al (2007) A probable causative factor for an old problem: Selenium and glutathione peroxidase appear to play important roles in epilepsy pathogenesis. Epilepsia 48:1750–1755. https://doi.org/10.1111/J.1528-1167.2007.01143.X
105. Shin EJ, Jeong JH, Chung YH et al (2011) Role of oxidative stress in epileptic seizures. Neurochem Int 59:122. https://doi.org/10.1016/J.NEUINT.2011.03.025
106. Abdel-Zaher AO, Abdel-Hady RH, Abdel Moneim WM, Salim SY (2011) Alpha-lipoic acid protects against potassium cyanide-induced seizures and mortality. Exp Toxicol Pathol 63:161–165. https://doi.org/10.1016/J.ETP.2009.11.004
107. Freitas RM (2009) The evaluation of effects of lipoic acid on the lipid peroxidation, nitrite formation and antioxidant enzymes in the hippocampus of rats after pilocarpine-induced seizures. Neurosci Lett 455:140–144. https://doi.org/10.1016/J.NEULET.2009.03.065
108. Banach M, Gurdziel E, Drych MJ$, Borowicz KK (2011) Melatonin in experimental seizures and epilepsy. Pharmacol Reports 63:1–11
109. Naziroğlu M (2015) Role of melatonin on calcium signaling and mitochondrial oxidativestress in epilepsy: focus on TRP channels. Turkish J Biol 39:813–821. https://doi.org/10.3906/biy-1505-43
110. Baluchnejadmojarad T, Roghani M (2013) Coenzyme Q10 ameliorates neurodegeneration, mossy fiber sprouting, and oxidative stress in intrahippocampal kainate model of temporal lobe epilepsy in rat. J Mol Neurosci 49:194–201. https://doi.org/10.1007/S12031-012-9886-2/FIGURES/5
111. Liao WC, Huang CW, Hsiao YH et al (2021) Association between the serum coenzyme Q10 level and seizure control in patients with drug-resistant epilepsy. Healthcare 9:1118. https://doi.org/10.3390/HEALTHCARE9091118
112. Borowicz-Reutt KK, Czuczwar SJ (2020) Role of oxidative stress in epileptogenesis and potential implications for therapy. Pharmacol Reports 72:1218–1226. https://doi.org/10.1007/S43440-020-00143-W/TABLES/1
113. Upaganlawar AB, Wankhede NL, Kale MB et al (2021) Interweaving epilepsy and neurodegeneration: Vitamin E as a treatment approach. Biomed Pharmacother 143:112146. https://doi.org/10.1016/J.BIOPHA.2021.112146
114. Sawicka-Glazer E, Czuczwar SJ (2014) Vitamin C: a new auxiliary treatment of epilepsy? Pharmacol Reports 66:529–533. https://doi.org/10.1016/J.PHAREP.2014.02.016
115. Martinc B, Grabnar I, Vovk T (2014) Antioxidants as a preventive treatment for epileptic process: a review of the current status. Curr Neuropharmacol 12:527. https://doi.org/10.2174/1570159X12666140923205715
116. Hamid NAK (2011) The effect of antioxidant supplementation in the treatment of epilepsy. Iraqi J Pharm 11:27–33. https://doi.org/10.33899/iphr.2011.49893
117. Diniz TC, Silva JC, Lima-Saraiva SRG De, et al (2015) The role of flavonoids on oxidative stress in epilepsy. Oxid Med Cell Longev 2015. https://doi.org/10.1155/2015/171756
118. Kwon JY, Jeon MT, Jung UJ et al (2019) Perspective: therapeutic potential of flavonoids as alternative medicines in epilepsy. Adv Nutr 10:778–790
119. Amin I, Majid S, Farooq A, et al (2020) Naringenin (4,5,7-trihydroxyflavanone) as a potent neuroprotective agent: From chemistry to medicine. 65:271–300. https://doi.org/10.1016/B978-0-12-817905-5.00008-1
120. Shakeel S, Rehman MU, Tabassum N et al (2017) Effect of naringenin (A naturally occurring flavanone) against pilocarpine-induced status epilepticus and oxidative stress in mice. Pharmacogn Mag 13:S154. https://doi.org/10.4103/0973-1296.203977
121. Szala-Rycaj J, Zagaja M, Szewczyk A, Andres-Mach M (2021) Selected flavonoids and their role in the treatment of epilepsy-a review of the latest reports from experimental studies. https://doi.org/10.21307/ane-2021-014

122. Dhir A (2020) Natural polyphenols in preclinical models of epilepsy. Phyther Res 34:1268–1281. https://doi.org/10.1002/PTR.6617
123. Kiasalari Z, Roghani M, Khalili M et al (2013) Antiepileptogenic effect of curcumin on kainate-induced model of temporal lobe epilepsy. http://dx.doi.org/103109/138802092013803128 51:1572–1578. https://doi.org/10.3109/13880209.2013.803128
124. Dhir A (2018) Curcumin in epilepsy disorders. Phyther Res 32:1865–1875. https://doi.org/10.1002/PTR.6125
125. Hamdy M, Antar A, El-Mesery M et al (2020) Curcumin offsets PTZ-induced epilepsy: involving inhibition of apoptosis, wnt/β-catenin, and autophagy pathways. 7:240–251. https://doi.org/10.1080/2314808X.2020.1809848
126. Bertoncello KT, Aguiar GPS, Oliveira JV, Siebel AM (2018) Micronization potentiates curcumin's anti-seizure effect and brings an important advance in epilepsy treatment. Sci Reports 8:1–9. https://doi.org/10.1038/s41598-018-20897-x
127. Sharma V, Nehru B, Munshi A, Jyothy A (2010) Antioxidant potential of curcumin against oxidative insult induced by pentylenetetrazol in epileptic rats. Methods Find Exp Clin Pharmacol 32:227–232. https://doi.org/10.1358/MF.2010.32.4.1452090
128. Valko K, Ciesla L (2019) Amyotrophic lateral sclerosis. Prog Med Chem 58:63–117. https://doi.org/10.1016/bs.pmch.2018.12.001
129. Henriques A, Croixmarie V, Priestman DA et al (2015) Amyotrophic lateral sclerosis and denervation alter sphingolipids and up-regulate glucosylceramide synthase. Hum Mol Genet 24:7390–7405. https://doi.org/10.1093/HMG/DDV439
130. Wier CG, Crum AE, Reynolds AB et al (2019) Muscle contractility dysfunction precedes loss of motor unit connectivity in SOD1(G93A) mice. Muscle and Nerve 59:254–262. https://doi.org/10.1002/mus.26365
131. van Es MA, Hardiman O, Chio A et al (2017) Amyotrophic lateral sclerosis. Lancet 390:2084–2098. https://doi.org/10.1016/S0140-6736(17)31287-4
132. Carrera-Juliá S, Moreno ML, Barrios C et al (2020) Antioxidant alternatives in the treatment of amyotrophic lateral sclerosis: a comprehensive review. Front Physiol 11:63. https://doi.org/10.3389/FPHYS.2020.00063
133. Park HR, Yang EJ (2021) Oxidative Stress as a Therapeutic Target in Amyotrophic Lateral Sclerosis: Opportunities and Limitations. Diagnostics 11. https://doi.org/10.3390/DIAGNOSTICS11091546
134. Lucchetti J, Marino M, Papa S, et al (2013) A mouse model of familial ALS has increased CNS levels of endogenous Ubiquinol9/10 and does not benefit from exogenous administration of Ubiquinol10. PLoS One 8:. https://doi.org/10.1371/journal.pone.0069540
135. Neves Carvalho A, Firuzi O, Joao Gama M et al (2016) Oxidative stress and antioxidants in neurological diseases: is there still hope? Curr Drug Targets 18:705–718. https://doi.org/10.2174/1389450117666160401120514
136. De Paula CZ, Gonçalves BDC (2015) Vieira LB (2015) An overview of potential targets for treating amyotrophic lateral sclerosis and huntington's disease. Biomed Res Int. https://doi.org/10.1155/2015/198612
137. Jaiswal MK (2017) Riluzole but not melatonin ameliorates acute motor neuron degeneration and moderately inhibits sod1-mediated excitotoxicity induced disrupted mitochondrial Ca^{2+} signaling in amyotrophic lateral sclerosis. Front Cell Neurosci:10. https://doi.org/10.3389/FNCEL.2016.00295/FULL
138. Pandi-Perumal SR, Bahammam AS, Brown GM et al (2013) Melatonin antioxidative defense: therapeutical implications for aging and neurodegenerative processes. Neurotox Res 23:267–300. https://doi.org/10.1007/S12640-012-9337-4
139. Dardiotis E, Panayiotou E, Feldman ML et al (2013) Intraperitoneal melatonin is not neuroprotective in the G93ASOD1 transgenic mouse model of familial ALS and may exacerbate neurodegeneration. Neurosci Lett 548:170–175. https://doi.org/10.1016/j.neulet.2013.05.058
140. Cunha-Oliveira T, Montezinho L, Mendes C, et al (2020) Oxidative stress in amyotrophic lateral sclerosis: pathophysiology and opportunities for pharmacological intervention. Oxid Med Cell Longev:2020. https://doi.org/10.1155/2020/5021694

141. Kim K (2021) Glutathione in the nervous system as a potential therapeutic target to control the development and progression of amyotrophic lateral sclerosis. Antioxidants 10. https://doi.org/10.3390/ANTIOX10071011
142. Chi L, Ke Y, Luo C et al (2007) Depletion of reduced glutathione enhances motor neuron degeneration in vitro and in vivo. Neuroscience 144:991–1003. https://doi.org/10.1016/J.NEUROSCIENCE.2006.09.064
143. Phatnani H, Maniatis T (2015) Astrocytes in neurodegenerative disease. Cold Spring Harb Perspect Biol 7:1–18. https://doi.org/10.1101/cshperspect.a020628
144. Chen T, Turner BJ, Beart PM et al (2018) Glutathione monoethyl ester prevents TDP-43 pathology in motor neuronal NSC-34 cells. Neurochem Int 112:278–287. https://doi.org/10.1016/J.NEUINT.2017.08.009
145. Vargas MR, Johnson DA, Sirkis DW et al (2008) Nrf2 activation in astrocytes protects against neurodegeneration in mouse models of familial amyotrophic lateral sclerosis. Soc Neurosci 28:13574–13581. https://doi.org/10.1523/JNEUROSCI.4099-08.2008
146. Kokić AN, Stević Z, Stojanović S et al (2013) Biotransformation of nitric oxide in the cerebrospinal fluid of amyotrophic lateral sclerosis patients. http://dx.doi.org/101179/135100005X70242 10:265–270. https://doi.org/10.1179/135100005X70242
147. Abati E, Bresolin N, Comi G, Corti S (2020) Silence superoxide dismutase 1 (SOD1): a promising therapeutic target for amyotrophic lateral sclerosis (ALS). 24:295–310. https://doi.org/10.1080/14728222.2020.1738390
148. Clara MC, Dennys CN, Ross HF, Estévez AG (2013) Superoxide dismutase and oxidative stress in amyotrophic lateral sclerosis. Curr Adv Amyotroph Lateral Scler. https://doi.org/10.5772/56488
149. Furukawa Y, Tokuda E (2020) Does wild-type Cu/Zn-superoxide dismutase have pathogenic roles in amyotrophic lateral sclerosis? Transl Neurodegener 9:1–16. https://doi.org/10.1186/S40035-020-00209-Y
150. Dixon SJ, Patel D, Welsch M et al (2014) Pharmacological inhibition of cystine-glutamate exchange induces endoplasmic reticulum stress and ferroptosis. Elife 2014. https://doi.org/10.7554/ELIFE.02523
151. Vargas MR, Johnson DA, Johnson JA (2011) Decreased glutathione accelerates neurologi-Cal deficit and mitochondrial pathology in familial ALS-linked hSOD1G93A mice model. Neurobiol Dis 43:543–551. https://doi.org/10.1016/j.nbd.2011.04.025
152. Wang M, Liu Z, Sun W et al (2020) Association between vitamins and amyotrophic lateral sclerosis: a center-based survey in Mainland China. Front Neurol 11:488. https://doi.org/10.3389/FNEUR.2020.00488/BIBTEX
153. Eom J, Son B, Kim SH (2022) Relationship between dietary total antioxidant capacity and the prognosis of amyotrophic lateral sclerosis. Nutrients 14:3264
154. Zoccolella S, Santamato A, Lamberti P (2009) Current and emerging treatments for amyotrophic lateral sclerosis. Neuropsychiatr Dis Treat 5:577–595. https://doi.org/10.2147/NDT.S7788
155. Guest J, Grant R (2016) Carotenoids and neurobiological health. Adv Neurobiol 12:199–228. https://doi.org/10.1007/978-3-319-28383-8_11
156. Nisar N, Li L, Lu S et al (2015) Carotenoid metabolism in plants. Mol Plant 8:68–82. https://doi.org/10.1016/J.MOLP.2014.12.007
157. Krishnaraj RN, Kumari SSS, Mukhopadhyay SS (2015) Antagonistic molecular interactions of photosynthetic pigments with molecular disease targets: a new approach to treat AD and ALS. 36:67–71. https://doi.org/10.3109/10799893.2015.1024851
158. He L, He T, Farrar S et al (2017) Antioxidants maintain cellular redox homeostasis by elimination of reactive oxygen species. Cell Physiol Biochem 44:532–553. https://doi.org/10.1159/000485089
159. Winter AN, Ross EK, Wilkins HM et al (2018) An anthocyanin-enriched extract from strawberries delays disease onset and extends survival in the hSOD1G93A mouse model

of amyotrophic lateral sclerosis. Nutr Neurosci 21:414–426. https://doi.org/10.1080/1028415X.2017.1297023
160. Novak V, Rogelj B, Župunski V (2021) Therapeutic potential of polyphenols in amyotrophic lateral sclerosis and frontotemporal dementia. Antioxidants 10:1328. https://doi.org/10.3390/ANTIOX10081328
161. Sharma DR, Wani WY, Sunkaria A et al (2016) Quercetin attenuates neuronal death against aluminum-induced neurodegeneration in the rat hippocampus. Neuroscience 324:163–176. https://doi.org/10.1016/J.NEUROSCIENCE.2016.02.055
162. Mancuso R, Del Valle J, Morell M, et al (2014) Lack of synergistic effect of resveratrol and sigma-1 receptor agonist (PRE-084) in SOD1G93A ALS mice: Overlapping effects or limited therapeutic opportunity? Orphanet J Rare Dis:9. https://doi.org/10.1186/1750-1172-9-78
163. Mancuso R, del Valle J, Modol L et al (2014) Resveratrol improves motoneuron function and extends survival in SOD1G93A ALS mice. Neurotherapeutics 11:419–432. https://doi.org/10.1007/s13311-013-0253-y
164. Koza LA, Winter AN, Holsopple J et al (2020) Protocatechuic acid extends survival, improves motor function, diminishes gliosis, and sustains neuromuscular junctions in the HSOD1g93a mouse model of amyotrophic lateral sclerosis. Nutrients 12:1–26. https://doi.org/10.3390/NU12061824
165. Maya S, Prakash T, Goli D (2018) Effect of wedelolactone and gallic acid on quinolinic acid-induced neurotoxicity and impaired motor function: significance to sporadic amyotrophic lateral sclerosis. Neurotoxicology 68:1–12. https://doi.org/10.1016/J.NEURO.2018.06.015
166. Shimojo Y, Kosaka K, Noda Y et al (2010) Effect of rosmarinic acid in motor dysfunction and life span in a mouse model of familial amyotrophic lateral sclerosis. Wiley Online Libr 88:896–904. https://doi.org/10.1002/jnr.22242
167. Fontanilla CV, Wei X, Zhao L et al (2012) Caffeic acid phenethyl ester extends survival of a mouse model of amyotrophic lateral sclerosis. Neuroscience 205:185–193. https://doi.org/10.1016/j.neuroscience.2011.12.025
168. Bagur MJ, Antonia Murcia M, Jiménez-Monreal AM et al (2017) Influence of diet in multiple sclerosis: a systematic review. Adv Nutr 8:463–472. https://doi.org/10.3945/AN.116.014191
169. Mirshafiey A, Mohsenzadegan M (2009) Antioxidant therapy in multiple sclerosis. 31:13–29. https://doi.org/10.1080/08923970802331943
170. Miller ED, Dziedzic A, Saluk-Bijak J, Bijak M (2019) A review of various antioxidant compounds and their potential utility as complementary therapy in multiple sclerosis. Nutrients 11. https://doi.org/10.3390/NU11071528
171. Miller E, Walczak A, Majsterek I, Kedziora J (2013) Melatonin reduces oxidative stress in the erythrocytes of multiple sclerosis patients with secondary progressive clinical course. J Neuroimmunol 257:97–101. https://doi.org/10.1016/j.jneuroim.2013.02.012
172. Garcia-Alloza M, Borrelli LA, Rozkalne A et al (2007) Curcumin labels amyloid pathology in vivo, disrupts existing plaques, and partially restores distorted neurites in an Alzheimer mouse model. J Neurochem 102:1095–1104. https://doi.org/10.1111/J.1471-4159.2007.04613.X
173. Miller E, Markiewicz Ł, Kabziński J et al (2017) Potential of redox therapies in neurodegenerative disorders. Front Biosci (Elite Ed) 9:214–234. https://doi.org/10.2741/E797
174. Plemel JR, Juzwik CA, Benson CA et al (2015) Over-the-counter anti-oxidant therapies for use in multiple sclerosis: a systematic review. Mult Scler 21:1485–1495. https://doi.org/10.1177/1352458515601513
175. Stoiloudis P, Kesidou E, Bakirtzis C et al (2022) The role of diet and interventions on multiple sclerosis: a review. Nutrients 14. https://doi.org/10.3390/nu14061150
176. Khalili M, Eghtesadi S, Mirshafiey A et al (2014) Effect of lipoic acid consumption on oxidative stress among multiple sclerosis patients: a randomized controlled clinical trial. Nutr Neurosci 17:16–20. https://doi.org/10.1179/1476830513Y.0000000060
177. Hussain G, Wang J, Rasul A et al (2019) Role of cholesterol and sphingolipids in brain development and neurological diseases. Lipids Health Dis 18:26

178. Esposito S, Bonavita S, Sparaco M et al (2018) The role of diet in multiple sclerosis: a review. Nutr Neurosci 21:377–390. https://doi.org/10.1080/1028415X.2017.1303016
179. AlAmmar WA, Albeesh FH, Ibrahim LM et al (2021) Effect of omega-3 fatty acids and fish oil supplementation on multiple sclerosis: a systematic review. Nutr Neurosci 24:569–579. https://doi.org/10.1080/1028415X.2019.1659560
180. Riccio P, Rossano R (2015) Nutrition facts in multiple sclerosis. ASN Neuro 7:1–20. https://doi.org/10.1177/1759091414568185
181. Murray AJ, Rogers JC, Katshu MZUH et al (2021) Oxidative stress and the pathophysiology and symptom profile of Schizophrenia spectrum disorders. Front Psychiatr 12:1235. https://doi.org/10.3389/FPSYT.2021.703452/BIBTEX
182. Madireddy S, Madireddy S (2020) Regulation of reactive oxygen species-mediated damage in the pathogenesis of Schizophrenia. Brain Sci 10:1–24. https://doi.org/10.3390/BRAINSCI10100742
183. Lu Z, Wen T, Wang Y et al (2020) Peripheral non-enzymatic antioxidants in patients with schizophrenia: a case-control study. BMC Psychiatr 20:241. https://doi.org/10.1186/S12888-020-02635-8/TABLES/5
184. Pandya CD, Howell KR, Pillai A (2013) Antioxidants as potential therapeutics for neuropsychiatric disorders. Prog Neuropsychopharmacol Biol Psychiatr 46:214. https://doi.org/10.1016/J.PNPBP.2012.10.017
185. Sacco RL, Kasner SE, Broderick JP et al (2013) An updated definition of stroke for the 21st century. Stroke 44:2064–2089. https://doi.org/10.1161/STR.0B013E318296AECA
186. Uttara B, Singh AV, Zamboni P, Mahajan R (2009) Oxidative stress and neurodegenerative diseases: a review of upstream and downstream antioxidant therapeutic options. Curr Neuropharmacol 7:65. https://doi.org/10.2174/157015909787602823
187. Thanoon IAJ, Abdul-Jabbar HAS, Taha DA (2012) Oxidative stress and C-reactive protein in patients with cerebrovascular accident (Ischaemic Stroke): the role of Ginkgo biloba extract. Sultan Qaboos Univ Med J 12:197. https://doi.org/10.12816/0003113
188. Halliwell B, Gutteridge JMC (2015) Free Radicals in biology and medicine. Google Books
189. Niizuma K, Yoshioka H, Chen H et al (2010) Mitochondrial and apoptotic neuronal death signaling pathways in cerebral ischemia. Biochim Biophys Acta Mol Basis Dis 1802:92–99. https://doi.org/10.1016/J.BBADIS.2009.09.002
190. Wang X, Wu D, Zhong P (2020) Serum bilirubin and ischaemic stroke: a review of literature. Stroke Vasc Neurol 5:198–204. https://doi.org/10.1136/SVN-2019-000289
191. Watchko JF, Tiribelli C (2013) Bilirubin-induced neurologic damage – mechanisms and management approaches. N Engl J Med 369:2021–2030. https://doi.org/10.1056/NEJMRA1308124/SUPPL_FILE/NEJMRA1308124_DISCLOSURES.PDF
192. Perlstein TS, Pande RL, Creager MA, et al (2008) Serum total bilirubin level, prevalent stroke, and stroke outcomes: NHANES 1999–2004. Am J Med 121:781–788.e1. https://doi.org/10.1016/J.AMJMED.2008.03.045
193. Thakkar M, Edelenbos J, Doré S (2019) Bilirubin and ischemic stroke: rendering the current paradigm to better understand the protective effects of bilirubin. Mol Neurobiol 56:5483–5496. https://doi.org/10.1007/S12035-018-1440-Y/TABLES/2
194. Justicia C, Salas-Perdomo A, Pérez-de-Puig I et al (2017) Uric acid is protective after cerebral ischemia/reperfusion in hyperglycemic mice. Transl Stroke Res 8:294–305. https://doi.org/10.1007/S12975-016-0515-1/FIGURES/6
195. Amaro S, Jiménez-Altayó F, Chamorro Á (2019) Uric acid therapy for vasculoprotection in acute ischemic stroke. Brain Circ 5:55. https://doi.org/10.4103/BC.BC_1_19
196. Clarke MW, Burnett JR, Croft KD (2008) Vitamin E in human health and disease. Crit Rev Clin Lab Sci 45:417–450
197. Cheng P, Wang L, Ning S et al (2018) Vitamin E intake and risk of stroke: a meta-analysis. Br J Nutr 120:1181–1188. https://doi.org/10.1017/S0007114518002647
198. Narasimhan S, Balasubramanian P (2017) Role of Vitamin D in the outcome of ischemic stroke – a randomized controlled trial. J Clin Diagn Res 11:CC06. https://doi.org/10.7860/JCDR/2017/24299.9346

199. Gupta A, Prabhakar S, Modi M et al (2014) Vitamin D status and risk of ischemic stroke in North Indian patients. Indian J Endocrinol Metab 18:721. https://doi.org/10.4103/2230-8210.139241
200. Chaudhuri JR, Mridula KR, Alladi S et al (2014) Serum 25-hydroxyvitamin D deficiency in Ischemic stroke and subtypes in Indian patients. J Stroke 16:44. https://doi.org/10.5853/JOS.2014.16.1.44
201. Myung SK, Ju W, Cho B et al (2013) Efficacy of vitamin and antioxidant supplements in prevention of cardiovascular disease: systematic review and meta-analysis of randomised controlled trials. BMJ:346. https://doi.org/10.1136/BMJ.F10
202. Myint PK, Luben RN, Welch AA et al (2008) Plasma vitamin C concentrations predict risk of incident stroke over 10 y in 20 649 participants of the European prospective investigation into Cancer–Norfolk prospective population study. Am J Clin Nutr 87:64–69. https://doi.org/10.1093/AJCN/87.1.64
203. Arteaga O, Álvarez A, Revuelta M et al (2017) Role of antioxidants in neonatal hypoxic–ischemic brain injury: new therapeutic approaches. Int J Mol Sci 18:265. https://doi.org/10.3390/IJMS18020265
204. Komsiiska D (2019) Oxidative stress and stroke: a review of upstream and downstream antioxidant therapeutic options. Comp Clin Path 28:915–926. https://doi.org/10.1007/S00580-019-02940-Z/FIGURES/5
205. Blondeau N, Tauskela JS (2013) A new future in brain preconditioning based on nutraceuticals: A focus on a-linolenic omega-3 fatty acid for stroke protection. Innate Toler CNS Transl Neuroprotection by Pre- Post-Conditioning 133–163. https://doi.org/10.1007/978-1-4419-9695-4_6/COVER
206. Abdel-Aleem GA, Khaleel EF, Mostafa DG, Elberier LK (2016) Neuroprotective effect of resveratrol against brain ischemia reperfusion injury in rats entails reduction of DJ-1 protein expression and activation of PI3K/Akt/GSK3b survival pathway. Arch Physiol Biochem 122:200–213. https://doi.org/10.1080/13813455.2016.1182190
207. Ritz MF, Ratajczak P, Curin Y et al (2008) Chronic treatment with red wine polyphenol compounds mediates neuroprotection in a rat model of ischemic cerebral stroke. J Nutr 138:519–525. https://doi.org/10.1093/JN/138.3.519
208. Cassidy A, Rimm EB, O'Reilly ÉJ et al (2012) Dietary flavonoids and risk of stroke in women. Stroke 43:946–951. https://doi.org/10.1161/STROKEAHA.111.637835/-/DC1
209. Wright B, Moraes LA, Kemp CF et al (2010) A structural basis for the inhibition of collagen-stimulated platelet function by quercetin and structurally related flavonoids. Br J Pharmacol 159:1312–1325. https://doi.org/10.1111/J.1476-5381.2009.00632.X
210. Hadjipanayis CG, Van Meir EG (2009) Brain cancer propagating cells: biology, genetics and targeted therapies. Trends Mol Med 15:519–530. https://doi.org/10.1016/J.MOLMED.2009.09.003
211. Kheirollahi M, Dashti S, Khalaj Z et al (2015) Brain tumors: special characters for research and banking. Adv Biomed Res 4:4. https://doi.org/10.4103/2277-9175.148261
212. Dandil E, Çakiroğlu M, Ekşi Z (2015) Computer-aided diagnosis of malign and benign brain tumors on MR images. Adv Intell Syst Comput 311:157–166. https://doi.org/10.1007/978-3-319-09879-1_16/COVER
213. Illán-Cabeza NA, García-García AR, Martínez-Martos JM et al (2013) A potential antitumor agent, (6-amino-1-methyl-5-nitrosouracilato-N3)-triphenylphosphine-gold(I): structural studies and in vivo biological effects against experimental glioma. Eur J Med Chem 64:260–272. https://doi.org/10.1016/J.EJMECH.2013.03.067
214. Ramírez-Expósito MJ, Martínez-Martos JM (2019) The delicate equilibrium between oxidants and antioxidants in brain glioma. Curr Neuropharmacol 17:342. https://doi.org/10.2174/1570159X16666180302120925
215. Hung YC, Pan TL, Hu WL (2016) Roles of reactive oxygen species in anticancer therapy with salvia miltiorrhiza Bunge. Oxid Med Cell Longev 2016:. https://doi.org/10.1155/2016/5293284

216. Dhar SK, Tangpong J, Chaiswing L et al (2011) Manganese superoxide dismutase is a p53-regulated gene that switches cancers between early and advanced stages. Cancer Res 71:6684–6695. https://doi.org/10.1158/0008-5472.CAN-11-1233/657103/P/MANGANESE-SUPEROXIDE-DISMUTASE-IS-A-P53-REGULATED
217. Atukeren P, Yigitoglu MR (2013) The stance of antioxidants in brain tumors. Clin Manag Evol Nov Ther Strateg Patients with Brain Tumors. https://doi.org/10.5772/54791
218. Miriyala S, Spasojevic I, Tovmasyan A et al (2012) Manganese superoxide dismutase, MnSOD and its mimics. Biochim Biophys Acta Mol Basis Dis 1822:794–814. https://doi.org/10.1016/J.BBADIS.2011.12.002
219. Backos DS, Franklin CC, Reigan P (2012) The role of glutathione in brain tumor drug resistance. Biochem Pharmacol 83:1005–1012. https://doi.org/10.1016/J.BCP.2011.11.016
220. Jeong C-H, Joo SH (2016) Downregulation of reactive oxygen species in apoptosis. J Cancer Prev 21:13. https://doi.org/10.15430/JCP.2016.21.1.13
221. Yilmaz N, Dulger H, Kiymaz N, et al (2009) Lipid peroxidation in patients with brain tumor. 116:937–943. https://doi.org/10.1080/00207450600553141
222. de Groot J, Sontheimer H (2011) Glutamate and the biology of gliomas. Glia 59:1181–1189. https://doi.org/10.1002/GLIA.21113
223. Nicoletti F, Arcella A, Iacovelli L et al (2007) Metabotropic glutamate receptors: new targets for the control of tumor growth? Trends Pharmacol Sci 28:206–213. https://doi.org/10.1016/J.TIPS.2007.03.008
224. Brocke KS, Staufner C, Luksch H, et al (2010) Glutamate receptors in pediatric tumors of the central nervous system. 9:455–468. https://doi.org/10.4161/CBT.9.6.10898
225. Sheweita SA, Sheikh BY (2013) Can dietary antioxidants reduce the incidence of brain tumors? Curr Drug Metab 12:587–593. https://doi.org/10.2174/138920011795713733
226. Mawson AR (2012) Retinoids in the treatment of glioma: a new perspective. Cancer Manag Res 4:233–241. https://doi.org/10.2147/CMAR.S32449
227. Doldo E, Costanza G, Agostinelli S, et al (2015) Vitamin A, cancer treatment and prevention: the new role of cellular retinol binding proteins. Biomed Res Int:2015. https://doi.org/10.1155/2015/624627
228. Liu RZ, Li S, Garcia E et al (2016) Association between cytoplasmic CRABP2, altered retinoic acid signaling, and poor prognosis in glioblastoma. Glia 64:963–976. https://doi.org/10.1002/GLIA.22976
229. Bielecka J, Markiewicz-żukowska R (2020) The influence of nutritional and lifestyle factors on glioma incidence. Nutritions 12:1812. https://doi.org/10.3390/NU12061812
230. Pawlowska E, Szczepanska J (2019) Blasiak J (2019) Pro- And antioxidant effects of Vitamin C in cancer in correspondence to its dietary and pharmacological concentrations. Oxid Med Cell Longev. https://doi.org/10.1155/2019/7286737
231. Zhang W, Jiang J, He Y, et al (2022) Association between vitamins and risk of brain tumors: A systematic review and dose-response meta-analysis of observational studies. Front Nutr 0:1720. https://doi.org/10.3389/FNUT.2022.935706
232. Hervouet E, Staehlin O, Pouliquen D, et al (2013) Antioxidants delay clinical signs and systemic effects of ENU induced brain tumors in rats 65:686–694. https://doi.org/10.1080/01635581.2013.789541

Chapter 9
Antioxidants as Adjuncts to Conventional Therapies Against Oxidative Stress

Aftab Ahmed, Muhammad Awaise Saleem, Muhammad Afzaal, Shinawar Waseem Ali, Muhammad Nadeem, and Noshiza Majeed

9.1 Introduction

Oxidative stress is a state that damages the proper function of the brain. The brain is vulnerable to oxidative stress since it exploits chemically varied reactive molecules for signal transmission [1]. More than 86 billion neurons and 250–300 billion glial cells make up the human brain, which uses more than 20% of the body's total basal oxygen [2]. The brain's mitochondria utilize the oxygen that is inspired to decrease O_2 to H_2O to promote the making of adenosine triphosphate (ATP) [3]. Reactive oxygen species release from mitochondria occurs in the presence of excess oxygen because in mitochondria electron transport chain (ETC) is inefficient. Redox signaling in the brain serves as an innate sensor in response to oxidative stress when a signal malfunction occurs due to disease conditions. The intense creation of reactive oxygen metabolites, low antioxidant levels, reduced ability for repair, non-replicating nerve cells, and high membrane-to-cytoplasm proportion are all characteristics of brain tissue [4].

A. Ahmed (✉) · M. A. Saleem · N. Majeed
Department of Nutritional Sciences, Government College University, Faisalabad, Pakistan
e-mail: aftabahmed@gcuf.edu.pk; masaleem.fhs@must.edu.pk

M. Afzaal
Department of Food Sciences, Government College University, Faisalabad, Pakistan
e-mail: muhammadafzaal@gcuf.edu.pk

S. W. Ali
Department of Food Sciences, Faculty of Agricultural Sciences, University of Punjab, Lahore, Pakistan
e-mail: shinawar.foodsciences@pu.edu.pk

M. Nadeem
Institute of Food Science and Nutrition, University of Sargodha, Sargodha, Pakistan
e-mail: nadeem.abdul@uos.edu.pk

A. Imran, G. Hussain (eds.), *The Role of Natural Antioxidants in Brain Disorders*, Food Bioactive Ingredients, https://doi.org/10.1007/978-3-031-41188-5_9

Free radicals are known as atoms or molecules with one or more than one unpaired electron that are not involved in chemical interactions and have the propensity to receive electrons from the other molecules, which results in their oxidation [5]. According to the severity of the oxidative stress status such as basal, low, intermediate& high [6], which results from an imbalance between oxidation and reduction in living things, reactive oxygen species and nitrogen species (RONS) are produced excessively [7].

Numerous antioxidant defense mechanisms exist against RONS. Enzymatic and non-enzymatic chemicals make up the two classes of antioxidant molecules. Superoxide dismutase (SOD), catalase (CAT), glutathione peroxidase (GPx), and glutathione reductase (GR) are among the enzymes in this group. One of the primary defenses against ROS is SOD, which catalyzes the transformation of O_2 into H_2O_2 and O_2, whereas CAT transforms the produced H_2O_2 into water and O_2 [8]. Glutathione (GSH), which is present in large amounts in the brain cells, thioredoxin (Trx), the vitamins A, C, and E, selenium, carotenoids, retinoic acid, and flavonoids are included in the non-enzymatic category. Together with GPx and GR, GSH interacts with Reactive oxygen species to produce glutathione disulfide (GSSG), which then initiates a cycle [9].

When polyunsaturated lipid levels are considerable and antioxidant defenses are inadequate, there may be a buildup of biomolecules that have been RONS-damaged [10]. Therefore, due to their high oxygen consumption, poor antioxidant defense, and high content of polyunsaturated fatty acids in their membranes, neuronal cells are particularly susceptible to oxidative damage. Eicosapentaenoic (C20:5) and docosahexaenoic (C22:6) acids are particularly susceptible to assault from free radicals[11].

Antioxidants are chemicals that can lessen or stop the cell damage often brought on by free radicals [11]. They are often referred to as scavengers of free radicals. There are numerous neurological illnesses based on oxidative damage or free radicals as the underlying mechanism. In general, free radical scavengers are utilized to postpone and prevent neurological diseases [12]. Scavenging activity, metal chelation, or lipid peroxidation are often some of these [13]. Antioxidants have mostly been explored for their physiological defense against oxidative stress by stopping the chain processes that spread many ailments and by neutralizing free radicals [14].

Oxidative stress is primarily a threat to all aerobic organisms. Combined antioxidant therapy has recently been investigated as a unique strategy for treating and preventing neurodegenerative disorders where oxidative stress is a contributing component to the disease's development and/or progression. The therapeutic effects of natural antioxidants as preventatives and/or treatments for neurodegenerative disorders are examined in this review, with a focus on natural antioxidant combinations and conjugates that are currently being studied in human clinical trials. It also examines the connection between oxidative stress and neurodegenerative disorders.

9.2 A Link Between Oxidative Stress and Neurodegenerative Disorders

As a result of mitochondrial malfunction, neuroinflammation, apoptosis, and tissue necrosis, cell damage sets off a series of degenerative events [15, 16]. Many neurodegenerative illnesses, including Alzheimer's disease (AD), Parkinson's disease (PD), and Amyotrophic Lateral Sclerosis (ALS), continue to have a key component of oxidative stress-induced homeostatic dysregulation [17]. Stroke, spinal cord injury (SCI), peripheral nerve injury (PNI), and other conditions are examples of neurodegenerative disorders that are produced by injury [19]as shown in Fig. 9.1.

9.2.1 Neurodegenerative Disorders That Worsen Over Time

9.2.1.1 Alzheimer's Disease (AD)

Progressive deterioration in cognitive function characterizes AD, the most frequent cause of dementia [15, 16]. Oxidative stress has been identified as the primary factor in the genesis of AD, leading to mitochondrial malfunction in synapses and neurons as well as the synthesis of A before the onset of plaque disease [18]. The "free radical hypothesis of aging," which claims that free radicals are crucial to the aging process, is where the idea of oxidative stress in AD originated. According to [19], impaired mitochondrial complexes, and harm to the promoter of the mitochondrial ATP synthase gene, which regulates ATP synthesis are examples of mitochondrial dysfunction in AD [20].

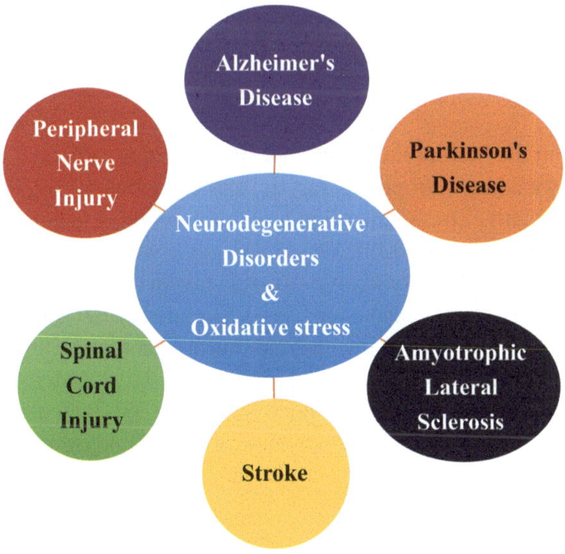

Fig. 9.1 Different types of neurological diseases and oxidative stress

Additionally, damaged mitochondria release 4-HNE, which activates the -secretase complex and encourages the breakdown of the amyloid precursor protein (APP), causing A to accumulate [21]. Additionally, elevated Ca2+ and ROS levels cause a buildup of toxic p-tau clumps, which are one of the pathogenic characteristics that characterize AD [22]. In stress kinases such as the phosphor-c-Jun N-terminal kinase 1 (p-JNK) pathway, which is connected to tau hyperphosphorylation and cell death in reaction to A buildup, ROS also plays a crucial role [23]. Additionally, oxidative stress decreases the activities of antioxidants including glutathione S-transferase (GST), catalase, and superoxide dismutase (SOD), weakening the CNS's natural antioxidant defense. As oxidative stress upregulates the expression of -secretase, it increases amyloidogenesis, which is strongly correlated with elevated levels of LPO and neurotoxicity in AD [24].

9.2.1.2 Parkinson's Disease (PD)

After AD, PD is the neurodegenerative condition that most frequently results in both motor and nonmotor symptoms [25]. Presynaptic neuronal protein α-synuclein, which accumulates and aggregates in the nervous system, is what causes PD [26]. Anomalous protein homeostasis, bioenergetic impairment, and oxidative stress are among the factors linked to the etiology of Parkinson's disease (PD) [27]. The protein aggregation of α-synuclein is related to oxidative stress. Oxidative stress is also connected to the series of events that cause dopaminergic neurons to degenerate in Parkinson's disease [28]. Increased amounts of oxidative stress indicators, including 4-HNE, protein carbonyl, 8-hydroxy-20-deoxyguanosine, and 8-hydroxyguanosine, have been found in the postmortem brain tissue of PD victims [29]. Lewy bodies, which are protein aggregates found in the brains of people with Parkinson's disease, are also linked to oxidative stress [30, 31].

9.2.1.3 Amyotrophic Lateral Sclerosis (ALS)

Motor neurons in the brain, brain stem, and spinal cord are affected in ALS, which causes muscle weakness, atrophy, paralysis, and early death [32]. The etiology of ALS includes oxidative stress, mitochondrial malfunction, and gene alterations that affect mitochondrial activities [33]. The SOD1 gene, which is a crucial constituent of the body's defense against oxidative stress, is mutated in the majority of familial ALS patients (15–20%) [34]. And over 150 ALS-related SOD1 gene mutations have been found in various regions of the enzyme. These mutations cause protein aggregation and misfolding, elevated ROS generation, and redox system disequilibrium, which ultimately leads to the death of nerve cells [35, 36]. Patients with ALS have extremely high levels of oxidative stress indicators [37]. Cysteine/glutamate antiporter overexpression, which results in elevated oxidative stress and extracellular glutamate buildup, has been seen in sporadic ALS patients. Additionally, motor neuron death is influenced by the dysregulation of a retinoic acid signaling pathway (RA), a vitamin A product [38].

9.2.2 *Injury-Induced Oxidative Stress*

Physical or ischemic neuronal tissue damage is known to cause oxidative stress, which sets off a subsequent injury and leads to progressive degeneration.

9.2.2.1 Stroke

A stroke occurs when a thrombus builds up in cerebral blood vessels, causing an ischemia situation that leads to the production of free radicals and tissue damage [39]. Resuming blood flow to the ischemic area worsens the disease because additional free radicals are created, which is known as "reperfusion injury or reoxygenation injury." It is known as ischemia/reperfusion (I/R) damage collectively [40]. Oxidative stress causes neuroinflammation, glutamate excitotoxicity, mitochondrial dysfunction, and damage to the blood-brain barrier (BBB), as well as apoptosis and necrosis in neurons and their supporting cellular components (glial cells and arteries) [41]). These are the salient signs of neurodegeneration in the brain pathology associated with stroke [42–44]. Furthermore, vasoconstriction enhanced platelet aggregation, and endothelial cell permeability is stimulated by excessive ROS generation or poor ROS degradation [45, 46], which affects cerebral blood flow. The disruption of the cerebral extracellular matrix (ECM) caused by the activation of matrix metalloproteinases (MMPs) results in immunocyte infiltration and neuroinflammation, which in turn causes the breakdown of the neurovascular unit (NVU), which causes bleeding and edema [47, 48].

9.2.2.2 Spinal Cord Injury (SCI)

Another frequent type of neuronal injury that results in neurological dysfunctions is SCI, which is distinguished by an initial primary insult and a secondary phase of injury [49, 50]. Disruption to the blood vessels and axons results from the first trauma right away, creating primary injury [51]. In comparison, secondary injury is the oxidative stress and inflammation caused by the main injury. With time, the situation worsens and becomes more incapacitating due to the secondary damage progression, which spreads throughout the entire spinal cord, including the distant spinal cord regions that have not been impacted. Due to trauma, the increased levels of ROS and the resulting oxidative stress are seen as important events linked to the development of secondary injury [52]. In an environment of oxidative stress, malfunctioning mitochondria produce ROs [53], which triggers a series of additional degenerative processes. In particular, this reduces the synthesis of ATP, which is necessary for normal cellular activity and encourages apoptosis [53]. By modifying ion channels, the extra ROS affects how cells operate. This is followed by an excessive buildup of intracellular calcium ions, which ultimately results in excitotoxicity. Ischemic injury results from oxidative stress damage to the microvascular endothelium, which lowers blood flow to the spinal cord's white matter [54].

9.2.2.3 Peripheral Nerve Injury (PNI)

The central nervous system is connected to various sections of the body by a network of long nerve fibers known as the peripheral nervous system (PNS). Communication between the brain and the target organs may be hampered as a result of impairment to the peripheral nerves caused by shock and compression. These wounds have an impact on consciousness, perception, motor, and sensory responses, as well as skin and joint feelings [55]. Malfunctioning of sensory and motor function, which could result in total paralysis of the damaged limb or perhaps the emergence of uncontrollable neuropathic discomfort, are the most typical signs of PNI. The peripheral nerves can potentially sustain damage during a variety of surgical procedures, including oral and maxillofacial surgery [56]. The main causes of PNI's pathogens and mechanism are oxidative stress and inflammation, which worsen neuronal damage and hinder the regeneration process. Inhibiting oxidative stress at the preclinical level has been shown to hasten the repair processes, which could enhance functional recovery [57–60]. Down syndrome, Vascular dementia, Autism, attention-deficit or hyperactivity disorder (ADHD), multiple sclerosis (MS), Huntington's disease (HD), epilepsy, and hopelessness are further neurodegenerative situations that have been accompanied by oxidative stress. Similar to this, gradual deterioration owing to an accumulation of too many free radicals, Ca2+ overload, glutamate release, and mitochondrial malfunction, which results in apoptosis/necrosis, happens in traumatic brain injury (TBI) [61–63].

9.3 Mechanism of Action of Antioxidants

Antioxidants' low quantity in the human body is crucial for reducing and preventing the oxidation of oxidizable substrates. Antioxidants that are both enzymatic and non-enzymatic collectively make up a complex system that collaborates to stop free radical cell damage and organs (Fig. 9.2) [64].

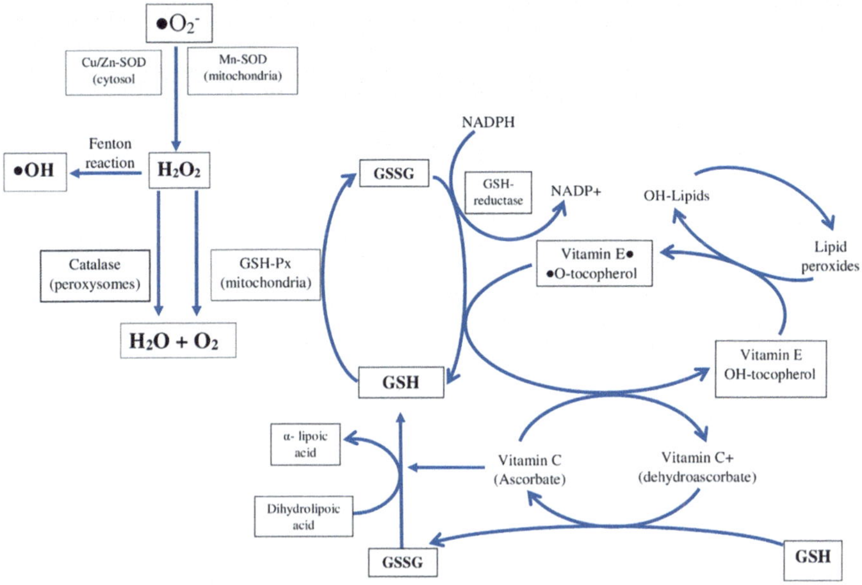

Fig. 9.2 The body's antioxidant defense system. To guard the body's cells and organ systems against free radical damage, enzymatic or non-enzymatic antioxidant systems collaborate and operate in harmony with one another [64]

9.3.1 Enzymatic Antioxidant Mechanism of Action

Different free radical-scavenging enzymes use cell signaling systems to keep ROS levels at optimal ranges. Simply an imbalance between ROS generation and degradation causes oxidative stress, which can disrupt biological processes. Elevated ROS levels can cause irreversible damage to cell macromolecules like DNA, proteins, and lipids, which can further start the carcinogenesis process. The oxidation of macromolecules is thus prevented and slowed down by several enzymatic antioxidants through a variety of defense mechanisms. Certain enzyme antioxidants, including catalases, SOD, and peroxidases, protect cells against harm and lysis through a variety of methods. Enzymatic antioxidants work in a variety of ways, including preventing the formation of free radicals, scavenging reactive oxygen species (ROS), converting more toxic substances into less toxic ones, preventing the production of secondary toxic metabolites and inflammatory mediators, stopping the chain propagation step of secondary oxidant reactions, repairing broken molecules, and boosting endogenous antioxidants. To defend the cell from harm, all of these processes cooperate (Fig. 9.3) [65].

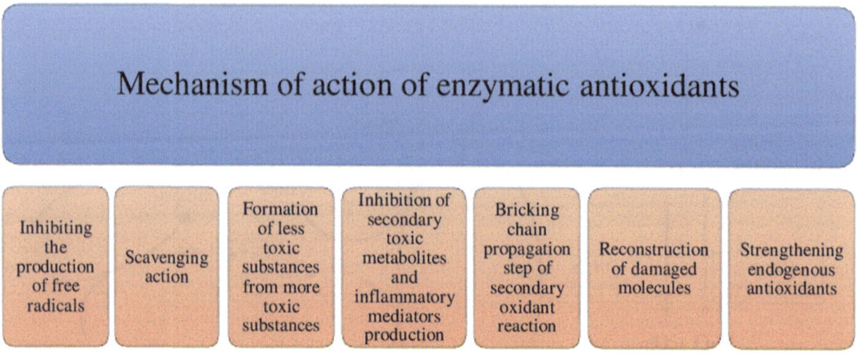

Fig. 9.3 The way enzymatic antioxidants work. The systems for preventing cell damage all operate simultaneously [66]

Catalases, SOD, and glutathione peroxidases are the three main groups of enzymatic antioxidants that keep most body cells in a state of homeostasis (GPx). These enzymes are induced as a result of excessive signaling brought on by increasing oxidative stress [67]. The scavenging activity of SOD causes the superoxide radical to be converted into hydroperoxide [68]. GPx completes further reduction of lipid hydroperoxide, hydrogen peroxide, and another organic hydroperoxide [69]. Superoxide dismutase triggers the breakdown of the superoxide ions into hydrogen and oxygen peroxidase [70].

9.3.2 Mechanism of Action of Nonenzymatic Antioxidants

Since both protein-bound and non-protein-bound thiols operate as cellular protection and a reducing agent through the -SH group against the majority of inorganic pollutants, they are typically the first line of defense against oxidative stress. In addition, severe oxidative stress also may result in lower levels of thiol due to loss in the adaptive mechanism, as illustrated in Fig. 9.4. The rise in thiol production levels may function as an adaptive response to oxidative stress.

As a cellular antioxidant, glutathione contributes to the preservation of the cellular redox state [71]. Reactive oxygen species (ROS) are reduced and neutralized by ascorbic acid [72]. Vitamin E reacts with the lipid radical to eliminate the free radical intermediate and prevent oxidation of the cell membrane [73]. In general, beta-carotene defends against the onslaught of free radicals by eliminating singlet oxygen.

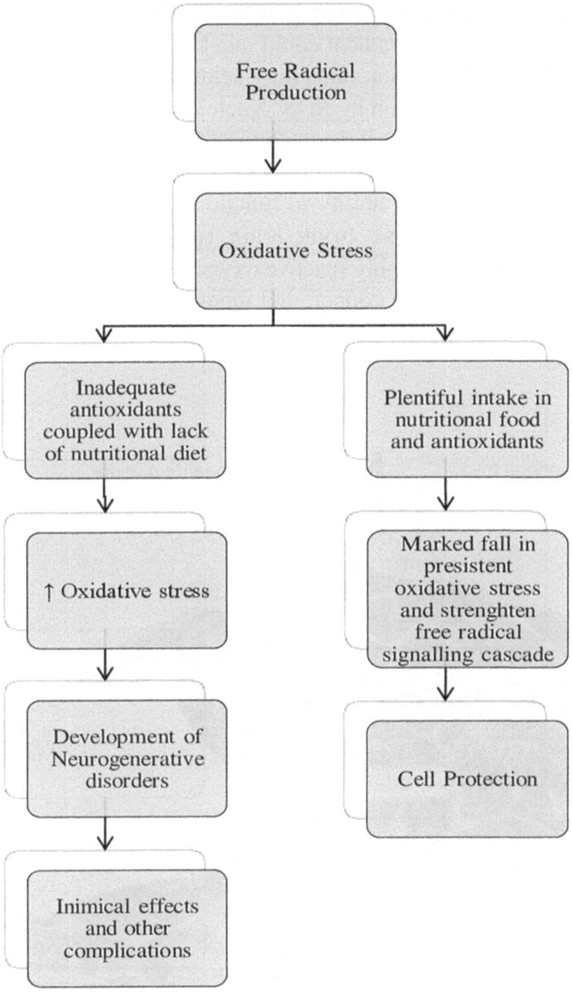

Fig. 9.4 Significance of antioxidants in neurodegenerative disorders [66]

9.4 Factors Contributing to the Vulnerability of the Brain to Oxidative Stress

Oxidative stress is brought on by free radicals and other reactive chemicals that disrupt cellular energy metabolism. Free radicals, which are crucial in oxidative stress (OS), cell death, and ultimately tissue damage, cause an imbalance in endogenous redox through pro-oxidant or antioxidant substances. Pro-oxidant mechanisms that produce too many free radicals can interact with fats, proteins, nucleic acids, as well as other biomolecules to change the structure and function of these molecules. In the membrane lipids of the brain, polyunsaturated fatty acids are highly concentrated. These are sources of lipid peroxidation, a type of breakdown reaction where one initial free radical could quickly cause oxidation of the nearby molecules. Particularly, polyunsaturated fatty acids are important biological targets for the oxidative harm brought on by ROS. Nucleic acids are another potential biological target for free radicals. Cell death and DNA breaks can be caused by altered nucleotides or breaks in DNA. Free radicals can also damage proteins' side chains and backbones, impairing their ability to function as enzymes, neurotransmitters, receptors, and structural proteins. In the brain, several metals and their reducing equivalents combine to form highly reactive oxygen radicals. By way of the Fenton reaction, metals like iron, zinc, copper, and manganese are linked to enhance the production of free radicals [4, 74] (Fig. 9.5).

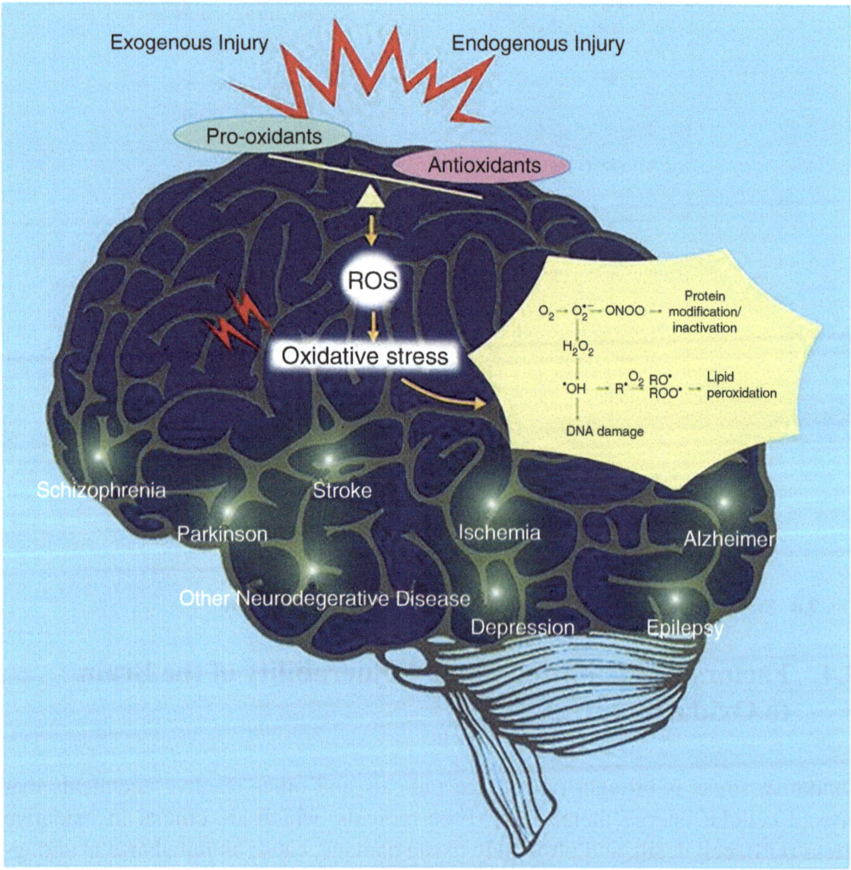

Fig. 9.5 The pathophysiology of oxidative stress in the brain leads to the development of numerous disorders. Pro-oxidants and antioxidants in balance or out of balance against the formation of reactive oxygen species cause oxidative stress, which then results in neuronal damage and neurodegenerative disorders [83]

In light of a thorough understanding of signaling, the following explanations explain why brain neurons are susceptible to oxidative stress: In redox signaling, the reduced form of NADPH oxidase 2 produces O2 /H2O2 which causes the activation of signaling proteins by the production of sulfenic acid [75]. Synaptic plasticity which is an essential brain function depends on Ca2+ signaling, and the brain uses a lot of Adenosine Triphosphate (ATP) to keep intracellular Ca2+ homeostasis in check. Furthermore, Ca2+ overload in mitochondria produces O2 /H2O2 creation, which might result in ONOO formation and excitotoxicity. Synaptic terminal glutamate-induced Ca2+ transients trigger neuronal nitric oxide synthase (nNOS)-mediated NO formation [76].

Particularly, the monoamine oxidase enzyme isoform catalyzes the production of H_2O_2in metabolism, and mitochondria produce O_2 in complex I and complex III [77]. To support neuronal activity, the human brain consumes about 25% of the blood's glucose. Though, protein inactivation caused by the development of advanced end glycation products (AGE) as a result of lower glycolytic rates might result in glucose-induced oxidative stress [78]. The brain is, however, more susceptible to oxidative stress when neurotransmitters having catechol groups are present. During the oxidation of neurotransmitters, redox-active transition metals promote auto-oxidation of the dopamine to a semiquinone radical [79].

The brain's mature neurons have a variety of processes that support long-term neuronal survival and guard against cell death, but in comparison to other tissues, the brain's endogenous antioxidant defense system is more underdeveloped, making it vulnerable to redox homeostasis imbalance. According to [80–82], neurons having fifty times less catalase than hepatocytes have restricted glutathione peroxidase 4 (GPX4) activity and a moderate antioxidant defense system.

9.5 The Neuroprotective Role of Natural Antioxidants

To combat oxidative stress, antioxidant defense systems are activated. Such systems stop the creation of Reactive Oxygen Species (ROS) and block and trap radicals that are produced. These systems, which can be enzymatic or non-enzymatic, are found in aqueous or membrane cell compartments. An enzyme antioxidant system composed of glutathione peroxidase, catalase, and superoxide dismutase has been named as the first line of defense. Reduced thiol and nonenzymatic antioxidants, such as hydro and liposoluble and metabolic substances, serve as the second line of defense [84]. Second, the removal of harmed biomolecules from the antioxidant repair process before they aggregate results in modifications to the cell metabolism. The interventions of the repair system consist of restoring oxidatively damaged nucleic acids by certain enzymes, eradicating oxidized proteins by the proteolytic systems, and restoring oxidized lipids by the peroxidases, phospholipases, or acryl transferases [85]. In addition, non-enzymatic antioxidants can be synthesized or regenerated with the aid of antioxidant enzymes [84].

9.5.1 *Enzymatic Antioxidants*

9.5.1.1 Superoxide Dismutase

The enzymatic defense system enzyme superoxide dismutase (SOD) converts superoxide radical anion to H_2O_2 through oxidative damage. Dehydratase protection from superoxide inactivation by free radicals is another function of SOD. Humans have three different types of SOD. The cytosol contains SOD-1, a Cu, Zn-SOD (Cu, Zn-SOD) that selectively catalyzes the dismutation in a pH-independent media. Eight antiparallel beta strands and two metal atoms, make up the homodimer SOD-1 protein, which catalyzes the transformation of harmful O_2 anions into O_2 and H_2O_2. Particularly, copper minerals are essential for this enzyme's catalytic activity, and zinc is necessary for its structural integrity. SOD-1 is a crucial initial line of defense because it detoxifies superoxide radicals.

Additionally, the absence of this enzyme showed a clear sensitivity to the toxicity of paraquat. The mitochondrial matrix contains SOD-2, a manganese-containing SOD (Mn-SOD), which lowers the superoxide radical anion produced by the electron transport chain (ETC) [86].

9.5.1.2 Catalase

When the body is affected by oxidative stress, it spontaneously produces the antioxidant enzyme catalase (CAT), which is a heme-containing tetrameric protein. In organisms, catalase and the activities of numerous peroxidases catabolize H_2O_2 produced inside the cell through enzymatic means. One of the most effective enzymes, CAT, is invulnerable to H2O2 saturation at any concentration. Using Fe as a cofactor, it interacts with H_2O_2 to produce water, alcohol, and oxygen [87]. CAT defends cells by eliminating the H_2O_2 that is produced, and it is crucial for the progress of oxidative stress tolerance as an adaptive reaction [88]. For recurrent cycles of chemical reduction or even for direct contact with toxins, CAT can keep the O2 content stable. Therefore, it is hypothesized that CAT shortage or dysfunction contributes to the etiology of several age-associated degenerative illnesses [89].

Subsequent Aβ-induced toxicity in the neuronal culture, CAT therapy lowers H_2O_2 levels and enhances neuronal survival. By lowering H_2O_2 levels, CAT-SKL (serine-lysine-leucine) treatment in A toxicity minimized the pathology of microglial activation and hence the brains of rats did not exhibit any deficits in long-term memory [90]. In the Parkinson's disease treatment model, mutant synuclein decreased CAT expression and activity, increased H_2O_2 generation, and decreased catalase activity [91]. This anti-oxidant enzyme is crucial for preserving the proper level of oxidative stress.

9.5.1.3 Glutathione Peroxidase

To shield mammalian cells from oxidative damage, glutathione peroxidase (GPx) uses GSH to catalyze the conversion of a range of hydroperoxides (ROOH and H2O2) to water or the equivalent alcohols. According to [80], GPx has four identical subunits that each contain one selenocysteine for crucial enzyme function. These subunits are categorized as selenium-containing GPxs (GPx1-4 and -6) and also their nonselenium congeners (GPx5, 7, and -8). Although CAT is comparatively important in defending against more severe oxidative stress, the glutathione redox cycle is more dynamic with low levels of oxidant stress. Despite being ubiquitously expressed, GPx isoenzymes appear to contain antioxidant roles at various sites including the cellular compartments, and levels of expression of each isoform fluctuate depending on the type of tissue.

Additionally, glutathione reductase (GR) converts oxidized glutathione disulfide (GSSG) from GSH back to GSH. GPx is the enzyme that converts GSH to GSSG [92].

While GPx1 and GPx4 are present in the majority of tissues, brain tissue contains the majority of GPx forms. GPx1 is expressed in astrocytes and neurons, and it lowers organic hydroperoxides and H2O2 [93]. The cytosolic enzyme GPx4, which may directly decrease phospholipid hydroperoxides, cholesterol hydroperoxides, and fatty acid hydroperoxides, is a phospholipid hydroperoxide glutathione peroxidase [94]. In the brain, neurons of the cerebral cortex, cerebellum, and hippocampus have been found to express the mitochondrial and cytosolic GPx4 isoforms. Hence, glial cells in contrast, scarcely express GPx4 under physiologically normal circumstances, while reactive astrocytes do so after a specific type of brain injury [94].

9.5.2 Non-Enzymatic Antioxidants

9.5.2.1 Antioxidant Enzyme Cofactors

Coenzyme Q10

Coenzyme Q10, also known as vitamin Q10, ubidecarenone, or ubiquinone. About 3-5 milligrams of an endogenous chemical are created daily by mitochondria. It functions as an antioxidant and is one of the key components in mitochondrial oxidative phosphorylation. CoQ10 has been proven to smoothly pass the blood-brain barrier in vitro tests [95]. In a pilot investigation on eleven children with Rett Syndrome, which is a severe neurodevelopmental illness in which hypoxia-induced oxidative stress is linked to the origin and progression of the disease, the antioxidant capacity of CoQ10 was further assessed [96]. The energy condition of red blood cells significantly improved after a 12-month CoQ10 administration (300 mg/day), indicating a reduction in oxidative stress [97].

Selenium

The recommended dietary range for selenium, an essential element, is quite constrained. Around 55 g/day is the RDA for selenium, which can be included in a particular dietary intake. 25 types of selenoproteins, which includes GPx, thioredoxins, and selenoproteins P, W, and R, and contain selenium in the form of selenocysteine (TrxR). As an antioxidant, it guards against cellular damage brought on by ROS [98]. Patients with Multiple Sclerosis and Alzheimer's disease have altered concentrations of this substance in their brains; as a result, it may have a significant protective function against neurodegeneration [99]. Given that aging increases the risk of selenium shortage due to changes in metabolism decreased bioavailability, and dietary alterations [100], Its exogenous assumption has been theorized as a possibility in various investigations to stop age-related disorders.

Zinc and Essential Metals

Despite not directly interacting with ROS, zinc, a redox inactive metal, is essential for maintaining redox equilibrium in the cell's antioxidant defense mechanism. Antioxidant enzymes like SOD, GPx, and CAT are more activated when zinc is present. Additionally, it functions as an indirect cofactor for GPx and a direct cofactor for SOD-1, SOD-3, and GPx [101]. The zinc-binding protein known as metallothionein, which releases the metal in oxidative circumstances and functions as a Se scavenging oxidant, is induced through its mediation. The modulation of protein thiol redox state and glutathione metabolism are both influenced by zinc [102]. With the release of glutamate, zinc in synaptic vesicles functions as a powerful extracellular modulator by interrelating with a variety of synaptic receptors throughout synaptic activity [103]. The preservation of life and the preservation of cell homeostasis are critical functions of other vital metals including copper, iron, and magnesium. They perform crucial structural, regulating, and catalytic roles in a variety of proteins, including transporters, receptors, and enzymes [104, 105]

9.5.2.2 Reactive Oxygen Species Scavengers (Vitamin C, E, and A)

Vitamin C and E

An aged person's chance of developing dementia is decreased and their cognitive function is favorably correlated with a diet strong in fruits and vegetables, often high in the vitamin E, carotenoids, and vitamin C. Chemically, Ascorbic Acid (AA) is also known as Vitamin C. It consists of a six-carbon molecule with two acidizing groups [106]. The area of the human body where AA is present in the highest concentration is the brain. This high level of concentration confirms that AA has a fundamental impact on how the brain functions. Indeed, numerous studies indicate that AA may play a neuroprotective effect via modulating antioxidant activity [107].

This regulation is connected to the buffer of the oxidizing species caused by methamphetamine, homocysteine, ethanol, and other compounds [108].

It is intriguing to observe that the Ascorbic acid activity is fairly broad, especially when taking into account how it interacts with vitamin E. Their cooperation in protecting membranes and other hydrophobic compartments is extraordinary [109]. According to [110] and Shen and colleagues (2012), a clinical investigation has shown a link between vitamin C and E intake and a delay in the start of AD in a group of older participants [111].

A lipophilic substance called vitamin E can be found in plants and many foods that are part of the Mediterranean diet [112]. Compounds known as tocopherols and tocotrienols are referred to as vitamin E [113]. These typically contain eight compounds (including tocotrienols and tocopherols) with high antioxidant potential [113, 114]. Regarding PD, research indicates that Vitamin E supplementation may help patients with their symptoms, functional skills, and inflammatory condition [115]. To prevent the emergence of neurodegenerative illnesses & associated complications, it is essential to supplement with vitamins E and C as antioxidants.

9.6 Factors Associated with the Lifestyle That Promotes Brain Health

With effective clinical medications that have a range of negative side effects, there are presently no disease-modifying therapeutic approaches. A load of symptoms develops with time, minimizing the quality of life. Optimistic lifestyle attitudes may motivate people, enhance their quality of life, lessen signs, and even decrease the course of diseases. Diet, cognitive stimulation, physical exercise, and stress management are examples of lifestyle habits that possess a positive influence on mental health and quality of life. Although there is some evidence that suggests a link between lifestyle and the incidence of neurological disorders, it is challenging to make clinical prescriptions due to the limited and sometimes contradictory data. Here, we showcase research demonstrating positive links between a healthy lifestyle and brain function.

9.6.1 Mental Wellness and Brain Function

This section aims to investigate the connection between brain health and mental wellness. Having positive feelings, performing well, and being able to deal with the difficulties of daily life are all characteristics of mental wellness. On the subject of mental wellness and brain health, a survey was undertaken. Based on the consequences of this research, it is predicted that mental wellness and brain health are positively correlated [116].

Sutin, Stephan, and Terracciano studied the effects of optimism and life goal on mental well-being [116]. They discovered that a lower incidence of dementia was connected with more optimism, a friendly mindset, an improved mood, satisfaction with life, and a sense of mission. They found that one of the best indicators of greater brain health was a sense of a life purpose when they integrated this data with socioeconomic, psychosocial, and generational characteristics. The study also discovered that possessing a mission in life was linked to a 20% lower incidence of dementia. Second research by Galderisi et al. supported these results and revealed that living a purposeful life also lowers the risk of dementia [117].

The inverse appears to be true as well. Early life stress seems to possess an influence on the brain and mental health, according to the investigation. Individuals are more susceptible to mental disease later in life when they experience significant negative circumstances at an early age [118]. Even during pregnancy, exposure to high amounts of cortisone brought on by anxiety could have long-lasting impacts by affecting brain growth and reducing levels of brain well-being in the future [119]. For instance, investigations have especially observed the significance of adversity in early life [120]. According to the research, people who experienced early adversity had smaller hippocampus and other cognition-related brain areas [121, 122].

9.6.2 Cognitively Stimulating Activities

Activities that stimulate a person's capacity to think and integrate information are referred to as cognitively stimulating activities. These consist of puzzle games, learning exercises, intellectual pursuits, and mental assessments. It is commonly known that as people age, their brain's architecture and functions fluctuate. But these modifications are not widespread. Some people, including those in their 30s, still retain the same volume of some brain regions connected to cognitive function, such as the hippocampus, at the age of 80. Others have significant volume loss [123]. The difficult part is figuring out what causes these significant disparities. It is well-recognized that education and learning improve cognitive reserves, rendering a person less vulnerable to the impacts of aging or disease-related brain impairments [124].

Cognitively stimulating actions may assist maintain cognitive function or prevent cognitive decline since research has revealed that brain plasticity endures throughout the aging process [125]. Numerous investigations have found a connection between engaging in self-initiated, intellectually stimulating activities and good brain health, which gives legitimacy to this concept [126–128]. However, no evidence has yet been shown to support this concept. According to the National Academies committee, "AHRQ's (the U.S. Agency for Healthcare Research and Quality) systematic evaluation did not identify any particular therapies to support launching an active public health strategy to promote their adoption for the aim of avoiding cognitive decline and dementia" [128]. This involves engaging in activities that may be categorized as intellectually stimulating.

9.6.3 Link Between Sleep and the Brain

As people grow older, their sleep habits vary, drastically altering both the pattern and length of their sleep [129]. Although the requirement for sleep does not change between the ages of 25 and 50, sleep habits do. According to recent research, most individuals need between 7 and 8 hours of sleep every night to maintain optimal physical and mental health. The length and consistency of sleep cycles are two of the most significant alterations that have been noticed. Whereas the amount of time it takes older persons to fall asleep does not alter, many nighttime interruptions and early morning awakenings are usual. These alterations are a typical aspect of aging and do not influence the well-being of the brain. However, less overall sleep duration does have a detrimental impact [130, 131].

Aside from changing regular sleep cycles, lack of light exposure can lead to hormonal imbalances that change sleep patterns [132]. Devastating sleep problems that are increasingly common as people become older and pharmacological regimens intended to treat a range of chronic diseases are additional variables that might disturb sleep. Luckily, a variety of lifestyle changes may be made to counteract these detrimental changes. Increasing time spent in the sun, maintaining regular sleep schedules, and learning about the possessions of drugs is all essential elements in maintaining sleep rhythms [133].

Light exposure has a significant impact on the hormone stability engaged in sleep regulation. To sustain sufficient melatonin production, exposure to light is essential [134]. Melatonin's regular production offers a crucial regulatory mechanism for other brain chemicals to adhere to in regulating sleep patterns. According to a study, older people frequently receive little daylight. Because of this, studies recommend allowing for more light exposure during daylight and spending time outdoors without sunglasses to optimize sleep patterns [135].

Furthermore, multi-drug prescriptions in elderly people can be quite alarming and tend to change regular sleep patterns [136, 137]. Insomnia represents the most often diagnosed sleep problem in older persons when it comes to abnormal sleep patterns [138]. When it comes to older people, insomnia is sometimes misdiagnosed as a problem with getting asleep when it is truly a problem with interrupted sleep and trouble staying asleep. Numerous studies have demonstrated the danger that interrupted sleep brings to brain function [139]. Poor intellectual and behavioral functioning as well as an elevated risk of cerebral small vessel disease both present in older persons who have interrupted sleep as compared to those who do not [140].

Another highly prevalent sleep condition in elderly persons is sleep apnea. It is characterized by airways compressing as you sleep, making it challenging to breathe. While up to two-thirds of those 65 and older would experience signs that might range from minor to severe, only around one-third of those with sleep apnea will have serious complications [141]. The brain stems and brain tissue has both been found to deteriorate as a result of sleep apnea [142]. Fortunately, research has shown that some of the injuries may be repaired with the best possible care, leading to better sleep and higher blood oxygen levels [143].

9.6.4 Brain Function and Social Interaction

Good social connections appear to be a significant feature in sustaining brain health, but isolation brought on by stressed relations raises the risk of psychological and behavioral diseases. The distinction between loneliness and isolation must be made; individuals can experience loneliness even while engaged by others[143]. Regrettably, social networks seem to become increasingly restricted as people age. They may start to disappear when illness, disease, or death, which will only exacerbate your sense of isolation. According to studies, severe loneliness is more harmful than occasional loneliness in terms of the likelihood of cognitive deterioration in older persons [144, 145]. It's interesting to note that a preliminary study indicates that the impact of digital involvement on a person's cognitive abilities later in life is comparable to that shown in face-to-face contact [146, 147].

Weddings and other close partnerships are important social links as well. Marriage relationships and their effects on people's bodies and emotions have both been examined. Engaging in a satisfying romantic relationship can be highly advantageous, however relying solely on one major connection for social contact can result in loneliness if the other person in this family circle were to pass away [148, 149].

In addition, researchers have exposed that Alzheimer's disease victims exhibit less verbal hostility and fear among animals [150]. Associations with neighbors and other members of the community can also have similar good impacts. Greater neighborhood social connectedness promotes a sense of community among inhabitants and encourages healthy lifestyle choices like greater exercise and efforts to quit smoker [151].

9.6.5 Diet and Nutrition for Brain Health

The Global Council on Brain Health (GCBH) also investigated the relationship between specific nutrients, diets, and nutritional behaviors that can improve cognitive achievement in elder persons [152, 153]. Some investigations highlight the significance of having a balanced diet by demonstrating how eating habits impact brain health through the progression of an individual's whole life [154, 155]. For instance, there is mounting evidence that vitamins and minerals, consumed as a portion of a well-balanced diet rather than as concentrated supplements, are very helpful.

9.6.5.1 The Mediterranean Diet

The eating habits of Mediterranean countries like Greece, Italy, and Spain are represented in the Mediterranean Diet (MD). Wide-ranging, it is considered by the high ingesting of fish, bread, pasta, whole grains, fruits, vegetables, dairy, poultry, eggs, seeds, nuts, and olive oil. In this diet, red meat, processed carbohydrates, and sweets are restricted. It frequently involves having a glass of wine, typically red, with one of the mealtimes. According to research [155, 156], the MD also has further health advantages such as a lower menace of AD, heart disease, and type 2 diabetes. For instance, research by Scarmeas et al. examined the impact of the MD on more than 2,000 New Yorkers over 1.5 years. According to the findings, those who moderately followed the MD had a 15–21% lower chance of developing AD, while those who firmly committed to the MD seemed to have a 39–40% lower risk [157].

9.6.5.2 Nordic Diet

The eating habits of Scandinavian countries notably Denmark, Finland, Iceland, Norway, and Sweden provide the basis for the Nordic Diet [158]. The focus is on non-animal nutrients such as fruits and vegetables, similar to the MD. Consuming fish, oils, and various sorts of meat is also part of the Nordic Diet. However, there are amendments amid the MD and the Nordic Diet in terms of the particular fruits and vegetables, cooking methods, and the quality and amount of oil used. For instance, the Nordic Diet substitutes rapeseed (also known as canola) oil to replace olive oil. To investigate the cross-sectional and longitudinal relationships between the Nordic Diet and mental performance, 4-year research of 1,135 randomly chosen men and women with average cognition was undertaken. Individuals who adhered to the Nordic Diet's recommendations showed improved mental abilities [159].

9.6.5.3 DASH (Dietary Approaches to Stop Hypertension)

Investigations have revealed that the DASH Diet has significant health advantages due to its minimal salt consumption and small portion sizes. Improvements in serum lipid, blood pressure, and the likelihood of long-term illness are among them [160]. The presence of nuts, low-fat and non-fat dairy products, lean meats, and whole cereals in the DASH Diet distinguishes it from the MD and Nordic diets. In research by Appel et al. 459 persons with systolic blood pressures of less than 160 mm Hg and diastolic blood pressures of 80 to 95 mm Hg was included, 133 of whom had high blood pressure. Following the DASH Diet for 3 weeks resulted in noteworthy drops in systolic and diastolic blood pressure in the 133 hypertensive patients than in the control group of subjects, by 11.2 and 5.6 mm Hg, accordingly [161].

9.6.5.4 Okinawan Diet

The Okinawan diet relies on the dietary patterns of the native Japanese Ryukyu Islanders. As researchers got conscious of the population's unusual lifespan, they began to investigate their dietary patterns. The Okinawan diet places a strong emphasis on eating green, yellow, and orange vegetables as well as soy and legumes.

It also restricts the use of dairy, sugar, sodium, and processed carbohydrates. Remarkably, the diet emphasizes the consumption of sweet potatoes together with relatively little seafood and grains. This group eats until they are no longer hungry rather than until they are full, which is another intriguing characteristic. According to research by Miyagi et al., Okinawans had the greatest life expectancy at birth (LEB) rates in the world in 1990, and their likelihood of developing cerebrovascular disease was at an all-time low. A 90-gram daily consumption of meat and pulses, which is roughly 20 and 30% greater than the national norm, was shown to be the cause of their excellent health. Their higher daily consumption of green and yellow vegetables than the national norm was another factor in their noteworthy LEB and lower risk of illness [162].

9.6.5.5 MIND (Mediterranean-DASH Intervention for Neurodegenerative Delay) Diet

The MD and DASH diets are both influences on the MIND diet. It advises eating two or more portions of berries each week, other vegetables at least once a day, and green leafy vegetables at least six times a week. On the MIND diet, nuts are recommended as a healthy snack in addition to beans every other day, chicken twice per week, and fish at least once. The diet also severely restricts the use of butter, cheese, fried foods, and fast food. The MIND diet was developed by Morris et al. using nutritional elements from the MD and the DASH diets that have been proven to be neuroprotective, such as a high intake of veggies and berries. Higher education, better engagement in mental and physical activities, and a lower frequency of cardiovascular diseases were all associated with a more promising risk profile for maintaining cognitive capacities in those who followed the MIND diet the most [163, 164].

These diets share certain similar patterns, which might be emphasized. They all advise staying away from trans-fats, which are included in processed foods like partly hydrogenated oils and are believed to increase LDL and reduce HDL (LDL) [165]. In contrast, omega-3 fatty acids are alternatives for healthy dietary fats. Given the extensive body of research on omega 3-fatty acids and cognitive performance, it is critical to draw attention to its advantages. Investigations on omega-3 fatty acids have looked at the entire spectrum of fatty acids rather than focusing on particular kinds. The utmost common omega-3 fatty acid in the brain is docosahexaenoic acid (DHA), which is intricate in the plasticity of neuronal membranes [166, 167].

The ingestion of caffeine and cocoa flavonoids are further interesting concerns. There is a clear link between drinking coffee and tea and preventing cognitive aging, according to several types of research [168]. These advantages have been linked to polyphenols, an important antioxidant present in plant-based foods. There is no set recommendation for the quality or quantity of food that should be consumed to maintain brain health across an individual's lifetime. However, improved awareness is a short-term advantage of coffee and tea use. Investigations on the impact of cocoa flavonoids, which are frequently present in dark chocolate, on mental well-being and cognition have been done [169, 170].

It is hard to debate nutrition without bringing up the topic of weight. Numerous studies have connected mid-life obesity to a higher risk of later cognitive deterioration in people [171]. However, whether weight loss at an advanced age is beneficial is up for debate. The best diet choices are nutritious ones with fewer calories to successfully optimize your diet for improved heart and brain health [172].

9.6.6 Physical Activity

Healthcare professionals have strongly supported physical activity because they believe that it will benefit brain health whether it arises from a physically energetic lifestyle (e.g., walking to the office or the shop instead of driving, using stairs, and participating in sports activities) or from a deliberate workout (brisk walking, strength training, as well as aerobic training). This idea is highly supported by recent studies. Researchers were able to conclude that exercise promotes neuroplasticity and enhances learning upshots by examining animal research. It appears that this is also true for people [173]. Coelho discovered that physical activity significantly raises peripheral concentrations of brain-derived neurotrophic factor (BDNF), a brain protein that supports the development and function of neurons in the senior citizen [174].

In addition, over six months, Baker et al. performed an experiment contrasting high-intensity aerobic exercise with stretching versus stretching alone [175]. The findings indicated that aerobic exercise had sex-specific benefits on cognition, glucose homeostasis, hypothalamic-pituitary-adrenal axis, and neurotrophic activity, as well as similar gains in cardiovascular endurance and body fat management. It's fascinating to see that men and women had different outcomes. Aerobic exercise improved performance in several tests for women. In addition to boosting glucose elimination during the metabolic clamp and decreasing fasting plasma levels of insulin, cortisol, and BDNF, their findings indicate better executive performance. Aerobicercise raised the plasma concentration of insulin-like growth factor 1 in males [176].

Other research has concentrated on showing the neurological links between aerobic exercise and good mental health. For instance, research by Colcombe et al. investigated the possibility of boosting brain volume in areas of the brain linked to age-related deterioration in both brain structure and cognition in older persons. In contrast to the older persons who contributed to the stretching and toning (non-aerobic) control group, they discovered a substantial rise in both grey and white matter areas of the brain in those who engaged in the aerobic workout program [176]. Ultimately, numerous meta-analyses have shown how important exercise is for maintaining brain function. These studies showed that exercise in middle age considerably lowers the risk of dementia and moderate cognitive decline [177, 178]

Skill-based, aerobic, strengthen, and cued training are standard physical activity therapies for persons with neurodegenerative illnesses. It is challenging to translate outcomes into patient recommendations since there is no standardization across outcome measurements, training methodologies, and intervention times. Yoga, dancing, and Tai Chi are a few examples of skill-based exercises [179, 180]. Balance and equilibrium are greatly improved by aerobic exercises, such as Nordic walking and elliptical training, although the effects on motor function and quality of life are unclear [181–183].

A minimum of 12 weeks are required for the benefits of progressive resistance training to take an effective quality of life, balance, and muscular strength [181]. It is uncertain whether resistance training correlates with better outcomes because there have been few studies that have evaluated muscular strength improvement [184]. It is also unknown whether resistance training is preferable to other forms of exercise or everyday activities [185]. Cued training is the process of learning movement assistance tactics by employing external visual, rhythmic auditory, or somatosensory signals. These methods increase walking speed, balance, and everyday life activities [186], but their effectiveness depends on cognitive function being retained. They may also provide similar benefits to those of traditional physical treatment [187].

9.6.7 Antioxidant Dietary Supplements for the Brain

In a study of women in their 70s and 80s, it was shown that supplementation with vitamins E and C together dramatically enhanced cognitive function compared to women who had never consumed either vitamin. The advantages for individuals taking vitamin E alone seemed to be less strong, and there was no indication of a progression with a time of usage. The obvious advantages were especially noticeable in women who consumed low -tocopherol in their diets; for these women, taking vitamins E and C combined was cognitively similar to being two years younger. However, there is no evidence to suggest that taking vitamin C supplements alone or having previously taken either supplement would have an impact on intellectual performance [188]. The stimulating impact of α-tocopherol on neuronal survival was demonstrated in rat brains by Sato et al. in 1993. When vitamin C was supplemented, this effect was more than doubled [189].

Two intervention studies examined the long-term and short-term effects of extracts comprising polyphenols on cognitive performance [190]. One cross-sectional study looked at the relationship between dietary phenolic acids and cognitive state in elderly people [191]. Regarding short-term memory, cognitive function, judicious and attentional control, and information processing, a crossover clinical study on healthy middle-aged volunteers supplemented with fruit- and vegetable-based extract containing polyphenols showed substantial perfections when equated to the placebo group [192]. A similar crossover clinical trial involving healthy college students found that the use of polyphenol-rich grape and blueberry extracts could be a secure substitute for immediately improving working memory and recognition during prolonged intellectual processing [190].

Ultimately, research carried out in a cohort of older Italian adults revealed a link between the usual dietary ingestion of phenolic acids and cognitive status, illustrating that people with reduced consumption of phenolic acids were more plausible to have impaired cognition. Phenolic acids are notably found in coffee, berries, nuts, artichokes, and olive oil [191]. Numerous studies have supported the effects of nutrition and dietary antioxidant supplements on the brain and mental health via a variety of processes, including the control of neuroinflammation, adult neurogenesis, synaptic plasticity, and mitophagy [193–195].

9.7 Conclusion

Antioxidant molecules, which shield brain cells from the harm caused by free radicals, are found in a variety of natural substances. Antioxidants have been shown to ameliorate the oxidative stress state of brain cells, cognitive abilities, and motor functions. Additional scientific experiments must be carried out to determine whether these natural ingredients, either singly or in conjunction with the proper pharmaceutical therapy, may successfully prevent the possible development of neurodegenerative disorders or improve brain functioning. Further, more research is needed to determine how well these natural antioxidants may cross the blood-brain barrier after being consumed orally, as well as their true bioavailability in the central nervous system. While research is still ongoing to understand the particular pathways by which adaptable lifestyle behaviors contribute to neuronal health in general, giving their known benefit on chronic condition preventive measures and treatment, as well as the quality of life. These must be recommended more consistently as fundamental abilities for health management and preventing illness.

References

1. Cobley JN, Fiorello ML, Bailey DM (2018) 13 reasons why the brain is susceptible to oxidative stress. Redox Biol 15:490–503. https://doi.org/10.1016/J.REDOX.2018.01.008
2. Magistretti PJ, Allaman I (2015) A cellular perspective on brain energy metabolism and functional imaging. Neuron 86:883–901. https://doi.org/10.1016/J.NEURON.2015.03.035
3. Bailey DM, Bärtsch P, Knauth M, Baumgartner RW (2009) Emerging concepts in acute mountain sickness and high-altitude cerebral edema: from the molecular to the morphological. Cell Mol Life Sci 66:3583–3594. https://doi.org/10.1007/S00018-009-0145-9
4. Niedzielska E, Smaga I, Gawlik M et al (2016) Oxidative stress in neurodegenerative diseases. Mol Neurobiol 53:4094–4125. https://doi.org/10.1007/S12035-015-9337-5
5. Valko M, Leibfritz D, Moncol J et al (2007) Free radicals and antioxidants in normal physiological functions and human disease. Int J Biochem Cell Biol 39:44–84. https://doi.org/10.1016/J.BIOCEL.2006.07.001

6. Sies H (2015) Oxidative stress: a concept in redox biology and medicine. Redox Biol 4:180–183. https://doi.org/10.1016/J.REDOX.2015.01.002
7. Lushchak VI (2014) Free radicals, reactive oxygen species, oxidative stress and its classification. Chem Biol Interact 224:164–175. https://doi.org/10.1016/J.CBI.2014.10.016
8. Rodriguez-Rocha H, Garcia-Garcia A, Pickett C et al (2013) Compartmentalized oxidative stress in dopaminergic cell death induced by pesticides and complex I inhibitors: distinct roles of superoxide anion and superoxide dismutases. Free Radic Biol Med 61:370–383. https://doi.org/10.1016/J.FREERADBIOMED.2013.04.021
9. Cenini G, Lloret A, Cascella R (2019) Oxidative stress in neurodegenerative diseases: from a mitochondrial point of view. Oxid Med Cell Longev 2019. https://doi.org/10.1155/2019/2105607
10. Wang Z, Ma C, Meng CJ et al (2012) Melatonin activates the Nrf2-ARE pathway when it protects against early brain injury in a subarachnoid hemorrhage model. J Pineal Res 53:129–137. https://doi.org/10.1111/J.1600-079X.2012.00978.X
11. Grewal AK, Singh TG, Sharma D et al (2021) Mechanistic insights and perspectives involved in neuroprotective action of quercetin. Biomed Pharmacother 140. https://doi.org/10.1016/J.BIOPHA.2021.111729
12. Sridevi P, Budde S, Raju BM, Adapa D (2018) Anti-oxidants and their role in disease management. Int J Med Res Heal Sci 7:175–190
13. Fadaka AO, Ajiboye BO, Adewale I et al (2019) Significance of antioxidants in the treatment and prevention of neurodegenerative diseases. J Phytopharm 8:75–83. https://doi.org/10.31254/phyto.2019.8210
14. Chang KH, Cheng ML, Chiang MC, Chen CM (2018) Lipophilic antioxidants in neurodegenerative diseases. Clin Chim Acta 485:79–87. https://doi.org/10.1016/J.CCA.2018.06.031
15. Eratne D, Loi SM, Farrand S et al (2018) Alzheimer's disease: clinical update on epidemiology, pathophysiology and diagnosis. Australas Psychiatr 26:347–357. https://doi.org/10.1177/1039856218762308
16. Deture MA (2019) Dickson DW (2019) The neuropathological diagnosis of Alzheimer's disease. Mol Neurodegener 141(14):1–18. https://doi.org/10.1186/S13024-019-0333-5
17. Cunha-Oliveira T, Montezinho L, Mendes C et al (2020) Oxidative stress in amyotrophic lateral sclerosis: pathophysiology and opportunities for pharmacological intervention. Oxid Med Cell Longev 2020. https://doi.org/10.1155/2020/5021694
18. Birben E, Sahiner UM, Sackesen C et al (2012) Oxidative stress and antioxidant defense. World Allergy Organ J 5:9–19. https://doi.org/10.1097/WOX.0B013E3182439613
19. Adav SS, Park JE, Sze SK (2019) Quantitative profiling brain proteomes revealed mitochondrial dysfunction in Alzheimer's disease. Mol Brain 12. https://doi.org/10.1186/S13041-019-0430-Y
20. Reed T, Perluigi M, Sultana R et al (2008) Redox proteomic identification of 4-hydroxy-2-nonenal-modified brain proteins in amnestic mild cognitive impairment: insight into the role of lipid peroxidation in the progression and pathogenesis of Alzheimer's disease. Neurobiol Dis 30:107–120. https://doi.org/10.1016/J.NBD.2007.12.007
21. Esposito L, Raber J, Kekonius L et al (2006) Reduction in mitochondrial superoxide dismutase modulates Alzheimer's disease-like pathology and accelerates the onset of behavioral changes in human amyloid precursor protein transgenic mice. J Neurosci 26:5167–5179. https://doi.org/10.1523/JNEUROSCI.0482-06.2006
22. Misrani A, Tabassum S, Yang L (2021) Mitochondrial dysfunction and oxidative stress in Alzheimer's disease. Front Aging Neurosci 13. https://doi.org/10.3389/FNAGI.2021.617588
23. Patten DA, Germain M, Kelly MA, Slack RS (2010) Reactive oxygen species: stuck in the middle of neurodegeneration. J Alzheimers Dis:20(Suppl 2). https://doi.org/10.3233/JAD-2010-100498
24. Chen L, Na R, Gu M et al (2008) Lipid peroxidation up-regulates BACE1 expression in vivo: a possible early event of amyloidogenesis in Alzheimer's disease. J Neurochem 107:197–207. https://doi.org/10.1111/J.1471-4159.2008.05603.X

25. Poewe W, Seppi K, Tanner CM et al (2017) Parkinson disease. Nat Rev Dis Prim 3:1–21. https://doi.org/10.1038/NRDP.2017.13
26. Simon DK, Tanner CM, Brundin P (2020) Parkinson disease epidemiology, pathology, genetics, and pathophysiology. Clin Geriatr Med 36:1–12. https://doi.org/10.1016/J.CGER.2019.08.002
27. Gómez-Benito M, Granado N, García-Sanz P et al (2020) Modeling Parkinson's disease with the alpha-synuclein protein. Front Pharmacol 11. https://doi.org/10.3389/FPHAR.2020.00356
28. Dias V, Junn E, Mouradian MM (2013) The role of oxidative stress in Parkinson's disease. J Parkinsons Dis 3:461–491. https://doi.org/10.3233/JPD-130230
29. Puspita L, Chung SY (2017) Shim JW (2017) Oxidative stress and cellular pathologies in Parkinson's disease. Mol Brain 101(10):1–12. https://doi.org/10.1186/S13041-017-0340-9
30. Trist BG, Hare DJ, Double KL (2019) Oxidative stress in the aging substantia nigra and the etiology of Parkinson's disease. Aging Cell 18:e13031. https://doi.org/10.1111/ACEL.13031
31. Tong H, Zhang X, Meng X et al (2018) Simvastatin inhibits activation of NADPH oxidase/p38 MAPK pathway and enhances expression of antioxidant protein in Parkinson disease models. Front Mol Neurosci 11. https://doi.org/10.3389/FNMOL.2018.00165
32. Park HR, Yang EJ (2021) Oxidative stress as a therapeutic target in amyotrophic lateral sclerosis: opportunities and limitations. Diagnostics 11. https://doi.org/10.3390/DIAGNOSTICS11091546
33. Petrov D, Daura X, Zagrovic B (2016) Effect of oxidative damage on the stability and dimerization of superoxide dismutase 1. Biophys J 110:1499–1509. https://doi.org/10.1016/J.BPJ.2016.02.037
34. Pansarasa O, Bordoni M, Diamanti L et al (2018) SOD1 in amyotrophic lateral sclerosis: "Ambivalent" behavior connected to the disease. Int J Mol Sci 19. https://doi.org/10.3390/IJMS19051345
35. Bellezza I, Grottelli S, Costanzi E et al (2018) Peroxynitrite activates the NLRP3 inflammasome cascade in SOD1(G93A) mouse model of amyotrophic lateral sclerosis. Mol Neurobiol 55:2350–2361. https://doi.org/10.1007/S12035-017-0502-X
36. Obrador E, Salvador R, Estrela JM et al (2020) Oxidative stress, neuroinflammation and mitochondria in the pathophysiology of amyotrophic lateral sclerosis. Antioxidants (Basel, Switzerland) 9:1–16. https://doi.org/10.3390/ANTIOX9090901
37. Wang Z, Bai Z, Qin X, Cheng Y (2019) Aberrations in oxidative stress markers in amyotrophic lateral sclerosis: a systematic review and meta-analysis. Oxid Med Cell Longev 2019. https://doi.org/10.1155/2019/1712323
38. Kazama M, Kato Y, Kakita A et al (2020) Astrocytes release glutamate via cystine/glutamate antiporter upregulated in response to increased oxidative stress related to sporadic amyotrophic lateral sclerosis. Neuropathology 40:587–598. https://doi.org/10.1111/NEUP.12716
39. Sun MS, Jin H, Sun X et al (2018) Free radical damage in ischemia-reperfusion injury: an obstacle in acute ischemic stroke after revascularization therapy. Oxid Med Cell Longev 2018. https://doi.org/10.1155/2018/3804979
40. Kalogeris T, Baines CP, Krenz M, Korthuis RJ (2012) Cell biology of ischemia/reperfusion injury. Int Rev Cell Mol Biol 298:229–317. https://doi.org/10.1016/B978-0-12-394309-5.00006-7
41. Wu L, Xiong X, Wu X et al (2020) Targeting oxidative stress and inflammation to prevent ischemia-reperfusion injury. Front Mol Neurosci 13. https://doi.org/10.3389/FNMOL.2020.00028
42. Nakano Y, Yamashita T, Li Q et al (2017) Time-dependent change of in vivo optical imaging of oxidative stress in a mouse stroke model. J Neurosci Res 95:2030–2039. https://doi.org/10.1002/JNR.24047
43. Lorenzano S, Rost NS, Khan M et al (2018) Oxidative stress biomarkers of brain damage: hyperacute plasma F2-isoprostane predicts infarct growth in stroke. Stroke 49:630–663. https://doi.org/10.1161/STROKEAHA.117.018440

44. Kishimoto M, Suenaga J, Takase H et al (2019) Oxidative stress-responsive apoptosis inducing protein (ORAIP) plays a critical role in cerebral ischemia/reperfusion injury. Sci Rep 9. https://doi.org/10.1038/S41598-019-50073-8
45. Lou Z, Wang AP, Duan XM et al (2018) Upregulation of NOX2 and NOX4 Mediated by TGF-β signaling pathway exacerbates cerebral ischemia/reperfusion oxidative stress injury. Cell Physiol Biochem 46:2103–2113. https://doi.org/10.1159/000489450
46. Xu N, Meng H, Liu T et al (2018) TRPC1 deficiency exacerbates cerebral ischemia/reperfusion-induced neurological injury by potentiating Nox4-derived reactive oxygen species generation. Cell Physiol Biochem 51:1723–1738. https://doi.org/10.1159/000495676
47. Rempe RG, Hartz AMS, Bauer B (2016) Matrix metalloproteinases in the brain and blood-brain barrier: Versatile breakers and makers. J Cereb Blood Flow Metab 36:1481–1507. https://doi.org/10.1177/0271678X16655551
48. Chen S, Shao L, Ma L (2021) Cerebral edema formation after stroke: emphasis on blood-brain barrier and the lymphatic drainage system of the brain. Front Cell Neurosci 15. https://doi.org/10.3389/FNCEL.2021.716825
49. Hodler J, Kubik-Huch RA, von Schulthess GK (2020) Diseases of the brain, head and neck, spine 2020–2023: Diagnostic Imaging [Internet]. https://doi.org/10.1007/978-3-030-38490-6
50. Eckert MJ, Martin MJ (2017) Trauma: Spinal Cord Injury. Surg Clin North Am 97:1031–1045. https://doi.org/10.1016/J.SUC.2017.06.008
51. Ahuja CS, Nori S, Tetreault L et al (2017) Traumatic spinal cord injury-repair and regeneration. Neurosurgery 80:S22–S90. https://doi.org/10.1093/NEUROS/NYW080
52. Lin J, Xiong Z, Gu J et al (2021) Sirtuins: potential therapeutic targets for defense against oxidative stress in spinal cord injury. Oxid Med Cell Longev 2021. https://doi.org/10.1155/2021/7207692
53. Bock FJ, Tait SWG (2020) Mitochondria as multifaceted regulators of cell death. Nat Rev Mol Cell Biol 21:85–100. https://doi.org/10.1038/S41580-019-0173-8
54. Rabchevsky AG, Michael FM, Patel SP (2020) Mitochondria focused neurotherapeutics for spinal cord injury. Exp Neurol 330. https://doi.org/10.1016/J.EXPNEUROL.2020.113332
55. Hussain G, Wang J, Rasul A et al (2020) Current status of therapeutic approaches against peripheral nerve injuries: a detailed story from injury to recovery. Int J Biol Sci 16:116. https://doi.org/10.7150/IJBS.35653
56. Agbaje JO, Van de Casteele E, Hiel M et al (2016) Neuropathy of trigeminal nerve branches after oral and maxillofacial treatment. J Maxillofac Oral Surg 15:321. https://doi.org/10.1007/S12663-015-0843-9
57. Khezri MK, Turkkan A, Koc C et al (2021) Anti-apoptotic and anti-oxidant effects of systemic uridine treatment in an experimental model of sciatic nerve injury. Turk Neurosurg 31:373–378. https://doi.org/10.5137/1019-5149.JTN.31127-20.3
58. Costa LS, Aidar FJ, de Matos DG et al (2020) Effects of resistance training and bowdichia virgilioides hydroethanolic extract on oxidative stress markers in rats submitted to peripheral nerve injury. Antioxidants (Basel, Switzerland) 9:1–17. https://doi.org/10.3390/ANTIOX9100941
59. Zafar S, Rasul A, Iqbal J et al (2021) Calotropis procera (leaves) supplementation exerts curative effects on promoting functional recovery in a mouse model of peripheral nerve injury. Food Sci Nutr 9:5016–5027. https://doi.org/10.1002/fsn3.2455
60. Tang C, Han R, Wu J, Fang F (2021) Effects of baicalin capsules combined with a-lipoic acid on nerve conduction velocity, oxidative stress and inflammatory injury in patients with diabetic peripheral neuropathy. Am J Transl Res 13:2774–2783
61. Pangrazzi L, Balasco L, Bozzi Y (2020) Oxidative stress and immune system dysfunction in autism spectrum disorders. Int J Mol Sci 21:3293. https://doi.org/10.3390/IJMS21093293
62. Geronzi U, Lotti F, Grosso S (2018) Oxidative stress in epilepsy. Expert Rev Neurother 18:427–434. https://doi.org/10.1080/14737175.2018.1465410
63. Bhatt S, Nagappa AN, Patil CR (2020) Role of oxidative stress in depression. Drug Discov Today 25:1270–1276. https://doi.org/10.1016/j.drudis.2020.05.001

64. Kurutas EB (2016) The importance of antioxidants which play the role in cellular response against oxidative/nitrosative stress: current state. Nutr J 15. https://doi.org/10.1186/S12937-016-0186-5
65. Harwell B (2007) Biochemistry of oxidative stress. Biochem Soc Trans 35:1147–1150. https://doi.org/10.1042/BST0351147
66. Elsayed Azab A, Adwas Almokhtar A, Ibrahim Elsayed AS et al (2019) Oxidative stress and antioxidant mechanisms in human body. J Appl Biotechnol Bioeng 6:43–47. https://doi.org/10.15406/JABB.2019.06.00173
67. Cheung CCC, Zheng GJ, Li AMY et al (2001) Relationships between tissue concentrations of polycyclic aromatic hydrocarbons and antioxidative responses of marine mussels, Perna viridis. Aquat Toxicol 52:189–203. https://doi.org/10.1016/S0166-445X(00)00145-4
68. McCord JM, Fridovich I (1969) Superoxide dismutase. An enzymic function for erythrocuprein (hemocuprein). J Biol Chem 244:6049–6055
69. Hayes PC, Bouchier IAD, Beckett GJ (1991) Glutathione S-transferase in humans in health and disease. Gut 32:813–818. https://doi.org/10.1136/GUT.32.7.813
70. Zelko IN, Mariani TJ, Folz RJ (2002) Superoxide dismutase multigene family: A comparison of the CuZn-SOD (SOD1), Mn-SOD (SOD2), and EC-SOD (SOD3) gene structures, evolution, and expression. Free Radic Biol Med 33:337–349. https://doi.org/10.1016/S0891-5849(02)00905-X
71. Ulusu NN, Tandoğan B (2007) Purification and kinetic properties of glutathione reductase from bovine liver. Mol Cell Biochem 303:45–51. https://doi.org/10.1007/S11010-007-9454-1
72. Linster CL, Van Schaftingen E (2007) Vitamin C. Biosynthesis, recycling and degradation in mammals. FEBS J 274:1–22. https://doi.org/10.1111/J.1742-4658.2006.05607.X
73. Sen CK, Khanna S, Roy S (2006) Tocotrienols: Vitamin E beyond tocopherols. Life Sci 78:2088–2098. https://doi.org/10.1016/J.LFS.2005.12.001
74. Li J, Wuliji O, Li W et al (2013) Oxidative stress and neurodegenerative disorders. Int J Mol Sci 14:24438–24475. https://doi.org/10.3390/IJMS141224438
75. Conway ME, Lee C (2015) The redox switch that regulates molecular chaperones. Biomol Concepts 6:269–284. https://doi.org/10.1515/BMC-2015-0015
76. Carballal S, Bartesaghi S, Radi R (2014) Kinetic and mechanistic considerations to assess the biological fate of peroxynitrite. Biochim Biophys Acta 1840:768–780. https://doi.org/10.1016/J.BBAGEN.2013.07.005
77. Brand MD, Goncalves RLS, Orr AL et al (2016) Suppressors of superoxide-H_2O_2 production at site IQ of mitochondrial complex I protect against stem cell hyperplasia and ischemia-reperfusion injury. Cell Metab 24:582–592. https://doi.org/10.1016/J.CMET.2016.08.012
78. Pun PBL, Logan A, Darley-Usmar V et al (2014) A mitochondria-targeted mass spectrometry probe to detect glyoxals: implications for diabetes. Free Radic Biol Med 67:437–450. https://doi.org/10.1016/J.FREERADBIOMED.2013.11.025
79. Przedborski S (2017) The two-century journey of Parkinson disease research. Nat Rev Neurosci 18:251–259. https://doi.org/10.1038/NRN.2017.25
80. Brigelius-Flohé R, Maiorino M (2013) Glutathione peroxidases. Biochim Biophys Acta 1830:3289–3303. https://doi.org/10.1016/J.BBAGEN.2012.11.020
81. Hong S, Dissing-Olesen L, Stevens B (2016) New insights on the role of microglia in synaptic pruning in health and disease. Curr Opin Neurobiol 36:128–134. https://doi.org/10.1016/J.CONB.2015.12.004
82. Thomas DD (2015) Breathing new life into nitric oxide signaling: A brief overview of the interplay between oxygen and nitric oxide. Redox Biol 5:225–233. https://doi.org/10.1016/J.REDOX.2015.05.002
83. Lee KH, Cha M, Lee BH (2020) Neuroprotective effect of antioxidants in the brain. Int J Mol Sci 21:1–29. https://doi.org/10.3390/IJMS21197152
84. Pisoschi AM, Pop A (2015) The role of antioxidants in the chemistry of oxidative stress: A review. Eur J Med Chem 97:55–74. https://doi.org/10.1016/J.EJMECH.2015.04.040

85. Cadet J, Davies KJA (2017) Oxidative DNA damage & repair: an introduction. Free Radic Biol Med 107:2–12. https://doi.org/10.1016/J.FREERADBIOMED.2017.03.030
86. Guan Y, Hickey MJ, Borgstahl GEO et al (1998) Crystal structure of Y34F mutant human mitochondrial manganese superoxide dismutase and the functional role of tyrosine 34. Biochemistry 37:4722–4730. https://doi.org/10.1021/BI972394L
87. Hunt CR, Sim JE, Sullivan SJ et al (1998) Genomic instability and catalase gene amplification induced by chronic exposure to oxidative stress. Cancer Res 58:3986–3992
88. Usui S, Komeima K, Lee SY et al (2009) Increased expression of catalase and superoxide dismutase 2 reduces cone cell death in retinitis pigmentosa. Mol Ther 17:778–786. https://doi.org/10.1038/MT.2009.47
89. Baxter PS, Bell KFS, Hasel P et al (2015) Synaptic NMDA receptor activity is coupled to the transcriptional control of the glutathione system. Nat Commun 61(6):1–13. https://doi.org/10.1038/ncomms7761
90. Nell HJ, Au JL, Giordano CR et al (2017) Targeted antioxidant, catalase-SKL, reduces beta-amyloid toxicity in the rat brain. Brain Pathol 27:86–94. https://doi.org/10.1111/BPA.12368
91. Yakunin E, Kisos H, Kulik W et al (2014) The regulation of catalase activity by PPAR γ is affected by α-synuclein. Ann Clin Transl Neurol 1:145–159. https://doi.org/10.1002/ACN3.38
92. Kemp M, Go YM, Jones DP (2008) Nonequilibrium thermodynamics of thiol/disulfide redox systems: a perspective on redox systems biology. Free Radic Biol Med 44:921–937. https://doi.org/10.1016/J.FREERADBIOMED.2007.11.008
93. Cardoso BR, Hare DJ, Bush AI, Roberts BR (2017) Glutathione peroxidase 4: a new player in neurodegeneration? Mol Psychiatry 22:328–335. https://doi.org/10.1038/MP.2016.196
94. Dixon SJ, Lemberg KM, Lamprecht MR et al (2012) Ferroptosis: an iron-dependent form of nonapoptotic cell death. Cell 149:1060–1072. https://doi.org/10.1016/J.CELL.2012.03.042
95. Sanoobar M, Eghtesadi S, Azimi A et al (2013) Coenzyme Q10 supplementation reduces oxidative stress and increases antioxidant enzyme activity in patients with relapsing-remitting multiple sclerosis. Int J Neurosci 123:776–782. https://doi.org/10.3109/00207454.2013.801844
96. Di Pierro D, Ciaccio C, Sbardella D et al (2020) Effects of oral administration of common antioxidant supplements on the energy metabolism of red blood cells. Attenuation of oxidative stress-induced changes in Rett syndrome erythrocytes by CoQ10. Mol Cell Biochem 463:101–113. https://doi.org/10.1007/S11010-019-03633-5
97. Biasini B, Marchi L, Angelino D et al (2018) Claimed effects, outcome variables and methods of measurement for health claims on foods related to the gastrointestinal tract proposed under regulation (EC) 1924/2006. Int J Food Sci Nutr 69:771–804. https://doi.org/10.1080/09637486.2018.1427220
98. Steinbrenner H, Sies H (2013) Selenium homeostasis and antioxidant selenoproteins in brain: implications for disorders in the central nervous system. Arch Biochem Biophys 536:152–157. https://doi.org/10.1016/J.ABB.2013.02.021
99. Cornett CR, Markesbery WR, Ehmann WD (1998) Imbalances of trace elements related to oxidative damage in Alzheimer's disease brain. Neurotoxicology 19:339–346
100. Letsiou S, Nomikos T, Panagiotakos D et al (2009) Serum total selenium status in Greek adults and its relation to age. The ATTICA study cohort. Biol Trace Elem Res 128:8–17. https://doi.org/10.1007/S12011-008-8252-2
101. Kloubert V, Rink L (2015) Zinc as a micronutrient and its preventive role of oxidative damage in cells. Food Funct 6:3195–3204. https://doi.org/10.1039/C5FO00630A
102. Oteiza PI (2012) Zinc and the modulation of redox homeostasis. Free Radic Biol Med 53:1748–1759. https://doi.org/10.1016/J.FREERADBIOMED.2012.08.568
103. Kawahara M, Tanaka KI, Kato-Negishi M (2018) Zinc, carnosine, and neurodegenerative diseases. Nutrients 10. https://doi.org/10.3390/NU10020147
104. Carocci A, Catalano A, Sinicropi MS, Genchi G (2018) Oxidative stress and neurodegeneration: the involvement of iron. Biometals 31:715–735. https://doi.org/10.1007/S10534-018-0126-2

105. Farina M, Avila DS, Da Rocha JBT, Aschner M (2013) Metals, oxidative stress and neurodegeneration: a focus on iron, manganese and mercury. Neurochem Int 62:575–594. https://doi.org/10.1016/J.NEUINT.2012.12.006
106. Ballaz SJ, Rebec GV (2019) Neurobiology of vitamin C: expanding the focus from antioxidant to endogenous neuromodulator. Pharmacol Res 146:104321. https://doi.org/10.1016/J.PHRS.2019.104321
107. Harrison FE, Meredith ME, Dawes SM et al (2010) Low ascorbic acid and increased oxidative stress in gulo(-/-) mice during development. Brain Res 1349:143–152. https://doi.org/10.1016/J.BRAINRES.2010.06.037
108. Stansley BJ, Yamamoto BK (2014) Chronic L-dopa decreases serotonin neurons in a subregion of the dorsal raphe nucleus. J Pharmacol Exp Ther 351:440–447. https://doi.org/10.1124/JPET.114.218966
109. Getoff N (2013) Vitamin c: Electron emission, free radicals and biological versatility. In Vivo (Brooklyn). 27:565–570
110. Shen L, Ji HF (2012) Vitamin E: supplement versus diet in neurodegenerative diseases. Trends Mol Med 18:443–445. https://doi.org/10.1016/J.MOLMED.2012.04.010
111. Kontush A, Schekatolina S (2004) Vitamin E in neurodegenerative disorders: Alzheimer's disease. Ann N Y Acad Sci 1031:249–262. https://doi.org/10.1196/ANNALS.1331.025
112. Schirinzi T, Martella G, Imbriani P et al (2019) Dietary Vitamin E as a protective factor for Parkinson's disease: clinical and experimental evidence. Front Neurol 10. https://doi.org/10.3389/FNEUR.2019.00148
113. Ulatowski LM, Manor D (2015) Vitamin E and neurodegeneration. Neurobiol Dis 84:78–83. https://doi.org/10.1016/J.NBD.2015.04.002
114. Jiang Q (2014) Natural forms of vitamin E: metabolism, antioxidant, and anti-inflammatory activities and their role in disease prevention and therapy. Free Radic Biol Med 72:76–90. https://doi.org/10.1016/J.FREERADBIOMED.2014.03.035
115. Simonetto M, Infante M, Sacco RL et al (2019) A novel anti-inflammatory role of omega-3 PUFAs in prevention and treatment of atherosclerosis and vascular cognitive impairment and dementia. Nutrients 11. https://doi.org/10.3390/NU11102279
116. Mehegan L, Rainville C (2018) 2018 AARP brain health and mental well-being survey. https://doi.org/10.26419/RES.00247.001
117. Galderisi S, Heinz A, Kastrup M et al (2015) Toward a new definition of mental health. World Psychiatr 14:231–233. https://doi.org/10.1002/WPS.20231
118. Brunson KL, Kramár E, Lin B et al (2005) Mechanisms of late-onset cognitive decline after early-life stress. J Neurosci 25:9328–9338. https://doi.org/10.1523/JNEUROSCI.2281-05.2005
119. Gulpers BJA, Oude Voshaar RC, van Boxtel MPJ et al (2019) Anxiety as a risk factor for cognitive decline: a 12-year follow-up cohort study. Am J Geriatr Psychiatr 27:42–52. https://doi.org/10.1016/J.JAGP.2018.09.006
120. Noble KG, Houston SM, Brito NH et al (2015) Family income, parental education and brain structure in children and adolescents. Nat Neurosci 18:773–778. https://doi.org/10.1038/NN.3983
121. Hardeman RR, Perry SP, Phelan SM et al (2016) Racial identity and mental well-being: the experience of African American medical students, a report from the medical student CHANGE study. J racial Ethn Heal disparities 3:250–258. https://doi.org/10.1007/S40615-015-0136-5
122. Mikolas P, Tozzi L, Doolin K et al (2019) Effects of early life adversity and FKBP5 genotype on hippocampal subfields volume in major depression. J Affect Disord 252:152–159. https://doi.org/10.1016/J.JAD.2019.04.054
123. Nyberg L, Lövdén M, Riklund K et al (2012) Memory aging and brain maintenance. Trends Cogn Sci 16:292–305. https://doi.org/10.1016/J.TICS.2012.04.005
124. Stern Y, Blumen HM, Rich LW et al (2011) Space Fortress game training and executive control in older adults: a pilot intervention. Neuropsychol Dev Cogn B Aging Neuropsychol Cogn 18:653–677. https://doi.org/10.1080/13825585.2011.613450

125. Belleville S, Clément F, Mellah S et al (2011) Training-related brain plasticity in subjects at risk of developing Alzheimer's disease. Brain 134:1623–1634. https://doi.org/10.1093/BRAIN/AWR037
126. Carlson MC, Kuo JH, Chuang YF et al (2015) Impact of the baltimore experience corps trial on cortical and hippocampal volumes. Alzheimers Dement 11:1340–1348. https://doi.org/10.1016/J.JALZ.2014.12.005
127. Cheng ST (2016) Cognitive reserve and the prevention of dementia: the role of physical and cognitive activities. Curr Psychiatry Rep 18. https://doi.org/10.1007/S11920-016-0721-2
128. Litwin H, Schwartz E, Damri N (2017) Cognitively stimulating leisure activity and subsequent cognitive function: a SHARE-based analysis. Gerontologist 57:940–948. https://doi.org/10.1093/GERONT/GNW084
129. Moraes W, Piovezan R, Poyares D et al (2014) Effects of aging on sleep structure throughout adulthood: a population-based study. Sleep Med 15:401–409. https://doi.org/10.1016/J.SLEEP.2013.11.791
130. Mintzer J, Donovan KA, Kindy AZ et al (2019) Lifestyle Choices and Brain Health. Front Med 6:204. https://doi.org/10.3389/FMED.2019.00204/BIBTEX
131. Watson NF, Badr MS, Belenky G et al (2015) Recommended amount of sleep for a healthy adult: a joint consensus statement of the American Academy of sleep medicine and sleep research society. Sleep 38:843–844. https://doi.org/10.5665/SLEEP.4716
132. Czeisler CA (2013) Perspective: casting light on sleep deficiency. Nature 497:S13. https://doi.org/10.1038/497S13A
133. Baron KG, Reid KJ, Zee PC (2013) Exercise to improve sleep in insomnia: exploration of the bidirectional effects. J Clin Sleep Med 9:819–824. https://doi.org/10.5664/JCSM.2930
134. Duffy JF, Czeisler CA (2009) Effect of light on human circadian physiology. Sleep Med Clin 4:165–177. https://doi.org/10.1016/J.JSMC.2009.01.004
135. Mead MN (2008) Benefits of sunlight: a bright spot for human health. Environ Health Perspect 116. https://doi.org/10.1289/EHP.116-A160
136. Gray SL, Anderson ML, Dublin S et al (2015) Cumulative use of strong anticholinergics and incident dementia: a prospective cohort study. JAMA Intern Med 175:401–407. https://doi.org/10.1001/JAMAINTERNMED.2014.7663
137. Milan R, Vasiliadis HM (2020) The association between side effects and adherence to antidepressants among primary care community-dwelling older adults. Aging Ment Health 24:1229–1236. https://doi.org/10.1080/13607863.2019.1594165
138. Jaussent I, Bouyer J, Ancelin ML et al (2011) Insomnia and daytime sleepiness are risk factors for depressive symptoms in the elderly. Sleep 34:1103–1110. https://doi.org/10.5665/SLEEP.1170
139. Lim ASP, Kowgier M, Yu L et al (2013) Sleep fragmentation and the risk of incident Alzheimer's disease and cognitive decline in older persons. Sleep 36:1027–1032. https://doi.org/10.5665/SLEEP.2802
140. Ju YES, Lucey BP, Holtzman DM (2014) Sleep and Alzheimer disease pathology – a bidirectional relationship. Nat Rev Neurol 10:115–119. https://doi.org/10.1038/NRNEUROL.2013.269
141. Pan W, Kastin AJ (2014) Can sleep apnea cause Alzheimer's disease? Neurosci Biobehav Rev 47:656–669. https://doi.org/10.1016/J.NEUBIOREV.2014.10.019
142. Baril AA, Carrier J, Lafrenière A et al (2018) Biomarkers of dementia in obstructive sleep apnea. Sleep Med Rev 42:139–148. https://doi.org/10.1016/J.SMRV.2018.08.001
143. Bliwise DL (2015) Never too old: beneficial neurobehavioural effects of continuous positive airway pressure in the elderly. Eur Respir J 46:13–15. https://doi.org/10.1183/09031936.00039915
144. Zhong BL, Chen SL, Conwell Y (2016) Effects of transient versus chronic loneliness on cognitive function in older adults: findings from the chinese longitudinal healthy longevity survey. Am J Geriatr Psychiatr 24:389–398. https://doi.org/10.1016/J.JAGP.2015.12.009

145. Donovan NJ, Wu Q, Rentz DM et al (2017) Loneliness, depression and cognitive function in older U.S. adults. Int J Geriatr Psychiatr 32:564–573. https://doi.org/10.1002/GPS.4495
146. Vaportzis E, Martin M, Gow AJ (2017) A tablet for healthy ageing: the effect of a tablet computer training intervention on cognitive abilities in older adults. Am J Geriatr Psychiatr 25:841–851. https://doi.org/10.1016/J.JAGP.2016.11.015
147. Lifshitz R, Nimrod G, Bachner YG (2018) Internet use and well-being in later life: a functional approach. Aging Ment Health 22:85–91. https://doi.org/10.1080/13607863.2016.1232370
148. DeLamater J, Koepsel E (2015) Relationships and sexual expression in later life: a biopsychosocial perspective. Sex Relatsh Ther 30:37–59. https://doi.org/10.1080/14681994.2014.939506
149. Forbes MK, Eaton NR, Krueger RF (2017) Sexual quality of life and aging: a prospective study of a nationally representative sample. J Sex Res 54:137–148. https://doi.org/10.1080/00224499.2016.1233315
150. Fogelman N, Canli T (2015) "Purpose in Life" as a psychosocial resource in healthy aging: an examination of cortisol baseline levels and response to the Trier Social Stress Test. NPJ aging Mech Dis 1. https://doi.org/10.1038/NPJAMD.2015.6
151. Kawachi I, Subramanian SV (2007) Neighbourhood influences on health. J Epidemiol Community Health 61:3–4. https://doi.org/10.1136/JECH.2005.045203
152. Shlisky J, Bloom DE, Beaudreault AR et al (2017) Nutritional considerations for healthy aging and reduction in age-related chronic disease. Adv Nutr 8:17–26. https://doi.org/10.3945/AN.116.013474
153. Solfrizzi V, Custodero C, Lozupone M et al (2017) Relationships of dietary patterns, foods, and micro- and macronutrients with Alzheimer's disease and late-life cognitive disorders: a systematic review. J Alzheimers Dis 59:815–849. https://doi.org/10.3233/JAD-170248
154. Veronese N, Stubbs B, Noale M et al (2016) Adherence to the Mediterranean diet is associated with better quality of life: data from the Osteoarthritis initiative. Am J Clin Nutr 104:1403–1409. https://doi.org/10.3945/AJCN.116.136390
155. Lauretti E, Iuliano L, Praticò D (2017) Extra-virgin olive oil ameliorates cognition and neuropathology of the 3xTg mice: role of autophagy. Ann Clin Transl Neurol 4:564–574. https://doi.org/10.1002/ACN3.431
156. McEvoy CT, Guyer H, Langa KM, Yaffe K (2017) Neuroprotective diets are associated with better cognitive function: the health and retirement study. J Am Geriatr Soc 65:1857–1862. https://doi.org/10.1111/JGS.14922
157. Scarmeas N, Stern Y, Tang MX et al (2006) Mediterranean diet and risk for Alzheimer's disease. Ann Neurol 59:912–921. https://doi.org/10.1002/ANA.20854
158. Morris MC (2016) Nutrition and risk of dementia: overview and methodological issues. Ann N Y Acad Sci 1367:31–37. https://doi.org/10.1111/NYAS.13047
159. Männikkö R, Komulainen P, Schwab U, et al (2015) The Nordic diet and cognition--The DR's EXTRA Study. Br J Nutr 114:231–239. https://doi.org/10.1017/S0007114515001890
160. Appel LJ, Moore TJ, Obarzanek E et al (1997) A clinical trial of the effects of dietary patterns on blood pressure. DASH Collaborative Research Group. N Engl J Med 336:1117–1124. https://doi.org/10.1056/NEJM199704173361601
161. Appel LJ, Espeland MA, Easter L et al (2001) Effects of reduced sodium intake on hypertension control in older individuals: results from the Trial of Nonpharmacologic Interventions in the Elderly (TONE). Arch Intern Med 161:685–693. https://doi.org/10.1001/ARCHINTE.161.5.685
162. Miyagi S, Iwama N, Kawabata T, Hasegawa K (2003) Longevity and diet in Okinawa, Japan: the past, present and future. Asia-Pacific J Public Heal 15(Suppl). https://doi.org/10.1177/101053950301500S03
163. Marcason W (2015) What are the components to the MIND diet? J Acad Nutr Diet 115:1744. https://doi.org/10.1016/J.JAND.2015.08.002
164. Morris MC, Tangney CC, Wang Y et al (2015) MIND diet slows cognitive decline with aging. Alzheimers Dement 11:1015–1022. https://doi.org/10.1016/J.JALZ.2015.04.011

165. Moubarac J-C, Parra DC, Cannon G, Monteiro CA (2014) Food classification systems based on food processing: significance and implications for policies and actions: a systematic literature review and assessment. Curr Obes Rep 3:256–272. https://doi.org/10.1007/S13679-014-0092-0
166. Davidson MH, Kling D, Maki KC (2011) Novel developments in omega-3 fatty acid-based strategies. Curr Opin Lipidol 22:437–444. https://doi.org/10.1097/MOL.0B013E32834BD642
167. Dyall SC (2015) Long-chain omega-3 fatty acids and the brain: a review of the independent and shared effects of EPA, DPA and DHA. Front Aging Neurosci 7. https://doi.org/10.3389/FNAGI.2015.00052
168. Panza F, Solfrizzi V, Barulli MR et al (2015) Coffee, tea, and caffeine consumption and prevention of late-life cognitive decline and dementia: a systematic review. J Nutr Health Aging 19:313–328. https://doi.org/10.1007/S12603-014-0563-8
169. Brickman AM, Khan UA, Provenzano FA et al (2014) Enhancing dentate gyrus function with dietary flavanols improves cognition in older adults. Nat Neurosci 17:1798–1803. https://doi.org/10.1038/NN.3850
170. Mastroiacovo D, Kwik-Uribe C, Grassi D, et al (2015) Cocoa flavanol consumption improves cognitive function, blood pressure control, and metabolic profile in elderly subjects: the Cocoa, Cognition, and Aging (CoCoA) Study--a randomized controlled trial. Am J Clin Nutr 101:538–548. https://doi.org/10.3945/AJCN.114.092189
171. Barnard ND, Bush AI, Ceccarelli A, et al (2014) Dietary and lifestyle guidelines for the prevention of Alzheimer's disease. Neurobiol Aging:35(Suppl 2). https://doi.org/10.1016/J.NEUROBIOLAGING.2014.03.033
172. Lehtisalo J, Lindström J, Ngandu T et al (2016) Association of long-term dietary fat intake, exercise, and weight with later cognitive function in the Finnish diabetes prevention study. J Nutr Health Aging 20:146–154. https://doi.org/10.1007/S12603-015-0565-1
173. Erickson KI, Gildengers AG, Butters MA (2013) Physical activity and brain plasticity in late adulthood. Dialogues Clin Neurosci 15:99–108. https://doi.org/10.31887/DCNS.2013.15.1/KERICKSON
174. de Melo Coelho FG, Gobbi S, CAA A et al (2013) Physical exercise modulates peripheral levels of brain-derived neurotrophic factor (BDNF): a systematic review of experimental studies in the elderly. Arch Gerontol Geriatr 56:10–15. https://doi.org/10.1016/J.ARCHGER.2012.06.003
175. Baker LD, Frank LL, Foster-Schubert K et al (2010) Effects of aerobic exercise on mild cognitive impairment: a controlled trial. Arch Neurol 67:71–79. https://doi.org/10.1001/ARCHNEUROL.2009.307
176. Colcombe SJ, Kramer AF, Erickson KI et al (2004) Cardiovascular fitness, cortical plasticity, and aging. Proc Natl Acad Sci U S A 101:3316–3321. https://doi.org/10.1073/PNAS.0400266101
177. Barnes DE, Santos-Modesitt W, Poelke G et al (2013) The Mental Activity and eXercise (MAX) trial: a randomized controlled trial to enhance cognitive function in older adults. JAMA Intern Med 173:797–804. https://doi.org/10.1001/JAMAINTERNMED.2013.189
178. Groot C, Hooghiemstra AM, Raijmakers PGHM et al (2016) The effect of physical activity on cognitive function in patients with dementia: a meta-analysis of randomized control trials. Ageing Res Rev 25:13–23. https://doi.org/10.1016/J.ARR.2015.11.005
179. Prodoehl J, Rafferty MR, David FJ et al (2015) Two-year exercise program improves physical function in Parkinson's disease: the PRET-PD randomized clinical trial. Neurorehabil Neural Repair 29:112–122. https://doi.org/10.1177/1545968314539732
180. Da Silva FC, Da Rosa IR, Dos Santos PD et al (2016) Effects of physical-exercise-based rehabilitation programs on the quality of life of patients with Parkinson's disease: a systematic review of randomized controlled trials. J Aging Phys Act 24:484–496. https://doi.org/10.1123/JAPA.2015-0162

181. Mak MK, Wong-Yu IS, Shen X, Chung CL (2017) Long-term effects of exercise and physical therapy in people with Parkinson disease. Nat Rev Neurol 13:689–703. https://doi.org/10.1038/NRNEUROL.2017.128
182. Nadeau A, Pourcher E, Corbeil P (2014) Effects of 24 wk of treadmill training on gait performance in Parkinson's disease. Med Sci Sports Exerc 46:645–655. https://doi.org/10.1249/MSS.0000000000000144
183. Bombieri F, Schena F, Pellegrini B et al (2017) Walking on four limbs: a systematic review of Nordic walking in Parkinson disease. Parkinsonism Relat Disord 38:8–12. https://doi.org/10.1016/J.PARKRELDIS.2017.02.004
184. Ramazzina I, Bernazzoli B, Costantino C (2017) Systematic review on strength training in Parkinson's disease: an unsolved question. Clin Interv Aging 12:619–628. https://doi.org/10.2147/CIA.S131903
185. Saltychev M, Bärlund E, Paltamaa J et al (2016) Progressive resistance training in Parkinson's disease: a systematic review and meta-analysis. BMJ Open 6. https://doi.org/10.1136/BMJOPEN-2015-008756
186. Ribas CG, Alves da Silva L, Corrêa MR et al (2017) Effectiveness of exergaming in improving functional balance, fatigue and quality of life in Parkinson's disease: a pilot randomized controlled trial. Parkinsonism Relat Disord 38:13–18. https://doi.org/10.1016/J.PARKRELDIS.2017.02.006
187. Yang WC, Wang HK, Wu RM et al (2016) Home-based virtual reality balance training and conventional balance training in Parkinson's disease: a randomized controlled trial. J Formos Med Assoc 115:734–743. https://doi.org/10.1016/J.JFMA.2015.07.012
188. Grodstein F, Chen J, Willett WC (2003) High-dose antioxidant supplements and cognitive function in community-dwelling elderly women. Am J Clin Nutr 77:975–984. https://doi.org/10.1093/AJCN/77.4.975
189. Sato K, Saito H, Katsuki H (1993) Synergism of tocopherol and ascorbate on the survival of cultured brain neurones. Neuroreport 4:1179–1182
190. Philip P, Sagaspe P, Taillard J et al (2019) Acute intake of a grape and blueberry polyphenol-rich extract ameliorates cognitive performance in healthy young adults during a sustained cognitive effort. Antioxidants (Basel, Switzerland) 8. https://doi.org/10.3390/ANTIOX8120650
191. Godos J, Caraci F, Micek A et al (2021) Dietary phenolic acids and their major food sources are associated with cognitive status in older italian adults. Antioxidants (Basel, Switzerland) 10. https://doi.org/10.3390/ANTIOX10050700
192. Carrillo JÁ, Arcusa R, Zafrilla MP, Marhuenda J (2021) Effects of fruit and vegetable-based nutraceutical on cognitive function in a healthy population: Placebo-controlled, double-blind, and randomized clinical trial. Antioxidants (Basel, Switzerland) 10:1–12. https://doi.org/10.3390/ANTIOX10010116
193. Godos J, Currenti W, Angelino D et al (2020) Diet and mental health: review of the recent updates on molecular mechanisms. Antioxidants (Basel, Switzerland) 9. https://doi.org/10.3390/ANTIOX9040346
194. Varghese N, Werner S, Grimm A, Eckert A (2020) Dietary mitophagy enhancer: a strategy for healthy brain aging? Antioxidants (Basel, Switzerland) 9:1–28. https://doi.org/10.3390/ANTIOX9100932
195. Limanaqi F, Biagioni F, Busceti CL et al (2020) Potential antidepressant effects of Scutellaria baicalensis, Hericium erinaceus and Rhodiola rosea. Antioxidants (Basel, Switzerland) 9. https://doi.org/10.3390/ANTIOX9030234

Chapter 10
Role of Antioxidants, and Lifestyle in Managing Brain Disorders Oxidative Stress Biomarkers and Antioxidant Treatments in Brain Diseases

Öznur Özge Özcan, Mesut Karahan, and Muhsin Konuk

10.1 Introduction

Alzheimer's Disease (AD), schizophrenia, and Autism Spectrum Disorder (ASD) are brain diseases that are the most current and popular research topics today. These three diseases occur in different periods of the life cycle. Investigation of the pathophysiological conditions of the brain continues with the emergence of these diseases especially in infancy, schizophrenia in youth, and AD in later ages. In addition, these three diseases fall into three different definitions in the field of brain diseases: neurodevelopmental disorders, neuropsychiatric diseases, and neurodegenerative disorders.

Oxidative stress can cause neurodegeneration by creating neurotrauma, especially when it occurs in neuron cells. For example, excessive reactive oxygen (ROS) and nitrogen (RNS) species, activity may impair neuronal signal transmission by myelin sheath damage by causing an imbalance of Ca^{2+} in the neuron cytoplasm [1]. Antioxidants are an innate class that regulates free or ROS depending on the damage caused by oxidative stress. While ROS belonging to our cells are produced by ETS in mitochondria, many enzymes, including superoxide dismutase (SOD) and

Ö. Ö. Özcan
Department of Molecular Neuroscience, Health Sciences Institute, Üsküdar University, Istanbul, Turkey

M. Karahan
Department of Biomedical Device Technology, Vocational School of Health Sciences, Üsküdar University, Istanbul, Turkey

M. Konuk (✉)
Department of Molecular Biology and Genetics, Faculty of Engineering and Natural Sciences, Üsküdar University, Istanbul, Turkey
e-mail: muhsin.konuk@uskudar.edu.tr

A. Imran, G. Hussain (eds.), *The Role of Natural Antioxidants in Brain Disorders*, Food Bioactive Ingredients, https://doi.org/10.1007/978-3-031-41188-5_10

glutathione peroxidase (GPx), which provide defense in times of stress and damage, act as ROS cleaning agents [2]. The brain is under more effective hyperbaric pressure than any other tissue in our body. This suggests the need for oxygen to produce adenosine triphosphate (ATP), which is required to continue oxidative phosphorylation in neural mitochondria and control vital mechanisms. The most of oxygen amount (almost %90) is utilized in oxidative phosphorylation in mitochondria and observed in the form of electron reactions in the electron transport chain (ETC). In some cases, however, oxiradicals from oxidation of molecules can leak into the cytoplasm, with electrons avoiding ETC, and may cause many types of peroxidative reactions that disrupt mitochondrial matrices and lipid membranes [3]. So, oxidative stress may play a crucial role in the pathophysiology of certain neuropsychiatric diseases, including AD, schizophrenia, and autism. Identification of readily accessible biomarkers is obligatory to let better diagnosis and control of brain disorders because much more effective therapies can be developed by looking at the effects of both drugs and many other alternative treatments with both cognitive and biomarkers levels. In this book chapter, we provided a deep insight into the main players in the neural mechanisms of antioxidants and particularly touched upon their effects on AD, schizophrenia, and autism, which cannot remove from the world of secrets. We tried to discuss updated and limiting aspects of oxidative stress biomarkers and the therapeutic effects of antioxidants on their potential in neuronal activities.

10.2 Schizophrenia

10.2.1 Oxidative Stress Biomarkers

The neuropathophysiology of schizophrenia was tried to be explained depending on the activity of dopamine pathways in the brain; In particular, the high activity of positive symptoms of the disease due to dopamine in the mesolimbic pathway and low activity in the frontotemporal region cause negative symptoms [4]. The diagnosis of schizophrenia is based on the clinician's subjective interpretations of the negative and positive symptoms of the disease. Investigation of molecular biomarkers of this condition has a very important place in the limited evaluation of clinical symptoms in many psychotic diseases, including schizophrenia [5]. The effects of neuronal oxidative stress in the pathophysiological studies of schizophrenia have begun to be clarified. In a systematic review of metabolite biomarkers of schizophrenia patients, elevated lipid peroxidation metabolites, glutamate, and decreased essential polyunsaturated fatty acids (EPUFAs), vitamin E, and creatinine levels were detected in blood serum samples [6]. In another systematic review and meta-analysis, TAS and docosahexaenoic acid levels were significantly lower, and in contrast, homocysteine, interleukin-6, and tumor necrosis factor-alpha levels were significantly higher in 3002 first-episode or early schizophrenia patients compared to 2806 control group [7].

Antioxidant enzymes have many complex and unclear functions in human nature, including SOD, GpX, and catalase (CAT). These enzymes suppress the initiation of many damaging chain reactions of reactive species [8]. Interestingly, Buosi et al. (2021) found lower SOD levels, higher malondialdehyde (MDA) and CAT levels, and the same GPx, total glutathione (GSH-t), and Trolox-equivalent antioxidant capacity (TEAC) levels in treatment-responsive and treatment-resistant schizophrenia patients compared to healthy group [9]. Cruz et al. (2021) found poor performance in working memory tests associated with higher serum levels of thiobarbituric acid reactive substances (TBARS) in 85 stable schizophrenia patients compared to a control group (n=75) [10].

Kim et al. (2019) measured the strong oxidant iridium (Ir) to examine the oxidative damage serum samples of 73 schizophrenia patients and 45 healthy people and found no significant differences. Likewise, the schizophrenia group treated with clozapine showed higher levels of oxidative stress and it was concluded that clozapine may cause this damage [11]. Another interesting finding was that oxidative stress levels in the serum could be related to oxidative stress in the cerebrospinal fluid (CSF), which may facilitate the diagnosis of schizophrenia in the future because neuroinflammatory and oxidative stress markers can cross the blood-brain barrier (BBB) [11, 12] GPx levels were low in the drug-free group, but total antioxidant capacity (T-AOC) and SOD levels were not significant in both groups in a total of 80 patients with schizophrenia (40 non-drug, and 40 medication use patients). In addition, GSH-Px and MDA were upregulated in the medication group, but SOD levels were reduced in patients Bai et al. (2018) and Juchnowicz et al. (2021) reported that GSH, total oxidant status (TOS) and GPx values are the most promising in the differential diagnosis of schizophrenia [13, 14]. On the other hand, the differential diagnosis of serum kynurenine (KYN) (lower in serum), enhanced oxidation protein products (AOPP), TAC, and nitric oxide (NO) levels among patients with chronic schizophrenia with the early diagnosis was ultimately promising. Importantly, KYN and TAC could be a biomarker for the treatment response. Also, other significant differences were found in terms of T-AOC, glutathione reductase (GR), total protein, enhanced glycation end products, the ferric reducing ability of plasma (FRAP), SOD, TOS, dityrosine, N-formylkynin urea levels [14]. In addition, the investigation of oxidative biomarkers in healthy and schizophrenia groups may provide an additional evaluation criterion for schizophrenia patients to diagnose and respond to treatment in the future. In a clinical study, the concentrations of monocyte chemotactic protein-1 and interleukin-8 were significantly higher and there was no significant difference in the serum concentrations of heme oxygenase-1 and 8-Hydroxydeoxyguanine in patients with schizophrenia compared to the control group [15]. For another oxidative stress marker study, oxidative stress index (OSI), TOS, myeloperoxidase (MPO) and Disulphide (DS) parameters were significantly higher and total antioxidant status (TAS), total thiol (TT), native thiol (NT) levels were significantly lower in Schizophrenia groups [16]. Grignon and Chianetta (2007) determined oxidative stress among patients with schizophrenia and reported a very high increase in MDA levels [17]. Zhang et al. (2010) reported upregulated NO and TBARS levels in schizophrenia patients' serum [18].

Oxidative stress also has many effects on DNA, because DNA has many negatively charged phosphate groups, it can bind many different cations such as $Fe^{+2/+3}$ and $Cu^{+1/+2}$, and these ions can be catalyzed by H_2O_2 if they bind to negatively charged DNA under oxidative stress [19]. For such a reason and beyond, bases and nucleic acid sugars may be modified due to oxidative stress, and damage due to single/double chain breaks may occur with the rupture of hydrogen and even covalent bonds[20]. Copoglu et al. (2015) found TAS, OSI, and 8-hydroxydeoxyguanosine (8-OhdG) serum levels were significantly elevated in without symptomatic group, TOS and OSI levels were significantly higher and TAS levels were significantly lower in the symptomatic schizophrenic patients, DNA damage was higher in only without symptomatic group compared to controls [21]. Animal studies have shown that some brain areas are affected differently by oxidative stress than others, and 8-OHdG levels were changed in different brain regions. A negative correlation was found between DNA damage in the detected regions and the effort of DNA to destroy 8-OhdG [22]. In a postmortem study, the 8-OHdG level was quite higher in the hippocampus of schizophrenia [23]. There may be situations where cells cannot cope with oxidative stress [24], in such cases, DNA cannot be repaired and cell death (apoptosis) occurs, especially in brain diseases [25]. This manifestation of apoptotic processes has been associated with many neuropsychiatric diseases, including schizophrenia, and has been associated with tumor suppressor elements. The study emphasized that about p53 for the regulation of apoptosis and explored the potential and sensitivity of nucleic acid in schizophrenia [26]. Damage in DNA or dysfunction of repair mechanisms may be a factor triggering apoptotic pathways [27] due to impaired intracellular signaling of hyperphosphorylated p53. Proving these theories, Levine (1997) reported that p53 is not active in healthy cell activity, but can be activated in DNA damage [28].

Level inconsistencies can be seen in many biomarkers attributable to oxidative stress. These may be caused by differences in the measurement techniques of these levels, differences in the tested material, medication, and different phases of the disease. Family history of the disease, geography, lifestyle, physical activity, and nutritional status may be the source of such problems.

10.2.2 Antioxidative Treatments

Drug therapy in the treatment of this disease is limited to antipsychotics. Although many drugs used in the clinic are important for the treatment of schizophrenia, they have not been satisfactory in managing the disease completely. Therefore, alternative treatment options are urgently needed. One possible way could be antioxidant therapy. For this reason, elucidating the antioxidant mechanism may suggest future treatment options and thus be more successful in the management of schizophrenia.

Also, findings regarding antioxidant enzyme levels in the earliest periods of the psychotic disorder may be contingent on the type of drug, the harshness of the psycho-pathology, or environmental circumstances. For example, studies are showing decreased GSH concentration in the CSF, prefrontal cortex, and postmortem caudate of schizophrenic patients [29–31]. Interestingly, most of the studies on antioxidant defense in patients reported a decrease in the defense system [8, 32–37], there are also studies in the literature that argue the opposite [38–41].

The most commonly used antioxidants in studies conducted with schizophrenia patients were vitamins C and E. As well known, Vitamin E is a fat-soluble vitamin that can keep from radicals that cause oxidative stress. However, cytosolic proteins, from which most harmful radicals are produced, have little potential to prevent oxidative damage to mitochondria and nuclei. Therefore, it makes sense to put together a water-soluble vitamin C along with the treatment [41].

Increasing research on the therapeutic potential effect of second-generation antipsychotics in the treatment of schizophrenia continues rapidly. Microglial cells in the brain show a protective effect against the production of ROS and RNS, which cause neuroinflammation against oxidative stress. ROS and RNS production cause neurodegeneration, white/grey matter abnormity, and decreased neurogenetic actions observed in schizophrenia [42] (in Fig. 10.1). Several studies have

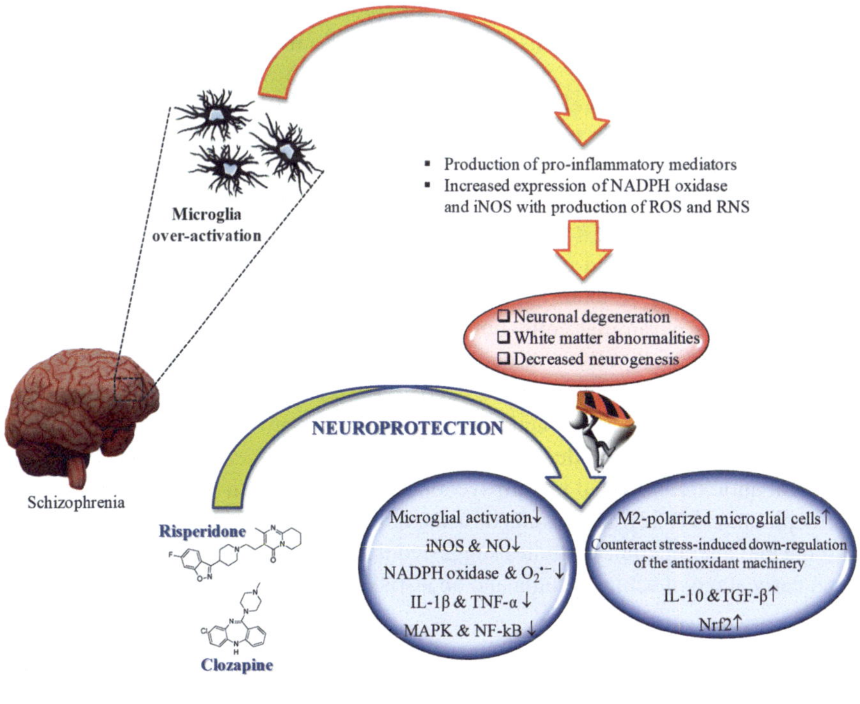

Fig. 10.1 Schematic representation of the oxidative effect of common second-generation antipsychotics used in schizophrenia on microglial cell activations, ↑ = increased; ↓ = decreased [50]

demonstrated the antioxidant and anti-inflammatory of second-generation antipsychotics on the regulation of microglia activity, specifically for proinflammatory cytokines, ROS and RNS. The experimental animal study reported that paliperidone (1 mg/kg i.p.) has the potential option to treat antioxidant and anti-inflammatory pathways in acute and chronic rat stress modeling [31]. Interestingly, Eneni et al. (2020) investigated the effects of haloperidol (1 mg/kg, i.p.), Disomine (25, 50, and 100 mg/kg, i.p.), and risperidone (0.5 mg/kg, i.p.) on schizophrenia-like behavior and the fundamental changes in oxidative stress biomarkers and acetylcholinesterase (AChE) activity in mice. While Disomine and Risperidone ameliorated acute and subacute ketamine-induced schizophrenia-like behaviors, cognitive status was better in the disomine-treated group, with high SOD and GSH levels and low levels of MDA and AChE [43].

As reported in the 12-week prospective longitudinal study results, the patients presented higher SOD, CAT activities, and TAS levels, but lower MDA levels and GPx activity after receiving risperidone monotherapy. The authors showed that the antioxidant defense enzymes and redox regulatory system may contribute to these values as a response to risperidone therapy in patients with schizophrenia [44]. In Yolland et al. (2020)'s study, a meta-analysis of randomized controlled experiments, total scores and cognitive status with working memory was significantly improved in the N-acetylcysteine group after 24 weeks of treatment [45]. Ermakov et al. (2021) suggested that not only antioxidants but also drugs targeting the redox-regulated transcription factor (including Nrf2 and FoxO activators or NF-kB inhibitors) have a distinguished promise in schizophrenia [46]. In an animal study, aripiprazole, ziprasidone, and olanzapine regulated ROS levels, SOD activity, and BCL2-related X protein (Bax) expression in mouse pheochromocytoma (PC12) cells to be protective against oxidative stress caused by 1-Methyl-4-*phenylpyridinium* (MPP+) ion [47, 48] In the ketamine-induced model, a permanent decrease was detected in the GSH/GSH disulfide ratio and parvalbumin expression in the medial prefrontal cortex. In addition, it caused a decrease in mitochondrial membrane potential while also increasing superoxide levels. In the synaptic examination, the excitatory and inhibitory effects were disrupted in the pyramidal cells, but the mitochondrial function returned to normal in the pyramidal cells with the applied NAC [49]. More studies on antioxidant treatments in schizophrenia are needed.

As Minarini et al. (2017), and Miyake and Miyamoto's (2016) papers showed that in randomized controlled trials, treatments with N-acetylcysteine (NAC), a powerful antioxidant added to antipsychotics, are effective in patients with chronic schizophrenia [51, 52]. The results of using a neurodevelopmental model of schizophrenia and also in clinical studies suggested that NAC may have promising effects in an early stage of schizophrenia and an at-risk mental state [51–54].

10.3 Autism

10.3.1 Oxidative Stress Biomarkers

ASD is defined as a neurodevelopmental disorder with a prevalence of at least 1%. Based on ASD, there are a series of brain studies that try to explain multigenetic, and epigenetic factors with environmental factors and the systems such as serotonergic and glutaminergic [55]. The symptoms of the disease include cognitive and social deficits as well as sensory disorders that begin in the developmental period of childhood [56]. Due to the multifactorial etiology of ASD, both the understanding of the mechanism of the disease and its treatment are complex. Haile et al. (2017) pointed out several mechanisms that may develop mitochondrial dysfunction due to oxidative stress in mitochondria. This situation can lead to misfolding of many proteins associated with mitochondrial membranes in the endoplasmic reticulum (ER) (e.g., guanosine triphosphatase (GTPase) Rab32), which may also be a precursor to many neurological disorders [57]. The cell also has an innate defense mechanism, particularly the expression of CYP cytochrome and many transcriptional factors (e.g., nuclear factor (erythroid-derived 2)-like 2 (Nrf2)) in response to oxidative stress [58].

Numerous studies on the pathogenesis of ASD have reported findings of increased oxidative stress levels which lead to DNA damage, neuroinflammation, weakened immune system and epigenetic impairment, and decreased antioxidant capacity in patients with ASD [59, 60]. In particular, factors such as oxidized biological markers, heavy metals, herbicides, pesticides, and UV light have been suggested for the relationship of ASD-related oxidative damage with environmental factors [61].

Two different studies reported that children with ASD were powerless against increased GSH in their plasma and decreased GSH levels in their neurons, and thus oxidative stress repair due to storage GSH deficiency [62, 63]. GSH, a thiol tripeptide class, is a powerful antioxidant for neurons (ie. effective for scavenging free radicals in dopamine neurons in the substantia nigra pars compacta). As a result, it is a powerful antioxidative agent with a neuroprotective effect by preventing neuroinflammatory reactions related to oxidative stress in ASD, especially Lipid peroxidation in newborns [64, 65]. Likewise, Rose et al. (2016) showed that plasma GSH storage of a child with autism decreased compared to his normally developing sibling, and this discrepancy would occur in cells sensitive to oxidative stress [62]. Interestingly, Burger et al. (2017), in their case study, underlined genetic biomarkers in the metabolic assessment of oxidative stress in children with ASD as a result of whole-exome sequencing (WES), a new c.795delT mutation in the WDR45 gene in a sibling with autism. They reported that this mutation has raised the mitochondrial activity of complex I + III in both muscle and fibroblasts due to elevating the respiration in peripheral blood mononuclear cells (PBMCs) [66]. Rose et al. (2016) associated this situation with stereotypical behaviors due to mitochondrial dysfunction in their study comparing sibling samples of ASD in vitro [62]. Mitochondrial dysfunction in ASD may be because redox inconsistency in this part results in oxidative stress and may be due to background gastrointestinal system (GIS) dysregulation [67].

Chauhan et al. (2004) concluded that oxidative stress may be higher by reporting that plasma lipid peroxidation and amino-glycerophospholipids level in the ASD group was more increased compared to their normal developing siblings and AGP levels could be a biomarker for ASD[68]. In Efe et al.'s (2021) study, of 60 children with autism, in which dynamic thiol/disulfide homeostasis (DTDH) levels could be a marker of oxidative stress, they reported lower plasma thiol levels in contrast to high disulfide and thiol oxidation-reduction ratio in plasma but showed that these oxidative stress biomarkers were not correlated with autism symptom severity [69]. Needham et al. (2021) examined plasma and stool samples from children with ASD and normal development. Differences were found in amino acid, lipid, and xenobiotic metabolism, oxidative stress, mitochondrial dysfunction, elevated hormone levels, changes in lipid profile, and levels of phenolic microbial metabolites [70].

James et al. (2004) reported that methionine, SAM, homocysteine, cysteine, and GSH-t levels were lower in children with ASD2 compared to healthy controls, while S-adenosylmethionine (SAM), S-adenosylhomocysteine (SAH), adenosine and oxidative glutathione (GSSG) levels were higher in blood plasma [71]. These values can be considered in the class of oxidative stress biomarkers. In a systematic review and meta-analysis of 87 studies (4928 ASD children and 4181 healthy control (HC) children), it was reported that blood nitric oxide, MDA, GSSG, homocysteine, S-adenosylhomocysteine, and copper levels were observed to be higher in children with ASD. However, the blood GSH-t, GSH/GSSG, tGSH/GSSG, GSH, vitamin B9, cysteine, methionine, vitamin B12, vitamin D, vitamin E, SAM/SAH, and calcium concentrations were significantly low in children with ASD [72]. In another systematic review and meta-analyses, although plasma GSH (27%), GPx (18%), methionine (13%), and cysteine (14%) levels were lower and oxidized glutathione (45%) concentrations were higher in ASDs, superoxide dismutase, homocysteine and cystathionine were significantly more no difference detected. The meta-OR of ASD exposure to homozygous mutant subjects (TT) relative to non-homozygous mutant (CC) C677T allele distribution in the methylene tetrahydrofolate reductase gene (MTHFR) was 2.26 [73]. Filippek et al. (2004) found that free carnitine and pyruvate levels were high at low ammonia and alanine levels in 100 ASD children [74], but there was no control group to compare and they only evaluated against reference ranges of values. Impaired ammonia, alanine, carnitine, and pyruvate levels could be an indicator of impaired mitochondrial energy production. Elevated lactate: pyruvate ratio in the cytoplasm of cells may also indicate an impaired cellular redox state resulting from mitochondrial dysfunction. Oliveira et al. (2005) found a higher ratio of lactate and lactate pyruvate in blood plasma in 20% of 69 children with ASD compared to healthy children [75].

Interestingly, in a study of parents of children with autism, the presence of impaired oxidative profiles was a possible factor and highlights the role of genetics in the development of autism [63]. Several oxidative biomarkers were significantly altered in persons with autism, strongly supporting the role of abnormalities in oxidative homeostasis in the pathophysiology of autism. Although valuable findings were found in all these studies, more comprehensive studies on biomarkers for oxidative stress are needed.

10.3.2 Antioxidative Treatments

Although there is no effective medicine for autism, current pharmacological treatments are used to improve some symptoms such as self-mutilation, aggression, repetitive and stereotyped behaviors, distraction, hyperactivity, and sleep disorders [76]. Pangrazzi et al. (2020) summarized the importance of the role of the Omega-3, dietary polyphenols, and vitamin E/vitamin C/GSH network in the clearance of intracellular ROS in the disorders observed with ASD in their published review [77]. There are limited clinical and preclinical studies on oxidative biomarkers or behavioral effects of antioxidants in autism. Zambrelli et al. (2021) searched more than 20 articles focalizing on the effects of antioxidant supplementation on sleep in ASD. Most of the studies were about melatonin and also TRY, and the remaining studies were about luteolin, Coenzyme Q10, and quercetin, which are known to have important antioxidative effects. Although antioxidants have been reported to be beneficial in sleep problems in ASD, more studies have been suggested in this direction due to the limited number of studies [60]. Therefore, there is an urgent need for more comprehensive studies with large samples in this area.

10.4 Alzheimer's Disease

10.4.1 Oxidative Stress Biomarkers

AD is an increasingly common neurodegenerative disease characterized by cognitive and cognitive damage and many brain pathophysiologies due to dementia. AD is clinically diagnosed by neuroimaging methods and some cognitive tests and is also characterized by neuronal loss and neuropathologic lesions that occur in many brain regions [78]. Clinical symptoms such as hippocampal type episodic memory loss are observed in patients in the Prodromal/Predementia AD stage, but their daily living activities are not affected yet and they are used for the early symptomatic, predementia stage because they cannot fully diagnose dementia. Detection of biomarkers at these stages is challenging, and even at earlier stages, it is difficult to detect clinical symptoms. However, in its later stages, the presence of biomarkers from CSF or imaging may prove the pathology of AD [79]. AD progression is dependent on the stage and age of the individual, with varying rates of decline in clinical markers such as cognitive, neuroimaging, and biological [80]. The most common biodiagnostic markers for AD are the 42 amino acids amyloid β (β-amyloid 1-42), neurofibrillary tangles, and hyperphosphorylated protein tau. However, these markers appear at later stages of the disease and are very costly [81]. The discovery of new biomarkers for AD with which we can understand the disease in its early stages is very important.

There is evidence that Amyloid plaques (Aβ) accumulate in the interneuronal space and cause oxidative damage. In addition, the accumulation of heavy metals in these regions may increase the effects of oxidative stress as the effect of plaques, especially metal ions such as copper and zinc. ROS production is triggered as the sensitivity to Aβ metal ions increases [82]. Oxidative stress damage may also occur in cases such as selenoproteins and Se accumulation. It can trigger the aggregation of Aβ plaques by stimulating hyperphosphorylation of the tau protein, resulting in neuronal toxicity and neurodegeneration (in Fig. 10.2) [83]. In the neural mechanism, redox metals such as Cu, Fe, and Hg can also cause such situations and can stimulate neuronal apolipoprotein E receptor-2 (ApoER2) by increasing signaling in AD [84]. The increase in microglial cells against interneuronal Aβ, which occurs in the pathogenesis of AD, unfortunately, affects the number of astrocytes that feed the neurons. In addition to this resulting inflammatory response, NO synthesis can be induced and cause a regional cytokine (eg TNF-a, IL-1p, IL-6) storm [85]. Meanwhile, Aβ causes dysfunction in mitochondria while triggering ROS

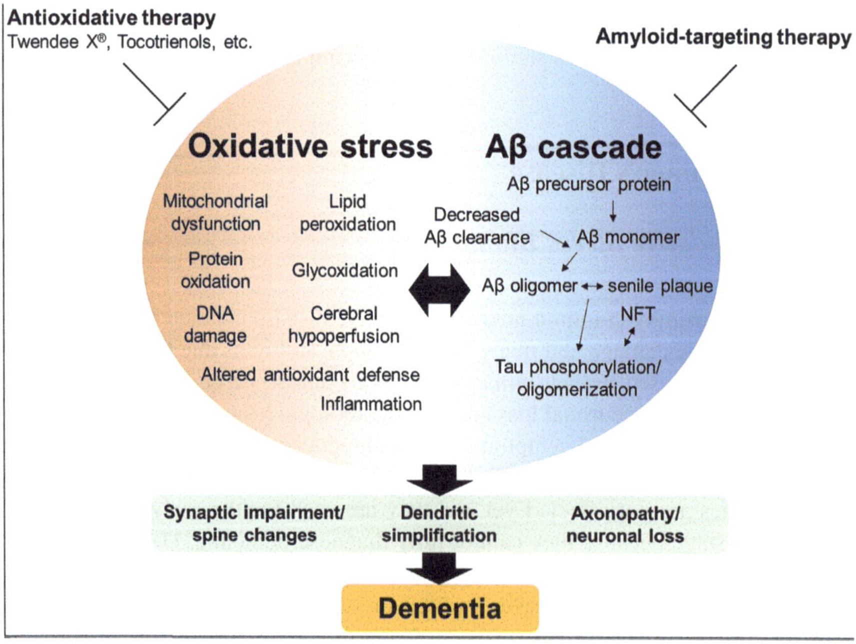

Fig. 10.2 Oxidative stress may develop in AD disease associated with Aβ accumulation. Mitochondrial dysfunction, protein oxidation, DNA damage, and lipid peroxidation develop as a cause/result of this situation, resulting in axonopathy, dendrite pruning, and synapse deficiency. Antioxidative agents such as Twendee X® and tocotrienols are currently available to prevent such situations [83, 89]

production via NADPH oxidase activation in microglia and astrocyte cells, causing oxidative damage [86, 87]. Oxidative damage in AD may have markers on cell membranes (lipid peroxidation), post-translational protein changes, and the production and function of genetic materials [88].

The current systematic review suggested that lipid peroxidation is the most potential oxidative stress marker in the diagnosis of AD due to its high lipid content and specificity in the brain [90]. Similarly, in another systematic review, although the included studies reported that the serum ratio of lipid peroxidation was high in the patients' sera, it was stated that there would not be sufficient evidence for the follow-up of treatment for AD and its evaluation as a complete biomarker [91]. Uruno et al. (2020) found that decreased GSH levels were increased by Nrf2 induction model Keap1$^{FA/FA}$ mice. The authors proposed a novel plasmalogen-phosphatidylethanolamine (PlsPE) level as the biomarker of AD [92]. These results suggest that Nrf2 overstimulation may ameliorate cognitive disorder in the AD mouse model by protecting them from both oxidative stress and neuroinflammation, and suggesting that Nrf2 is indeed an important therapeutic target of AD. The levels of 8-OHdG in the CSF of AD patients were found to be quite high compared to healthy controls, associated with impaired DNA repair [93]. In general, commonly reported oxidative stress biomarkers of AD are ApoE genotype, GSH/GSSG, MDA, coenzyme Q10, 8-OHdG, SOD, H_2O_2, GPx, and isoprostanes [90]. Consequently, as metalized AB is one of the major drivers of ROS production in the brain, the peptide itself is often attacked and oxidized by ROS activity, so it may appear as a very specific oxidative stress biomarker for AD [94].

Finally, there is a new research area that some microRNAs (miR) may relate to oxidative stress in AD. Kou et al. (2017) reported that miR-34a can increase APP accumulation in AD pathophysiology by suppressing oxidative stress associated with autophagy inhibition followed by mitochondrial damage [95]. miR-141-3p is detected at low levels in plasma exosomes of AD patients and mouse memory loss due to miR125b-5p overexpression [96] likewise miR-141-3p was found at high levels in mediator exosomes of astrocyte cells in the brain [97] and a recent study found that miR-125b-5p can weaken oxidative stress after Aβ-inducement [33]. However, another study reported that miR-125b-5p diminished the ROS levels and decreased mitochondrial membrane potential, and showed a neuroprotective effect against oxidative stress [98]. Although exosome-mediated or free miRNA levels have been investigated in many neurodegenerative diseases until now, it can be clearly said that especially for the oxidative process, many mysteries that need to be clarified in this area continue.

10.4.2 Antioxidative Treatments

Clinical trials based on the hypothesis that AD pathology is neurodegeneration as a result of amyloid-beta (Aβ) and progresses from hippocampal destruction to the brainstem, with dire outcomes initially led to new therapeutic research for disease control, but unsuccessful trials proved that this hypothesis was the result, not the cause of the disease [99]. Currently, the applied treatments only ease the symptoms and temporarily lessen the cognitive progression rate of AD symptoms. Considering the situations given in the section on antioxidant biomarkers, the importance of antioxidative approaches cannot be ignored. Naturally-sourced approaches are promising. In many reviews, Brahmi (Bacopa monnieri), Quercetin and *Ginkgo biloba* are among the traditional "anti-dementia treatments" and the importance of their neuroprotective and antioxidant effect is emphasized, Dubey ve Chinnathambi (2019), Khan et al. (2019), Singh et al. (2019) and Noori et al. (2021) published a review for many natural products that have antioxidant and neuroprotective effects on AD [100–103].

There is a lack of evidence for the use of probiotics. Athari Nik Azm et al. (2018) reported a decline in the accumulation of Aβ, neuroinflammation, and oxidative parameters in an Alzheimer's-probiotic supplementation (Lactobacilli and Bifidobacteria) group which was an β-amyloid (1–42) injected rats [104]. Kobayashi et al. (2017) also demonstrated that the introduction of *Bifidobacterium breve* strains A1 in an AD mouse model has inhibited hippocampal inflammation and oxidative stress-related gene expression [105]. Den et al. (2020) also supported the previous studies' reports with randomized controlled trials. They found that probiotics have improved cognitive performance in AD or Mild cognitive impairment (MCI) patients, maybe because of reducing both inflammatory and oxidative chemical levels [106]. These results suggest that probiotics may have many advantages because of their anti-inflammatory and antioxidant effects.

There are some studies on antioxidant therapy. These antioxidants mainly contain their substrates or coenzymes, various endogenous antioxidant enzymes, and non-enzymatic antioxidants, as well as natural and synthetic antioxidant sources that maintain the redox balance in the biological system [107]. Chen et al. (2020) reported a synthetic chalcone derivative and 2-Hydroxy-40-methoxychalcone decreased ROS activity, induced Nrf2 pathways, enhanced GSH levels, and anti-inflammatory agents, thus causing healing [108].

In another example, the authors showed the importance of the antioxidant effects of Proxison (belong to synthetic flavonoid groups) with improved cellular uptake, free ROS and RNS scavenging power, and neuroprotective action against the cell in a zebrafish animal model [109]. In the results of Systematic Review and Meta-Analysis on vitamins, other antioxidant components including a single antioxidant supplement such as vitamin C, vitamin E, or mixtures thereof did not clearly show a therapeutic effect on cognitive decline in AD [110, 111]. Twendee X (TwX) is an important supplement in AD disease containing 8 antioxidants. A multicenter, randomized, double-blind, and placebo-controlled prospective interventional study, reported that the use of TwX for 6 months increased cognitive functions in MCI patients, but did not affect their daily living activities [83]. In general, it has been seen that such antioxidants are emphasized in the literature and that biomarker studies should be increased for new therapies to be developed.

10.5 Conclusion

Recently, extensive clinical and preclinical studies of oxidative stress biomarkers of AD, ASD, and schizophrenia have provided strong evidence that they play a role in both the etiology and course of these diseases. Lipid peroxidation levels were particularly promising for the potential oxidative stress biomarker and Probiotic supplement has positive contributions to both cognitive and oxidative stress in AD. Although the results of the levels of oxidative stress biomarkers related to schizophrenia and ASD are controversial, more studies are needed in the future. However, the effects of NAC supplementation on schizophrenia symptoms and markers cannot be ignored. The findings of regulating the antioxidant balance of antidepressant, anxiolytic or antipsychotic drugs used in the clinic for AD and schizophrenia have been reported, and the promising aspects of supplements/combined therapies with antioxidant effects in all three diseases are highlighted. However, many important issues remain to be fully elucidated, and more preclinical and clinical studies are needed to evaluate the precise contribution of oxidative stress in psychiatric disorders.

References

1. Milbourn HR, Fitzgerald M (2017) Limiting oxidative stress following neurotrauma with a combination of ion channel inhibitors. Discov Med 23
2. Fransen M, Nordgren M, Wang B, Apanasets O (2012) Role of peroxisomes in ROS/RNS-metabolism: implications for human disease. Biochim Biophys Acta 1822:1363–1373. https://doi.org/10.1016/J.BBADIS.2011.12.001
3. Mahadik SP, Scheffer RE (1996) Oxidative injury and potential use of antioxidants in schizophrenia. Prostaglandins Leukot Essent Fatty Acids 55:45–54. https://doi.org/10.1016/S0952-3278(96)90144-1
4. Mhillaj E, Morgese M, Trabace L (2015) Early life and oxidative stress in psychiatric disorders: what can we learn from animal models? Curr Pharm Des 21:1396–1403. https://doi.org/10.2174/1381612821666150105122422
5. Cannon TD, Yu C, Addington J et al (2016) An individualized risk calculator for research in prodromal psychosis. Am J Psychiatr 173:980–988. https://doi.org/10.1176/APPI.AJP.2016.15070890
6. Davison J, O'Gorman A, Brennan L, Cotter DR (2018) A systematic review of metabolite biomarkers of schizophrenia. Schizophr Res 195:32–50. https://doi.org/10.1016/J.SCHRES.2017.09.021
7. Fraguas D, Díaz-Caneja CM, Ayora M et al (2019) Oxidative stress and inflammation in first-episode psychosis: a systematic review and meta-analysis. Schizophr Bull 45:742–751. https://doi.org/10.1093/SCHBUL/SBY125
8. Mahadik SP, Mukherjee S (1996) Free radical pathology and antioxidant defense in schizophrenia: a review. Schizophr Res 19:1–17. https://doi.org/10.1016/0920-9964(95)00049-6
9. Buosi P, Borghi FA, Lopes AM et al (2021) Oxidative stress biomarkers in treatment-responsive and treatment-resistant schizophrenia patients. Trends psychiatry Psychother 43:278–285. https://doi.org/10.47626/2237-6089-2020-0078

10. Cruz BF, de Campos-Carli SM, de Oliveira AM et al (2021) Investigating potential associations between neurocognition/social cognition and oxidative stress in schizophrenia. Psychiatry Res 298. https://doi.org/10.1016/J.PSYCHRES.2021.113832
11. Kim E, Keskey Z, Kang M et al (2019) Validation of oxidative stress assay for schizophrenia. Schizophr Res 212:126–133. https://doi.org/10.1016/J.SCHRES.2019.07.057
12. Najjar S, Pahlajani S, De Sanctis V et al (2017) Neurovascular unit dysfunction and blood-brain barrier hyperpermeability contribute to schizophrenia neurobiology: a theoretical integration of clinical and experimental evidence. Front Psychiatr 8. https://doi.org/10.3389/FPSYT.2017.00083
13. Le Bai Z, Li XS, Chen GY et al (2018) Serum oxidative stress marker levels in unmedicated and medicated patients with schizophrenia. J Mol Neurosci 66:428–436. https://doi.org/10.1007/S12031-018-1165-4
14. Juchnowicz D, Dzikowski M, Rog J et al (2021) Pro/antioxidant state as a potential biomarker of Schizophrenia. J Clin Med 10. https://doi.org/10.3390/JCM10184156
15. Ma J, Yan L, Guo T et al (2020) A pilot study of biomarkers of oxidative stress in serum and schizophrenia. Psychiatry Res 284. https://doi.org/10.1016/J.PSYCHRES.2020.112757
16. Guler EM, Kurtulmus A, Gul AZ et al (2021) Oxidative stress and schizophrenia: A comparative cross-sectional study of multiple oxidative markers in patients and their first-degree relatives. Int J Clin Pract 75. https://doi.org/10.1111/IJCP.14711
17. Grignon S, Chianetta JM (2007) Assessment of malondialdehyde levels in schizophrenia: a meta-analysis and some methodological considerations. Prog Neuropsychopharmacol Biol Psychiatr 31:365–369. https://doi.org/10.1016/J.PNPBP.2006.09.012
18. Zhang M, Zhao ZM, He L, Wan CL (2010) A meta-analysis of oxidative stress markers in schizophrenia. Sci China Life Sci 53:112–124. https://doi.org/10.1007/S11427-010-0013-8
19. Sharpe MA, Robb SJ, Clark JB (2003) Nitric oxide and Fenton/Haber-Weiss chemistry: nitric oxide is a potent antioxidant at physiological concentrations. J Neurochem 87:386–394. https://doi.org/10.1046/J.1471-4159.2003.02001.X
20. Prabhulkar S, Li CZ (2010) Assessment of oxidative DNA damage and repair at single cellular level via real-time monitoring of 8-OHdG biomarker. Biosens Bioelectron 26:1743–1749. https://doi.org/10.1016/J.BIOS.2010.08.029
21. Markkanen E, Meyer U, Dianov GL (2016) DNA damage and repair in Schizophrenia and Autism: implications for cancer comorbidity and beyond. Int J Mol Sci 17. https://doi.org/10.3390/IJMS17060856
22. Cardozo-Pelaez F, Brooks PJ, Stedeford T et al (2000) DNA damage, repair, and antioxidant systems in brain regions: a correlative study. Free Radic Biol Med 28:779–785. https://doi.org/10.1016/S0891-5849(00)00172-6
23. Nishioka N, Arnold SE (2004) Evidence for oxidative DNA damage in the hippocampus of elderly patients with chronic schizophrenia. Am J Geriatr Psychiatr 12:167–175. https://doi.org/10.1097/00019442-200403000-00008
24. Esteve JM, Mompo J, Asuncion JG et al (1999) Oxidative damage to mitochondrial DNA and glutathione oxidation in apoptosis: studies in vivo and in vitro. FASEB J 13:1055–1064. https://doi.org/10.1096/FASEBJ.13.9.1055
25. Aizenman E, Stout AK, Hartnett KA et al (2000) Induction of neuronal apoptosis by thiol oxidation: putative role of intracellular zinc release. J Neurochem 75:1878–1888. https://doi.org/10.1046/J.1471-4159.2000.0751878.X
26. Catts VS, Catts SV (2000) Apoptosis and schizophrenia: is the tumour suppressor gene, p53, a candidate susceptibility gene? Schizophr Res 41:405–415. https://doi.org/10.1016/S0920-9964(99)00077-8
27. Kagan VE, Gleiss B, Tyurina YY et al (2002) A role for oxidative stress in apoptosis: oxidation and externalization of phosphatidylserine is required for macrophage clearance of cells undergoing Fas-mediated apoptosis. J Immunol 169:487–499. https://doi.org/10.4049/JIMMUNOL.169.1.487

28. Levine AJ (1997) p53, the cellular gatekeeper for growth and division. Cell 88:323–331. https://doi.org/10.1016/S0092-8674(00)81871-1
29. Do KQ, Trabesinger AH, Kirsten-Krüger M et al (2000) Schizophrenia: glutathione deficit in cerebrospinal fluid and prefrontal cortex in vivo. Eur J Neurosci 12:3721–3728. https://doi.org/10.1046/J.1460-9568.2000.00229.X
30. Lavoie S, Murray MM, Deppen P et al (2008) Glutathione precursor, N-acetyl-cysteine, improves mismatch negativity in schizophrenia patients. Neuropsychopharmacology 33:2187–2199. https://doi.org/10.1038/SJ.NPP.1301624
31. MacDowell KS, Caso JR, Martín-Hernández D et al (2016) The atypical antipsychotic paliperidone regulates endogenous antioxidant/anti-inflammatory pathways in rat models of acute and chronic restraint stress. Neurotherapeutics 13:833–843. https://doi.org/10.1007/S13311-016-0438-2
32. García S, Alberich S, Martínez-Cengotitabengoa M et al (2018) The complex association between the antioxidant defense system and clinical status in early psychosis. PLoS One 13. https://doi.org/10.1371/JOURNAL.PONE.0194685
33. Li P, Xu Y, Wang B, et al (2020) miR-34a-5p and miR-125b-5p attenuate Aβ-induced neurotoxicity through targeting BACE1. J Neurol Sci:413. https://doi.org/10.1016/J.JNS.2020.116793
34. Micó JA, Rojas-Corrales MO, Gibert-Rahola J, et al (2011) Reduced antioxidant defense in early onset first-episode psychosis: a case-control study. BMC Psychiatr:11. https://doi.org/10.1186/1471-244X-11-26
35. Raffa M, Atig F, Mhalla A, et al (2011) Decreased glutathione levels and impaired antioxidant enzyme activities in drug-naive first-episode schizophrenic patients. BMC Psychiatr:11. https://doi.org/10.1186/1471-244X-11-124
36. Ranjekar PK, Hinge A, Hegde MV et al (2003) Decreased antioxidant enzymes and membrane essential polyunsaturated fatty acids in schizophrenic and bipolar mood disorder patients. Psychiatr Res 121:109–122. https://doi.org/10.1016/s0165-1781(03)00220-8
37. Schiavone S, Trabace L (2018) The use of antioxidant compounds in the treatment of first psychotic episode: highlights from preclinical studies. CNS Neurosci Ther 24:465–472. https://doi.org/10.1111/CNS.12847
38. Altuntas I, Aksoy H, Coskun I et al (2000) Erythrocyte superoxide dismutase and glutathione peroxidase activities, and malondialdehyde and reduced glutathione levels in schizophrenic patients. Clin Chem Lab Med 38:1277–1281. https://doi.org/10.1515/CCLM.2000.201
39. Dakhale G, Khanzode S, Khanzode S et al (2004) Oxidative damage and schizophrenia: the potential benefit by atypical antipsychotics. Neuropsychobiology 49:205–209. https://doi.org/10.1159/000077368
40. Kuloglu M, Ustundag B, Atmaca M et al (2002) Lipid peroxidation and antioxidant enzyme levels in patients with schizophrenia and bipolar disorder. Cell Biochem Funct 20:171–175. https://doi.org/10.1002/CBF.940
41. Michel TM, Thome J, Martin D et al (2004) Cu, Zn- and Mn-superoxide dismutase levels in brains of patients with schizophrenic psychosis. J Neural Transm 111:1191–1201. https://doi.org/10.1007/S00702-004-0160-9
42. Monji A, Kato T, Kanba S (2009) Cytokines and schizophrenia: Microglia hypothesis of schizophrenia. Psychiatry Clin Neurosci 63:257–265. https://doi.org/10.1111/J.1440-1819.2009.01945.X
43. Okubo Eneni AE, Ben-Azu B, Mayowa Ajayi A, Oladele Aderibigbe A (2020) Diosmin attenuates schizophrenia-like behavior, oxidative stress, and acetylcholinesterase activity in mice. Drug Metab Pers Ther. https://doi.org/10.1515/DMDI-2020-0119
44. Li XR, Xiu MH, Guan XN et al (2021) Altered antioxidant defenses in drug-naive first episode patients with schizophrenia are associated with poor treatment response to risperidone: 12-week results from a prospective longitudinal study. Neurotherapeutics 18:1316–1324. https://doi.org/10.1007/S13311-021-01036-3

45. Yolland COB, Hanratty D, Neill E et al (2020) Meta-analysis of randomised controlled trials with N-acetylcysteine in the treatment of schizophrenia. Aust N Z J Psychiatr 54:453–466. https://doi.org/10.1177/0004867419893439
46. Ermakov EA, Dmitrieva EM, Parshukova DA, et al (2021) Oxidative stress-related mechanisms in schizophrenia pathogenesis and new treatment perspectives. Oxid Med Cell Longev:2021. https://doi.org/10.1155/2021/8881770
47. Park SW, Lee CH, Lee JG et al (2011) Protective effects of atypical antipsychotic drugs against MPP(+)-induced oxidative stress in PC12 cells. Neurosci Res 69:283–290. https://doi.org/10.1016/J.NEURES.2011.01.004
48. Sadowska-Bartosz I, Galiniak S, Bartosz G et al (2016) Antioxidant properties of atypical antipsychotic drugs used in the treatment of schizophrenia. Schizophr Res 176:245–251. https://doi.org/10.1016/J.SCHRES.2016.07.010
49. Phensy A, Driskill C, Lindquist K et al (2017) Antioxidant treatment in male mice prevents mitochondrial and synaptic changes in an NMDA receptor dysfunction model of schizophrenia. eNeuro 4. https://doi.org/10.1523/ENEURO.0081-17.2017
50. Caruso G, Grasso M, Fidilio A et al (2020) Antioxidant properties of second-generation antipsychotics: focus on microglia. Pharmaceuticals (Basel) 13:1–20. https://doi.org/10.3390/PH13120457
51. Dean O, Giorlando F, Berk M (2011) N-acetylcysteine in psychiatry: current therapeutic evidence and potential mechanisms of action. J Psychiatr Neurosci 36:78. https://doi.org/10.1503/JPN.100057
52. Miyake N, Miyamoto S (2016) Effectiveness of N-acetylcysteine in the treatment of schizophrenia. Japanese J Neuropsychopharmacol 36:29–35
53. Chauhan A, Chauhan V, Brown WT, Cohen I (2004) Oxidative stress in autism: Increased lipid peroxidation and reduced serum levels of ceruloplasmin and transferrin – the antioxidant proteins. Life Sci 75:2539–2549. https://doi.org/10.1016/J.LFS.2004.04.038
54. Raghu G, Berk M, Campochiaro PA et al (2021) The multifaceted therapeutic role of N-acetylcysteine (NAC) in disorders characterized by oxidative stress. Curr Neuropharmacol 19:1202–1224. https://doi.org/10.2174/1570159X19666201230144109
55. Bourgeron T (2015) From the genetic architecture to synaptic plasticity in autism spectrum disorder. Nat Rev Neurosci 16:551–563. https://doi.org/10.1038/NRN3992
56. Percário S, Da A, Barbosa S et al (2020) No Title. https://doi.org/10.1155/2020/2360872
57. Haile Y, Deng X, Ortiz-Sandoval C et al (2017) Rab32 connects ER stress to mitochondrial defects in multiple sclerosis. J Neuroinflammation 14. https://doi.org/10.1186/S12974-016-0788-Z
58. dos Santos AA, Ferrer B, Gonçalves FM, et al (2018) Oxidative stress in methylmercury-induced cell toxicity. Toxics:6. https://doi.org/10.3390/TOXICS6030047
59. Fatemi SH, Aldinger KA, Ashwood P et al (2012) Consensus paper: pathological role of the cerebellum in autism. Cerebellum 11:777–807. https://doi.org/10.1007/S12311-012-0355-9
60. Zambrelli E, Lividini A, Spadavecchia S et al (2021) Effects of supplementation with antioxidant agents on sleep in autism spectrum disorder: a review. Front Psychiatr 12. https://doi.org/10.3389/FPSYT.2021.689277
61. Skalny AV, Simashkova NV, Skalnaya AA et al (2017) Assessment of gender and age effects on serum and hair trace element levels in children with autism spectrum disorder. Metab Brain Dis 32:1675–1684. https://doi.org/10.1007/S11011-017-0056-7
62. Rose S, Bennuri SC, Wynne R et al (2017) Mitochondrial and redox abnormalities in autism lymphoblastoid cells: a sibling control study. FASEB J 31:904–909. https://doi.org/10.1096/FJ.201601004R
63. James SJ, Melnyk S, Fuchs G et al (2009) Efficacy of methylcobalamin and folinic acid treatment on glutathione redox status in children with autism. Am J Clin Nutr 89:425–430. https://doi.org/10.3945/AJCN.2008.26615
64. Poggi C, Dani C (2018) Sepsis and oxidative stress in the newborn: from pathogenesis to novel therapeutic targets. Oxid Med Cell Longev:2018. https://doi.org/10.1155/2018/9390140

65. Ghanizadeh A, Akhondzadeh S, Hormozi M et al (2012) Glutathione-related factors and oxidative stress in autism, a review. Curr Med Chem 19:4000–4005. https://doi.org/10.2174/092986712802002572
66. Burger BJ, Rose S, Bennuri SC et al (2017) Autistic siblings with novel mutations in two different genes: insight for genetic workups of autistic siblings and connection to mitochondrial dysfunction. Front Pediatr 5. https://doi.org/10.3389/FPED.2017.00219
67. Kang DW, Ilhan ZE, Isern NG et al (2018) Differences in fecal microbial metabolites and microbiota of children with autism spectrum disorders. Anaerobe 49:121–131. https://doi.org/10.1016/J.ANAEROBE.2017.12.007
68. Chauhan A, Chauhan V, Brown WT, Cohen I (2004) Oxidative stress in autism: increased lipid peroxidation and reduced serum levels of ceruloplasmin and transferrin – the antioxidant proteins. Life Sci 75:2539–2549. https://doi.org/10.1016/J.LFS.2004.04.038
69. Efe A, Neşelioğlu S, Soykan A (2021) An investigation of the dynamic thiol/disulfide homeostasis, as a novel oxidative stress plasma biomarker, in children with autism spectrum disorders. Autism Res 14:473–487. https://doi.org/10.1002/AUR.2436
70. Needham BD, Adame MD, Serena G et al (2021) Plasma and fecal metabolite profiles in autism spectrum disorder. Biol Psychiatr 89:451–462. https://doi.org/10.1016/J.BIOPSYCH.2020.09.025
71. James SJ, Cutler P, Melnyk S et al (2004) Metabolic biomarkers of increased oxidative stress and impaired methylation capacity in children with autism. Am J Clin Nutr 80:1611–1617. https://doi.org/10.1093/AJCN/80.6.1611
72. Chen L, Shi XJ, Liu H et al (2021) Oxidative stress marker aberrations in children with autism spectrum disorder: a systematic review and meta-analysis of 87 studies (N = 9109). Transl Psychiatry 11. https://doi.org/10.1038/S41398-020-01135-3
73. Frustaci A, Neri M, Cesario A et al (2012) Oxidative stress-related biomarkers in autism: systematic review and meta-analyses. Free Radic Biol Med 52:2128–2141. https://doi.org/10.1016/J.FREERADBIOMED.2012.03.011
74. Filipek PA, Juranek J, Nguyen MT et al (2004) Relative carnitine deficiency in autism. J Autism Dev Disord 34:615–623. https://doi.org/10.1007/S10803-004-5283-1
75. Oliveira G, Diogo L, Grazina M et al (2005) Mitochondrial dysfunction in autism spectrum disorders: a population-based study. Dev Med Child Neurol 47:185–189. https://doi.org/10.1017/S0012162205000332
76. Kardani A, Soltani A, Sewell RDE et al (2019) Neurotransmitter, antioxidant and anti-neuroinflammatory mechanistic potentials of herbal medicines in ameliorating autism spectrum disorder. Curr Pharm Des 25:4421–4429. https://doi.org/10.2174/1381612825666191112143940
77. Pangrazzi L, Balasco L, Bozzi Y (2020) Natural antioxidants: a novel therapeutic approach to autism spectrum disorders? Antioxidants (Basel, Switzerland) 9:1–16. https://doi.org/10.3390/ANTIOX9121186
78. Kukull WA, Bowen JD (2002) Dementia epidemiology. Med Clin North Am 86:573–590. https://doi.org/10.1016/S0025-7125(02)00010-X
79. Albert MS, DeKosky ST, Dickson D et al (2011) The diagnosis of mild cognitive impairment due to Alzheimer's disease: recommendations from the National Institute on Aging-Alzheimer's Association workgroups on diagnostic guidelines for Alzheimer's disease. Alzheimers Dement 7:270–279. https://doi.org/10.1016/J.JALZ.2011.03.008
80. Bertens D, Knol DL, Scheltens P, Visser PJ (2015) Temporal evolution of biomarkers and cognitive markers in the asymptomatic, MCI, and dementia stage of Alzheimer's disease. Alzheimers Dement 11:511–522. https://doi.org/10.1016/J.JALZ.2014.05.1754
81. Leung YY, Toledo JB, Nefedov A et al (2015) Identifying amyloid pathology-related cerebrospinal fluid biomarkers for Alzheimer's disease in a multicohort study. Alzheimer's Dement (Amsterdam, Netherlands) 1:339–348. https://doi.org/10.1016/J.DADM.2015.06.008

82. Praticò D, Clark CM, Liun F et al (2002) Increase of brain oxidative stress in mild cognitive impairment: a possible predictor of Alzheimer disease. Arch Neurol 59:972–976. https://doi.org/10.1001/ARCHNEUR.59.6.972
83. Tadokoro K, Morihara R, Ohta Y et al (2019) Clinical benefits of antioxidative supplement Twendee X for mild cognitive impairment: a multicenter, randomized, double-blind, and placebo-controlled prospective interventional study. J Alzheimers Dis 71:1063–1069. https://doi.org/10.3233/JAD-190644
84. Solovyev N, Drobyshev E, Bjørklund G et al (2018) Selenium, selenoprotein P, and Alzheimer's disease: is there a link? Free Radic Biol Med 127:124–133. https://doi.org/10.1016/J.FREERADBIOMED.2018.02.030
85. McGeer EG, McGeer PL (2010) Neuroinflammation in Alzheimer's disease and mild cognitive impairment: a field in its infancy. J Alzheimers Dis 19:355–361. https://doi.org/10.3233/JAD-2010-1219
86. Brown GC, Bal-Price A (2003) Inflammatory neurodegeneration mediated by nitric oxide, glutamate, and mitochondria. Mol Neurobiol 27:325–355. https://doi.org/10.1385/MN:27:3:325
87. Matos M, Augusto E, Oliveira CR, Agostinho P (2008) Amyloid-beta peptide decreases glutamate uptake in cultured astrocytes: involvement of oxidative stress and mitogen-activated protein kinase cascades. Neuroscience 156:898–910. https://doi.org/10.1016/J.NEUROSCIENCE.2008.08.022
88. Galasko D, Montine TJ (2010) Biomarkers of oxidative damage and inflammation in Alzheimer's disease. Biomark Med 4:27–36. https://doi.org/10.2217/BMM.09.89
89. Tadokoro K, Ohta Y, Inufusa H et al (2020) Prevention of cognitive decline in alzheimer's disease by novel antioxidative supplements. Int J Mol Sci 21. https://doi.org/10.3390/IJMS21061974
90. García-Blanco A, Baquero M, Vento M et al (2017) Potential oxidative stress biomarkers of mild cognitive impairment due to Alzheimer disease. J Neurol Sci 373:295–302. https://doi.org/10.1016/J.JNS.2017.01.020
91. Chang YT, Chang WN, Tsai NW, et al (2014) The roles of biomarkers of oxidative stress and antioxidant in Alzheimer's disease: a systematic review. Biomed Res Int:2014. https://doi.org/10.1155/2014/182303
92. Uruno A, Matsumaru D, Ryoke R et al (2020) Nrf2 suppresses oxidative stress and inflammation in app knock-in Alzheimer's disease model mice. Mol Cell Biol 40. https://doi.org/10.1128/MCB.00467-19
93. Cioffi F, Adam RHI, Bansal R, Broersen K (2021) A review of oxidative stress products and related genes in early Alzheimer's disease. J Alzheimers Dis 83:977–1001. https://doi.org/10.3233/JAD-210497
94. Collin F, Cheignon C, Hureau C (2018) Oxidative stress as a biomarker for Alzheimer's disease. Biomark Med 12:201–203. https://doi.org/10.2217/BMM-2017-0456
95. Kou X, Li J, Liu X et al (2017) Swimming attenuates d-galactose-induced brain aging via suppressing miR-34a-mediated autophagy impairment and abnormal mitochondrial dynamics. J Appl Physiol 122:1462–1469. https://doi.org/10.1152/JAPPLPHYSIOL.00018.2017
96. Lugli G, Cohen AM, Bennett DA, et al (2015) Plasma exosomal miRNAs in persons with and without Alzheimer disease: altered expression and prospects for biomarkers. PLoS One:10. https://doi.org/10.1371/JOURNAL.PONE.0139233
97. Gayen M, Bhomia M, Balakathiresan N, Knollmann-Ritschel B (2020) Exosomal MicroRNAs released by activated astrocytes as potential neuroinflammatory biomarkers. Int J Mol Sci:21. https://doi.org/10.3390/IJMS21072312
98. Shen Y, Shen Z, Guo L et al (2018) MiR-125b-5p is involved in oxygen and glucose deprivation injury in PC-12 cells via CBS/H2S pathway. Nitric Oxide Biol Chem 78:11–21. https://doi.org/10.1016/J.NIOX.2018.05.004
99. Tcw J, Goate AM (2017) Genetics of β-amyloid precursor protein in Alzheimer's disease. Cold Spring Harb Perspect Med 7. https://doi.org/10.1101/CSHPERSPECT.A024539

100. Dubey T, Chinnathambi S (2019) Brahmi (Bacopa monnieri): an ayurvedic herb against the Alzheimer's disease. Arch Biochem Biophys 676. https://doi.org/10.1016/J.ABB.2019.108153
101. Khan H, Ullah H, Aschner M et al (2019) Neuroprotective effects of quercetin in Alzheimer's disease. Biomolecules 10. https://doi.org/10.3390/BIOM10010059
102. Singh SK, Srivastav S, Castellani RJ et al (2019) Neuroprotective and antioxidant effect of Ginkgo biloba extract against AD and other neurological disorders. Neurotherapeutics 16:666–674. https://doi.org/10.1007/S13311-019-00767-8
103. Noori T, Dehpour AR, Sureda A et al (2021) Role of natural products for the treatment of Alzheimer's disease. Eur J Pharmacol 898. https://doi.org/10.1016/J.EJPHAR.2021.173974
104. Azm SAN, Djazayeri A, Safa M et al (2018) Lactobacilli and bifidobacteria ameliorate memory and learning deficits and oxidative stress in β-amyloid (1-42) injected rats. Appl Physiol Nutr Metab 43:718–726. https://doi.org/10.1139/APNM-2017-0648
105. Kobayashi Y, Sugahara H, Shimada K et al (2017) Therapeutic potential of Bifidobacterium breve strain A1 for preventing cognitive impairment in Alzheimer's disease. Sci Rep 7. https://doi.org/10.1038/S41598-017-13368-2
106. Deng H, Dong X, Chen M, Zou Z (2020) Efficacy of probiotics on cognition, and biomarkers of inflammation and oxidative stress in adults with Alzheimer's disease or mild cognitive impairment – a meta-analysis of randomized controlled trials. Aging (Albany NY) 12:4010–4039. https://doi.org/10.18632/AGING.102810
107. He L, He T, Farrar S et al (2017) Antioxidants maintain cellular redox homeostasis by elimination of reactive oxygen species. Cell Physiol Biochem 44:532–553. https://doi.org/10.1159/000485089
108. Chen YF, Wu SN, Gao JM et al (2020) The antioxidant, anti-inflammatory, and neuroprotective properties of the synthetic chalcone derivative AN07. Molecules 25. https://doi.org/10.3390/MOLECULES25122907
109. Drummond NJ, Davies NO, Lovett JE et al (2017) A synthetic cell permeable antioxidant protects neurons against acute oxidative stress. Sci Rep 7. https://doi.org/10.1038/S41598-017-12072-5
110. Butler M, Nelson VA, Davila H et al (2018) Over-the-counter supplement interventions to prevent cognitive decline, mild cognitive impairment, and clinical Alzheimer-type dementia: a systematic review. Ann Intern Med 168:52–62. https://doi.org/10.7326/M17-1530
111. Forbes SC, Holroyd-Leduc JM, Poulin PhD MJ, Hogan DB (2015) Effect of nutrients, dietary supplements and vitamins on cognition: a systematic review and meta-analysis of randomized controlled trials. Can Geriatr J 18:231–245. https://doi.org/10.5770/CGJ.18.189

Chapter 11
Clinical Use of Antioxidants for the Treatment of Brain Disorders

Bilal Ahmad, Muhammad Shareef Masoud, Usman Ali Ashfaq, Mahmood-ur-Rahman Ansari, Nazia Nahid, and Muhammad Qasim

11.1 What Are Antioxidants?

Antioxidants are defined as molecules that can minimize or stop cell damage triggered by free radicals [1] These are sometimes referred to as free radical scavengers. Free radicals are molecular entities made up of one or more unpaired electrons. Many biological alterations can be caused by free radicals, including DNA mutations, changes in enzyme function, cell membrane lipid peroxidation, and even may induce apoptosis [2]. Hydroxyl radicals, reactive oxygen species, and superoxides are a few examples of oxygen free radicals that are produced in the body as a consequence of aerobic metabolism [3] (Fig. 11.1).

Depending on how oxidative damage, there are different neurological ailments. In general, neurological diseases are prevented and delayed by the use of free radical scavengers. Antioxidants have mostly been explored for their physiological defense against oxidative stress by stopping the chain processes that spread various diseases and by eliminating free radicals [5]. Oxidative stress is primarily a threat to all aerobic organisms. The brain's membrane peroxidation is a result of oxidative damage, which primarily impacts the brain and other essential organs. For the therapy of neurological disorders, various antioxidants like pyrrolopyrimidine that can pass the blood-brain barrier (BBB) are often recommended [4]. Some chemicals' combined effects are more effective than their individual effects, like combining vitamin E with vitamin C is more effective than administering either one separately in promoting free radical scavenging activity. For disorders like Parkinson's disease,

B. Ahmad · M. S. Masoud · U. A. Ashfaq · Mahmood-ur-Rahman Ansari · N. Nahid · M. Qasim (✉)
Faculty of life sciences, Department of Bioinformatics and Biotechnology, Government College University Faisalabad, Faisalabad, Pakistan
e-mail: qasimawan@gcuf.edu.pk

A. Imran, G. Hussain (eds.), *The Role of Natural Antioxidants in Brain Disorders*, Food Bioactive Ingredients, https://doi.org/10.1007/978-3-031-41188-5_11

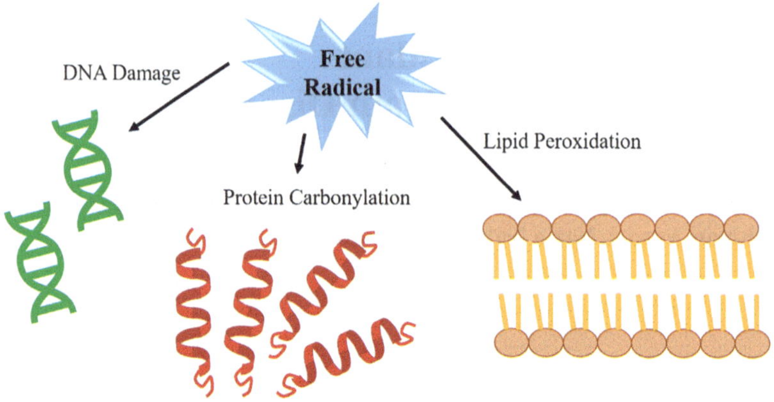

Fig. 11.1 The increased production of free radicals can result in oxidative damage to macromolecules including DNA, lipids, and proteins. This might potentially increase the chance of developing a variety of medical problems, including arthritis, tumor, diabetes, and neurological disorders [4]

additional combined medicines include pyrrolopyrimidine, deferoxamine, and coenzyme Q10 [6]. The multifunctional role of antioxidants in treating neurological disorders has been discussed in this chapter.

11.2 Antioxidants' Mode of Action

For therapeutic applications, antioxidant status assessment is receiving more and more attention. Due to the intricate processes of action for each specific antioxidant, it is challenging to determine the antioxidative potential in this situation. A few of them work by scavenging free radicals, others by stopping the production of reactive oxygen species (ROS), activating signaling pathways, or yet others by restoring oxidative stress. Catalase, glutathione peroxidase, and SOD (Enzymatic radicals) are the primary mechanisms by which cells are protected. Nonenzymatic radicals work in plasma. Furthermore, there are large differences in redox equilibrium state across people; as a result, control parameters have not yet been developed [7, 8].

Additionally, no direct approach is currently available for the precise evaluation of oxidative stress under in vivo environments. Because of this, oxidative stress is assessed using a variety of in vitro assays that can directly identify free radicals, such as fluorescent probes, electron spin resonance (ESR) spectroscopy, or indirect methods that can identify the stable products produced as a result of the free radical spell, such as enzymatic tests, chromatography, immune, or colorimetry [9].

11.3 Types of Antioxidants

In human cells, there are many antioxidant systems with an enzymatic or non-enzymatic activity that cooperate to defend the organism from free radical-induced damage.

11.3.1 Enzymatic Antioxidants

Only a few numbers of proteins make up enzymatic antioxidants, such as Peroxidase enzyme (POD), catalase, superoxide dismutase (SOD), and a few other auxiliary enzymes. Such enzymes all have distinct reactions and exhibit strong antioxidant activity in the body.

11.3.1.1 Superoxide Dismutase (SOD)

Having a widespread existence, superoxide dismutase (SOD) works by catalyzing superoxide dismutation in the body. As a result of this process, hydrogen peroxide is created. The human body produces an incredible number of reactive oxidants, including hydrogen peroxide, hydroxyl radicals, and superoxide [10] [11].

11.3.1.2 Peroxidase Enzyme (POD)

The oxidoreductase enzyme peroxidase (POD) is frequently present in both plants and animals, as well as in some microbe [12]. In the presence of hydrogen peroxide functioning as a hydrogen receiver, it promotes the oxidation of various hydrogen donor molecules including, hydroquinone, aromatic amines, hydroquinone amines, and phenol compounds. To protect plants from oxidative stress, peroxidase can stimulate the conversion of hydrogen peroxide to water [13].

11.3.1.3 Glutathione

The two enzymes glutathione peroxidase and glutathione reductase are the most common forms of glutathione. Only glutathione in its reduced form demonstrates a protection system; glutathione in its oxidized state does not. To neutralize the hydrogen peroxide that is created inside the cell, reduced glutathione is crucial [14]. As a result, glutathione's continual reduction and oxidation contribute to making it a free radical scavenger [15].

11.3.1.4 Catalase

One more antioxidant enzyme called catalase serves as a catalyst in the breakdown of hydrogen peroxide into water and oxygen. Because most of it is lost while manipulating tissue, the true quantity of catalase cannot be estimated [16].

11.3.1.5 Hydroxyl Radical

The hydroxyl radical is thought to be the most effective of them and works by demolishing nearby cells. For SOD, there are three possibilities. Normally, the cytoplasm houses the copper-zinc-containing enzymes, the mitochondria house manganese SOD, and the extracellular space houses the third enzyme [10].

11.3.2 Non-enzymatic Antioxidants

Antioxidants exist in a variety of types and are used to cure free radical damage or oxidative stress. These include vitamin C, carotenoids, vitamin E, and thiol antioxidants.

11.3.2.1 Vitamin E (α-Tocopherol)

Typically, α-tocopherol (vitamin E) is a fat-soluble antioxidant. The cell uses this powerful antioxidant to defend itself from lipid peroxidation [17].By supplying relevant hydrogen to lipid peroxyl radical during the antioxidant process, vitamin E is transformed into α-tocopherol radical [18].

11.3.2.2 Carotenoids (β-Carotene)

Plants and microbes both contain colorful pigments called carotenoids. According to a research survey, eating a diet high in carotenoids reduces the chance of developing age-related diseases [19] carotenoids work as antioxidant properties by decreasing singlet oxygen without breakage and delocalizing the valence electron [20].

11.3.2.3 Ascorbic Acid (Vitamin C)

Ascorbic acid (vitamin C) is often water soluble, it primarily interacts with enzymatic antioxidants in the body's moist environment. Together with vitamin E, it aids in the regeneration of α-tocopherol from the radical form found in membranes and lipoproteins [18].Increasing intracellular glutathione levels also have a significant impact on defending the protein thiol group from oxidation [21].

11.3.2.4 Proline

A particular amine acid called proline takes a role in the production of proteins. Because a plant cannot produce the necessary proteins to continue the deconstruction process, proline keeps building up. It is distinguished by its capacity to suppress lipid oxidation [22].

11.3.2.5 Melatonin

Tryptophan is converted into melatonin in the pineal gland. It primarily functions as a source of electrons in the metabolism of oxygen, preventing free radical damage to DNA, proteins, and membranes [23].

11.4 Antioxidants vs Oxidative Stress

Oxidative stress is the imbalance between the generation of highly reactive oxygen species (ROS) and the antioxidant defense against these radicals. It is well recognized to have significant roles in the development of diseases, immunological defense, as well as aging. Highly reactive entities with an unpaired electron in the valence shell are known as free radicals (Fig. 11.2). They can operate as an oxidant or a reducing agent, and they can also donate electron or take it from other molecules [24].In the human being, reactive forms are produced by metabolic procedures connected to prostaglandin production, respiratory chain, and phagocytosis. The hydroxyl radical, which is created when an oxygen molecule receives three electrons, for example through the Fenton reaction, and the superoxide radical, which is primarily produced in mitochondria as a byproduct of electron transport in the respiratory chain, are the two most reactive species that can be found in biological systems. Some other forms like reactive nitrogen, oxygen, and chlorine molecules that exist as free radicals can also readily become radicals when exposed to oxidizing agents [25].

External factors including exposure to industrial toxins, X-rays, air pollution, and ozone can also produce free radicals. The cell maintains a balance between the creation of a reactive form and the defensive systems' ability to neutralize it. This equilibrium is gradually altered in favor of pro-oxidative states during physiological circumstances, resulting in moderate oxidative stress [26].

Free radicals may build up in the body over time, accelerating the aging process and contributing to several neurological disorders like Parkinson's disease, Alzheimer's disease, stroke, etc [27]. The defense mechanisms of diverse species

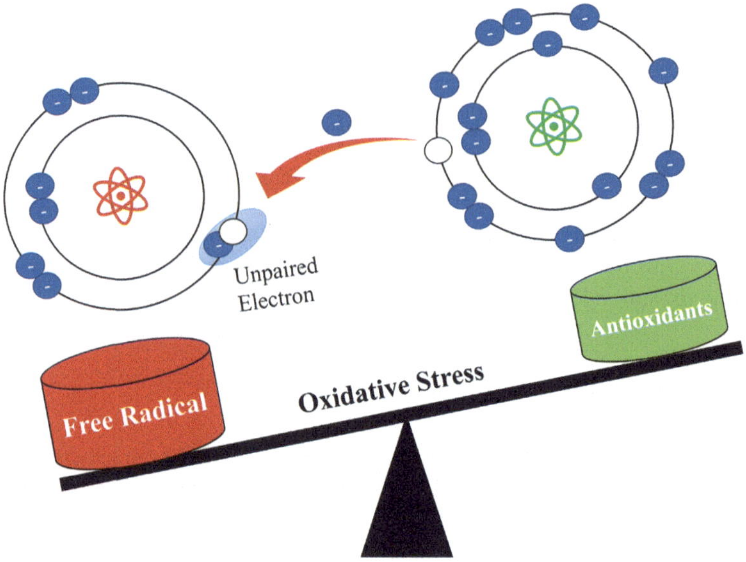

Fig. 11.2 Free radicals are extremely reactive and unstable molecules. They develop as a result of electron gains or losses in molecules or atoms. Because these unpaired electrons dislike being alone, they look for an electron to couple with throughout the system. Oxidative stress occurs when the ratio of free radicals to antioxidants is imbalanced

create antioxidants to counteract the harmful effects of free radicals. Reactive forms of nitrogen, oxygen, and even chlorine damage can be lessened by antioxidants. The protective system's function may reduce the adverse effects of free radicals by inhibiting the creation of reactive radicals [28].

11.5 Clinical Applications of Antioxidants in the Treatment of Brain Disorders

Neurological disorders are caused by the loss of nerve cells in the brain and spinal cord. It has been suggested that mitochondrial malfunction, which then triggers apoptosis, is a major cause of aging and several neurodegenerative ailments, including amyotrophic lateral sclerosis. Alzheimer's disease, multiple sclerosis, and Parkinson's disease. Such neurological ailments have been treated with a variety of antioxidant treatments. Eating a variety of nutritional supplements, vegetables, fruits, and different herbs can help to repair damaged brain cells and act as antioxidant medicine.

11.5.1 Ischemia Strokes

Several enzymes are in control of the redox balance in neurons under healthy and pathological circumstances. Some of these enzymes maintain proper expression of the antioxidant glutathione, which shields neurons from oxidative damage. An oligopeptide with a cysteine residue called glutathione functions as a nucleophile by providing electrons to help oxidized proteins dissolve their disulfide bonds. Although glutathione concentrations in neurons are modest, astrocytes produce and secrete significant quantities of glutathione in vitro, which may serve to protect neurons. Furthermore, the release of a cysteine-glycine dipeptide from astrocytes is a key component in neuronal glutathione production [29].

Hydrogen peroxide is transformed into water and molecular oxygen by the enzyme catalase, a typical part of cellular peroxisomes. Catalase is one of the most effective enzymes known to exist in nature and is widely expressed by neurons and glia of the Central nervous system [30]. According to a study that examined the antioxidant enzymes in human patients with neurological diseases Parkinson's disease or Ischemic stroke patients exhibited reduced levels of catalase activity in the brain. Increasing activity of the enzyme has thus been investigated as a possible treatment for ischemic stroke [31].

SOD1, SOD2, and SOD3 are the three Superoxide dismutase (SOD) isoforms that are expressed in mammalian cells. In animal models of stroke, targeting Superoxide dismutase (SOD) enzyme expression has also demonstrated significant efficacy [32]. In contrast to the vector alone, Davis and his colleagues demonstrated that SOD1 gene delivery via herpes simplex viral vectors protected mice against striatal damage [33]. Since SOD1 overexpressing animals failed to significantly reduce infarct volume when compared to wild-type mice, it is unfortunate that the neuroprotective impact of SOD1 overexpression appears to be restricted to models of cerebral ischemia damage [34]. According to several independent studies, both permanent and temporary middle cerebral artery occlusion increases the vulnerability of SOD2 mutant mice to oxidative stress [35, 36]. In models of ischemic stroke, increasing SOD3 is also a successful technique for preventing oxidative damage [37].

11.5.2 Alzheimer's Disease (AD)

The World Health Organization (WHO) estimates that globally there are approximately fifty million (~50 million) people suffering from dementia, with ten million new cases added every year [38]. A research survey reveals that sixty to seventy percent of all instances of dementia are caused by Alzheimer's disease (AD), making it the most common cause of dementia. Alzheimer's disease is a permanent, advancing, and accelerating brain ailment that causes memory loss as well as loss of the ability to do simple tasks [39]. Some antioxidants are discussed in the following as therapeutic agents to recover this disorder.

11.5.2.1 Ascorbyl Palmitate

It is a kind of vitamin C that preserves the entire vitamin C action without causing issues with ascorbic acids, such as decreased viability in vivo and less α-tocopherol recycling capability in the lipid bilayer [40]. Furthermore, it has been claimed that the lipophilic form of vitamin C can better meet hydrophilic variant demand. Ascorbyl palmitate has been found to have a substantial role in the treatment of Alzheimer's disease and can be passed through the blood-brain barrier [41]. It can hasten the formation of vitamin E since it is found in the cell membrane. Although it is yet unclear whether vitamin C works alone or in conjunction with other treatments to treat Alzheimer's disease, thus its preventive effect is under debate.

5.2.1 Astaxanthin and Quercetin

Astaxanthin is a type of carotenoid that can defend against the neurotoxic effects of Alzheimer's disease and prevent oxidative stress, memory loss, and inflammation [42]. The most notable and major dietary antioxidant that has been proven to be beneficial to health is quercetin, which stops serious disorders like cardiovascular disease, lung cancer, osteoporosis, etc. Clinical trials are still being conducted to determine its precise impact [43].

11.5.2.2 Catechins

Tea contains four different forms of catechins: epigallocatechin gallate, epicatechin, epigallocatechin, and epicatechin gallate [44].Green tea has the most catechins per cup (green tea catechins). By chelating metal ions like zinc, copper, iron, and reactive oxygen species, catechins have antioxidative actions that prevent reactive oxygen species from building up in the brains of Alzheimer's disease [45]. In the brain of the rat model, epigallocatechin gallate was found to lower lipid peroxidation, caspase levels, and oxidative stress. Catechins were said to have anti-inflammatory activity-inhibiting qualities in addition to their antioxidant capabilities [46]. Additionally, rat models have revealed that catechins are blood-brain barrier permeable, making them a viable therapeutic option for the treatment of Alzheimer's disease [45].

11.5.2.3 Estrogen and Gintonin

Estrogen functions as an antioxidant to shield neurons from the toxicity of Aβ. It is thought to have a therapeutic impact on Alzheimer's patients without enhancing memory [47, 48]. A glycol-lipoprotein called gintonin can support blood-brain barrier integrity maintenance [47] In the brains of mice that have received an injection of Aβ. it can also decrease active inflammatory mediators and microglial cells. Recent research suggests that gintonin therapy for Alzheimer's disease improves brain synapses and memory processes. In addition to having the capacity to control autophagy in primary cortical astrocytes, it shows a growing function as a regulator of synaptic transmission and neurogenesis [49]. However, further research is still necessary to fully comprehend gintonin's underlying mechanism of action in Alzheimer's disease.

11.5.2.4 Glutathione and Lipoic Acid

Glutathione is important for the production of DNA and proteins, for controlling the cell cycle, and for storing and transporting cysteine. Acrolein, 4-hydroxy-2-nonenal, and other lipid peroxidation byproducts may be scavenged by it. It is utilized to defend against metals, detox electrophiles, and oxidative stress, and maintain the thiol redox of cells [50]. It can also create metal complexes, which help the body remove the metals from the body and lessen their toxicity. A recent study has shown the redox mechanism of glutathione antioxidant, which controls the dynamics of the mitochondria in the axons [51]. But it's still unknown how glutathione specifically plays a part in Alzheimer's disease. The mitochondria contain the therapeutic antioxidant lipoic acid. It serves as a cofactor for both α-ketoglutarate dehydrogenase and pyruvate dehydrogenase. It is also engaged in the recycling of other antioxidants, including, vitamins C, E, and glutathione [52]. In some redox-active chelating metals, lipoic acid helps to stop the accumulation of lipid peroxidation. As a powerful antioxidant that may pass across the Blood-brain barrier, lipoic acid is perfect for therapeutic uses in Alzheimer's disease.

Box 11.1: List of Antioxidants Used for the Treatment of Alzheimer's Disease (AD)

- Ascorbyl Palmitate
- Astaxanthin
- Catechins
- Estrogen
- Gintonin
- Glutathione
- Lipoic Acid
- Melatonin
- Molecular Hydrogen
- Palmatine
- Quercetin
- Resveratrol
- Silibinin
- Vitamin E

Additional anti-oxidants used to treat Alzheimer's disease include melatonin, molecular hydrogen, palmatine, resveratrol, silibinin, and vitamin E (Box 11.1). Numerous studies and pieces of evidence suggest that oxidative stress or damage through a variety of mechanisms and pathways contributes significantly to the development of Alzheimer's disease. To prevent or minimize oxidative damage, new treatment methods are needed, and they may also be therapeutically effective.

11.5.3 Huntington's Disease (HD)

A cell that is under oxidative stress will have abnormally high amounts of oxidants compared to antioxidants. This will impair the cell's metabolism and redox balance, which will cause the cell to malfunction and die. Numerous studies have demonstrated that oxidative stress is one of the primary factors influencing Huntington's disease development, indicating that it may be a fundamental pathogenic process driving neuronal death and damage [53]. Current studies on antioxidant therapy employing Huntington's disease animal models have produced encouraging outcomes, including improvements in behavior and motor function. Antioxidants, which can stop, slow down, or mitigate the course of neurodegenerative diseases are frequently employed as free radical scavengers in these conditions.

Enzymatic and non-enzymatic kinds of antioxidants are generally utilized to treat neurological diseases. Catalase, Superoxide dismutase, and, peroxidase are examples of enzyme-based antioxidants. Vitamin C, vitamin E, retinoic acid, flavonoids and carotenoids, and other antioxidants are examples of non-enzymatic antioxidants [54]. Some antioxidants used for the treatment of Huntington's disease was listed in Box 11.2. Antioxidants were discovered to be helpful and effective in reducing Huntington's disease development using animal models. Yet antioxidant treatments have had conflicting effects in human clinical studies. Antioxidants are often utilized as adjuvant therapies or in conjunction with other treatments for Huntington's disease.

Box 11.2: List of Antioxidants Used for the Treatment of Huntington's Disease

- Ascorbic acid
- Creatine
- Curcumin
- Extra Virgin Olive Oil
- Grape Seed Phenolic Extract
- Lycopene
- Melatonin
- N-acetylcysteine
- Rutin
- Selenium
- Synthetic triterpenoid

11.5.4 Parkinson's Disease (PD)

Parkinson's disease, among the most prevalent forms of neurological disease, is characterized by persistent stiffness, tremor, and the death of pigment neurons [55]. Dopaminergic neurons often die preferentially in this condition [56]. It has been noted that in 2006, four million people with an average age of sixty years were living with Parkinson's disease, with fewer women than men being affected by the condition [57]. The estrogen level content is the cause of the disease's absence in the majority of females. Multiple sources of data from post-mortem studies indicated that the various processes are mostly caused by things like neuroinflammation, mitochondrial failure, and oxidative stress [58].

To assess the effectiveness of therapeutic agents for Parkinson's disease, various neurotoxin-based animal models are frequently used. These animal models, which go through a generation of dopamine neurons, exhibit sensory and motor dysfunction identical to Parkinson's disease patients. Vitamin E, Vitamin C, iron chelators, and creatine are a few antioxidants that can be used to treat Parkinson's disease (PD). In essence, "vitamin E" serves as a scavenger for different reactive oxygen species (ROS), which prevents lipid peroxidation. Results from the vitamin E supplementation provided to diseased animal models were examined, and the findings demonstrated that patients' ailment development was slowed down by vitamin E therapy [59]. The use of several antioxidants as recovery agents for this condition is explored in the below section.

11.5.4.1 Iron Chelators

Iron chelators can be used to reduce iron development in the Substantia Nigra of Parkinson's disease patients. Iron chelators have been effectively used in several preclinical investigations as therapeutic agents [60].

11.5.4.2 Melatonin

Studies conducted in vivo convincingly show melatonin's antioxidant abilities. Melatonin was administered to mice with multiple MPTP-based Parkinson's disease models to stop the neuronal degeneration in the nigrostriatal pathway. The neuroprotective effects of 6-OHDA in Parkinson's disease animal models have also been demonstrated [61].

11.5.4.3 Creatine

These are solid pieces of proof that creatine has antioxidant properties. In general, it helps to lessen mitochondrial dysfunction and has therapeutic properties. In vitro models of neurological disorders like Parkinson's disease utilize it [62].

11.6 Conclusion

Clinically, there is a need for effective treatment of neurological diseases. Oxidative stress is a significant factor which is involved in the development of disease, and anti-oxidant treatment may be beneficial. However, several obstacles, such as insufficient dosage, poor cytotoxicity, restricted transport to the central nervous system and temporary storage, and minimal antioxidant activity to fully detoxify the impact of free radicals, might have restricted the application in clinical research. Due to the strong catalytic activity of antioxidant enzymes, there has been a lot of effort done recently to create a Nano therapy-based method of administering these antioxidant enzymes. Many clinical situations might benefit therapeutically from an efficient antioxidant system.

References

1. Grewal AK, Singh TG, Sharma D et al (2021) Mechanistic insights and perspectives involved in neuroprotective action of quercetin. Biomed Pharmacother 140. https://doi.org/10.1016/J.BIOPHA.2021.111729
2. Arya A, Chahal R, Rao R et al (2021) Acetylcholinesterase inhibitory potential of various sesquiterpene analogues for Alzheimer's disease therapy. Biomolecules 11:1–30. https://doi.org/10.3390/BIOM11030350
3. Sharma VK, Singh TG, Garg N et al (2021) Dysbiosis and Alzheimer's disease: a role for chronic stress? Biomolecules 11. https://doi.org/10.3390/BIOM11050678
4. Akter R, Rahman H, Behl T et al (2021) Prospective role of polyphenolic compounds in the treatment of neurodegenerative diseases. CNS Neurol Disord Drug Targets 20:430–450. https://doi.org/10.2174/1871527320666210218084444
5. Galimberti D, Scarpini E (2018) Neurodegenerative diseases: clinical aspects, molecular genetics and biomarkers. Neurodegener Dis Clin Asp Mol Genet Biomarkers 1–407. https://doi.org/10.1007/978-3-319-72938-1/COVER
6. Delanty N, Dichter MA (2000) Antioxidant therapy in neurologic disease. Arch Neurol 57:1265–1270. https://doi.org/10.1001/ARCHNEUR.57.9.1265
7. Argüelles S, Gómez A, Machado A, Ayala A (2007) A preliminary analysis of within-subject variation in human serum oxidative stress parameters as a function of time. Rejuvenation Res 10:621–636. https://doi.org/10.1089/REJ.2006.0528
8. Poljšak B, Jamnik P (2010) Methodology for oxidative state detection in biological systems. Handb Free Radicals Form Types Eff:421–448
9. Sahnoun Z, Jamoussi K, Zeghal KM [Free radicals and antioxidants: physiology, human pathology and therapeutic aspects (part II)]. Therapie 53:315–39

10. Jeeva JS, Sunitha J, Ananthalakshmi R et al (2015) Enzymatic antioxidants and its role in oral diseases. J Pharm Bioallied Sci 7:S331–S333. https://doi.org/10.4103/0975-7406.163438
11. Many faces of superoxide dismutase, originally known as erythrocuprein on JSTOR. https://www.jstor.org/stable/24096687. Accessed 29 Mar 2023
12. Ahmad Mahmood Y, Ahmad Mahmood Al-Aloosy Y, Al-Tameemi AJH, Sameer Jumaa S (2019) The role of enzymatic and non-enzymatic antioxidants in facing the environmental stresses on plant: a review. Plant Archives:19
13. Kawano T (2003) Roles of the reactive oxygen species-generating peroxidase reactions in plant defense and growth induction. Plant Cell Rep 21:829–837. https://doi.org/10.1007/S00299-003-0591-Z
14. Sindhu RK, Kaur P, Kaur P et al (2022) Exploring multifunctional antioxidants as potential agents for management of neurological disorders. Environ Sci Pollut Res Int 29:24458–24477. https://doi.org/10.1007/S11356-021-17667-0
15. Maritim AC, Sanders RA, Watkins JB (2003) Diabetes, oxidative stress, and antioxidants: a review. J Biochem Mol Toxicol 17:24–38. https://doi.org/10.1002/JBT.10058
16. Aly DG, Shahin RS (2010) Oxidative stress in lichen planus. Acta Dermatovenerologica Alpina, Pannonica Adriat 19:3–11
17. New perspectives on vitamin E: gamma-tocopherol and carboxyelthylhydroxychroman metabolites in biology and medicine – PubMed. https://pubmed.ncbi.nlm.nih.gov/14732286/. Accessed 28 Mar 2023
18. Pryor WA (2000) Vitamin E and heart disease: basic science to clinical intervention trials. Free Radic Biol Med 28:141–164. https://doi.org/10.1016/S0891-5849(99)00224-5
19. Full article: Studies on free radicals, antioxidants, and co-factors
20. Mortensen A, Skibsted LH, Truscott TG (2001) The interaction of dietary carotenoids with radical species. Arch Biochem Biophys 385:13–19. https://doi.org/10.1006/ABBI.2000.2172
21. Naziroğlu M, Butterworth PJ (2005) Protective effects of moderate exercise with dietary vitamin C and E on blood antioxidative defense mechanism in rats with streptozotocin-induced diabetes. Can J Appl Physiol 30:172–185. https://doi.org/10.1139/H05-113
22. Trovato M, Mattioli R, Costantino P (2008) Multiple roles of proline in plant stress tolerance and development. Rend Lincei 19:325–346. https://doi.org/10.1007/s12210-008-0022-8
23. Rahimi R, Nikfar S, Larijani B, Abdollahi M (2005) A review on the role of antioxidants in the management of diabetes and its complications. Biomed Pharmacother 59:365–373. https://doi.org/10.1016/J.BIOPHA.2005.07.002
24. Cheeseman KH, Slater TF (1993) An introduction to free radical biochemistry. Br Med Bull 49:481–493. https://doi.org/10.1093/OXFORDJOURNALS.BMB.A072625
25. Apak R, Özyürek M, Güçlü K, Çapanoɋlu E (2016) Antioxidant activity/capacity measurement. 1. Classification, physicochemical principles, mechanisms, and electron transfer (ET)-based assays. J Agric Food Chem 64:997–1027. https://doi.org/10.1021/ACS.JAFC.5B04739
26. Poljšak B, Jamnik P, Kozyrev D, Slutsky V (2009) Methodology for oxidative state detection in biological systems. Handbook of Free Radicals: Formation, Types and Effects ISBN: 978-1-60876-101-2
27. Li Q, Zhao W, Zeng X, Hao Z (2018) Ursolic Acid Attenuates Atherosclerosis in ApoE-/- Mice: Role of LOX-1 Mediated by ROS/NF-κB Pathway. Molecules 23. https://doi.org/10.3390/MOLECULES23051101
28. Fang YZ, Yang S, Wu G (2002) Free radicals, antioxidants, and nutrition. Nutrition 18: 872–879. https://doi.org/10.1016/S0899-9007(02)00916-4
29. Pompella A, Visvikis A, Paolicchi A et al (2003) The changing faces of glutathione, a cellular protagonist. Biochem Pharmacol 66:1499–1503. https://doi.org/10.1016/S0006-2952(03)00504-5
30. Vainshtein BK, Melik-Adamyan WR, Barynin VV et al (1981) Three-dimensional structure of the enzyme catalase. Nature 293:411–412. https://doi.org/10.1038/293411A0
31. De La Torre MR, Casado A, López-Fernández ME et al (1996) Human aging brain disorders: role of antioxidant enzymes. Neurochem Res 21:885–888. https://doi.org/10.1007/BF02532336

32. Carlsson LM, Marklund SL, Edlund T (1996) The rat extracellular superoxide dismutase dimer is converted to a tetramer by the exchange of a single amino acid. Proc Natl Acad Sci U S A 93:5219–5222. https://doi.org/10.1073/PNAS.93.11.5219
33. Davis AS, Zhao H, Sun GH et al (2007) Gene therapy using SOD1 protects striatal neurons from experimental stroke. Neurosci Lett 411:32–36. https://doi.org/10.1016/J.NEULET.2006.08.089
34. Cloning, sequencing, and mutation of thiol-specific antioxidant gene of Saccharomyces cerevisiae – PubMed. https://pubmed.ncbi.nlm.nih.gov/8344960/. Accessed 29 Mar 2023
35. Macdonald RL, Stoodley M (1995) Cerebral infarction is exacerbated in mitochondrial manganese superoxide dismutase (SOD-2) knockout mutant mice after focal cerebral ischemia and reperfusion. Soc Neurosci Abstr 21:1268-. https://doi.org/10.2176/NMC.38.1
36. Kim GW, Kondo T, Noshita N, Chan PH (2002) Manganese superoxide dismutase deficiency exacerbates cerebral infarction after focal cerebral ischemia/reperfusion in mice: implications for the production and role of superoxide radicals. Stroke 33:809–815. https://doi.org/10.1161/HS0302.103745
37. Sheng H, Brady TC, Pearlstein RD et al (1999) Extracellular superoxide dismutase deficiency worsens outcome from focal cerebral ischemia in the mouse. Neurosci Lett 267:13–16. https://doi.org/10.1016/S0304-3940(99)00316-X
38. Qassem T, Khater MS, Emara T et al (2020) Validation of the Egyptian-Arabic Version of the Addenbrooke's Cognitive Examination III (ACE-III) in Diagnosing Dementia. Dement Geriatr Cogn Disord 49:179–184. https://doi.org/10.1159/000507758
39. Pyenson B, Sawhney TG, Steffens C et al (2019) The real-world medicare costs of alzheimer disease: considerations for policy and care. J Manag care Spec Pharm 25:800–809c. https://doi.org/10.18553/JMCP.2019.25.7.800
40. Ross D, Mendiratta S, Qu ZC et al (1999) Ascorbate 6-palmitate protects human erythrocytes from oxidative damage. Free Radic Biol Med 26:81–89. https://doi.org/10.1016/S0891-5849(98)00198-1
41. Bradford F, Sanjay G (2005) A review of antioxidants and Alzheimer's disease: Annals Clin Psychiatr:17(4)
42. Liu Q, Tang GY, Zhao CN et al (2019) Antioxidant activities, phenolic profiles, and organic acid contents of fruit vinegars. Antioxidants (Basel, Switzerland) 8. https://doi.org/10.3390/ANTIOX8040078
43. Gonzales MM, Garbarino VR, Marques Zilli E et al (2022) Senolytic therapy to modulate the progression of Alzheimer's disease (SToMP-AD): a pilot clinical trial. J Prev Alzheimer's Dis 9:22–29. https://doi.org/10.14283/JPAD.2021.62
44. Al-Gubory KH, Laher I (2018) Nutritional antioxidant therapies: treatments and perspectives. Nutr Antioxid Ther Treat Perspect 1–553. https://doi.org/10.1007/978-3-319-67625-8/COVER
45. Ide K, Matsuoka N, Yamada H et al (2018) Effects of tea catechins on Alzheimer's disease: recent updates and perspectives. Molecules 23. https://doi.org/10.3390/MOLECULES23092357
46. Kaur T, Pathak CM, Pandhi P, Khanduja KL (2008) Effects of green tea extract on learning, memory, behavior and acetylcholinesterase activity in young and old male rats. Brain Cogn 67:25–30. https://doi.org/10.1016/J.BANDC.2007.10.003
47. Ikram M, Jo MG, Park TJ et al (2021) Oral administration of gintonin protects the brains of mice against A β-induced Alzheimer disease pathology: antioxidant and anti-inflammatory effects. Oxid Med Cell Longev 2021. https://doi.org/10.1155/2021/6635552
48. Mulnard RA, Cotman CW, Kawas C et al (2000) Estrogen replacement therapy for treatment of mild to moderate Alzheimer disease: a randomized controlled trial. Alzheimer's disease cooperative study. JAMA 283:1007–1015. https://doi.org/10.1001/JAMA.283.8.1007
49. Jang M, Choi S-H, Choi JH et al (2021) Ginseng gintonin attenuates the disruptions of brain microvascular permeability and microvascular endothelium junctional proteins in an APPswe/PSEN-1 double-transgenic mouse model of Alzheimer's disease. Exp Ther Med 21. https://doi.org/10.3892/ETM.2021.9741

50. Butterfield DA, Castegna A, Pocernich CB et al (2002) Nutritional approaches to combat oxidative stress in Alzheimer's disease. J Nutr Biochem 13:444–461. https://doi.org/10.1016/S0955-2863(02)00205-X
51. Marí M, de Gregorio E, de Dios C et al (2020) Mitochondrial glutathione: recent insights and role in disease. Antioxidants (Basel, Switzerland) 9:1–21. https://doi.org/10.3390/ANTIOX9100909
52. Feng Y, Wang X (2012) Antioxidant therapies for Alzheimer's disease. Oxid Med Cell Longev. Oxid Med Cell Longev. 2012;2012:472932. doi: 10.1155/2012/472932. Epub 2012 Jul 25.
53. Kumar A, Ratan RR (2016) Oxidative stress and huntington's disease: the good, the bad, and the ugly. J Huntingtons Dis 5:217–237. https://doi.org/10.3233/JHD-160205
54. Johri A, Beal MF (2012) Mitochondrial dysfunction in neurodegenerative diseases. J Pharmacol Exp Ther 342:619–630. https://doi.org/10.1124/JPET.112.192138
55. Lin MT, Beal MF (2006) Mitochondrial dysfunction and oxidative stress in neurodegenerative diseases. Nature (443):787–795. https://doi.org/10.1038/nature05292
56. Filograna R, Beltramini M, Bubacco L, Bisaglia M (2016) Anti-oxidants in Parkinson's disease therapy: a critical point of view. Curr Neuropharmacol 14:260. https://doi.org/10.2174/1570159X13666151030102718
57. Bose A, Beal MF (2016) Mitochondrial dysfunction in Parkinson's disease. J Neurochem 139:216–231. https://doi.org/10.1111/JNC.13731
58. Esteves AR, Arduíno DM, Swerdlow RH et al (2009) Oxidative stress involvement in alpha-synuclein oligomerization in Parkinson's disease cybrids. Antioxid Redox Signal 11:439–448. https://doi.org/10.1089/ARS.2008.2247
59. Bové J, Prou D, Perier C, Przedborski S (2005) Toxin-induced models of Parkinson's disease. NeuroRx 2:484–494. https://doi.org/10.1602/NEURORX.2.3.484
60. Weinreb O, Mandel S, Youdim MBH, Amit T (2013) Targeting dysregulation of brain iron homeostasis in Parkinson's disease by iron chelators. Free Radic Biol Med 62:52–64. https://doi.org/10.1016/J.FREERADBIOMED.2013.01.017
61. Dabbeni-Sala F, Santo S, Franceschini D et al (2001) Melatonin protects against 6-OHDA-induced neurotoxicity in rats: a role for mitochondrial complex I activity. FASEB J 15:164–170. https://doi.org/10.1096/FJ.00-0129COM
62. Beal MF (2011) Neuroprotective effects of creatine. Amino Acids 40:1305–1313. https://doi.org/10.1007/S00726-011-0851-0

Index

A. Imran, G. Hussain (eds.), *The Role of Natural Antioxidants in Brain Disorders*, Food Bioactive Ingredients, https://doi.org/10.1007/978-3-031-41188-5

The manufacturer's authorised representative in the EU is Springer
Nature Customer Service Centre GmbH, Europaplatz 3, 69115 Heidelberg,
Germany. If you have any concerns regarding our products, please
contact ProductSafety@springernature.com

Printed and bound by CPI Group (UK) Ltd, Croydon, CR0 4YY
01/10/2026
02225896-0006